SEISMIC PROSPECTING FOR OIL

SEISMIC PROSPECTING FOR OIL

By C. HEWITT DIX

Professor Emeritus,
California Institute of Technology

INTERNATIONAL HUMAN RESOURCES DEVELOPMENT CORPORATION
Boston

SEISMIC PROSPECTING FOR OIL

 For information address: IHRDC, Publishers, 137 Newbury Street, Boston MA 02116.

ISBN: 0-934634-06-8

Library of Congress Catalog Card Number: 80-84573

Printed in the United States of America

Reprinted by arrangement with Harper & Row, Publishers, Inc.

CONTENTS

III. INTERPRETATIONS

V. THE PHYSICAL PROCESSES INVOLVED

EDITOR'S INTRODUCTION TO THE FIRST EDITION

Seismic Prospecting for Oil is a scholarly volume but a highly practical one as well. The second sentence in the book succinctly states its purpose and epitomizes its author's direct style. It reads, "The principal problem is to find oil."

The process of finding petroleum in practically all cases is, as Dr. Dix points out, an indirect one. In the past it has involved, and commonly still requires, library and laboratory studies, field surveys of surface geology, a methodical program of core drilling and subsurface analysis, and considerable aerial mapping. On the other hand, the finding process sometimes has consisted merely of a successful "scouting" program for gathering pertinent gossip, and "inside information" from persons who acquired their data in a more scientific fashion.

During the past three decades, however, the indirect approach to the problem of finding oil has been characterized chiefly by the rapid development of geophysical methods of exploration. The purpose of all such methods is not to discover petroleum directly but to reveal stratigraphic and structural data which are otherwise unavailable or cannot be obtained cheaply by simpler methods. The facts acquired through geophysics must then be given the correct geological *interpretation* before a successful drilling program can be developed.

Of great use in the ever-expanding, and now world-wide, geophysical program have been ground and, more recently, air-borne magnetometer surveys, torsion balance or gravity surveys, and geological exploration by means of the seismograph. The earlier seismic surveys employed the refraction method which was gradually supplanted in large part by the somewhat more accurate reflection surveys. A detailed description of the latter type of oil-finding process is the main topic of the present volume.

The seismic method of prospecting for oil is relatively new. It was first demonstrated to the public generally in 1933 in a series of electrically animated cross sections and models displayed at the Hall of Science at the Chicago World's Fair. Even the *Bulletin of the American Association of Petroleum Geologists* did not begin to carry scientific articles on the subject until 1931, when a paper on "Seismic Interpretation" appeared in Volume XIV of that publication. Then followed an article on "The Application of Seismography to Geological Problems" in the next volume of that journal; and, finally, in Volume XVI there appeared a paper on the "Applications of the Reflection Seismograph."

These are, of course, by no means the only papers, or the earliest, on the general subject, but it is significant that petroleum geologists did not begin to give space in their principal journal until some score of years ago. As a matter of fact, they could not have been expected to do so, for *The Journal of the Society of Petroleum Geophysicists* did not start publication until 1930; and Beno Gutenberg published his paper, "Geophysics as a Science," as recently as 1937.

A "Note on the Theory of Seismic Prospecting," by the author of the present volume in Harper's Geoscience Series, appeared in 1935, and he has been increasingly active in the field ever since. Inasmuch as a scientific book can reveal only a portion of its writer's background, training, and experience, there is here included a brief biographical note.

Charles Hewitt Dix is a native of Los Angeles who took his undergraduate training in the physical sciences at California Institute of Technology. His graduate work was done at Rice Institute, which granted him his doctorate in mathematics in

1931. After several years' experience as an instructor in mathematics at Rice, Dr. Dix became research geophysicist for the Humble Oil and Refining Company, and he served in a similar capacity for the Socony Vacuum Oil Company from 1939 to 1941. He then became Chief Seismologist and Vice President of the United Geophysical Company of Pasadena, California.

Since 1948 he has been a member of the geophysics staff at California Institute of Technology. He also served as geophysical consultant for several oil companies from 1948 through 1950; since then he has been an exclusive consultant for the California Research Corporation (Standard Oil Company of California.

With his mathematical training, teaching experience, and rich industrial and practical background, Dr. Dix has been able to approach a rather complex subject from particularly fortunate vantage points. As a consequence, *Seismic Prospecting for Oil* should be useful to teachers as well as students, to physicists and mathematicians as well as geologists, to the scientists, technicians, engineers, and practical field workers of the great petroleum industry generally as well as to the academicians.

Carey Croneis

EDITOR'S INTRODUCTION TO THE SECOND EDITION

This new edition of *Seismic Prospecting for Oil* was first conceived in the summer of 1979 as a response to the need expressed by many geophysicists for having this text back in print. This need was reiterated in the program at the Awards Luncheon for the 49th Annual International Meeting of the Society of Exploration Geophysicists when C. Hewitt Dix received the Maurice Ewing Award, the society's highest honor. Raymond A. Peterson in writing about this award notes that C. Hewitt Dix's "1952 book, *Seismic Prospecting for Oil* is still one of the best references for practicing exploration geophysicists."

In the ensuing months, IHRDC made arrangements with Harper & Row, Publishers for a new edition. It was felt by Dr. Dix that a new edition would be impractical without some updating. Accordingly, Dr. Dix prepared comments to each new chapter evaluating events that have occurred since 1952. These are not to be considered encyclopedic nor comprehensive but rather insights from the perspective of a lifelong and dedicated geophysicist. In addition, Dr. Dix made minor corrections in the text itself.

We are extremely honored to be able to bring this edition back into print.

Michael R. Hays

PREFACE TO THE FIRST EDITION

The aim of this book is entirely educational. The purpose is to train younger men in the interpretive techniques and in the backgrounds necessary for these techniques. The book incidentally will also serve as a kind of check review for the older members of the profession. By going over some of the items covered, they may decide whether or not their view of the science is more or less coincident with the one here presented. If they agree, nothing further need be said. If they disagree, they may do so either directly to me or in technical publications which, of course, will tend to improve this science.

In connection with the educational aim of this book, the exploration geologists have often been uppermost in my mind. The reason is that the geologists have the necessary broader view that must, in the final analysis, be applied to the results obtained by the seismologists. It is therefore desirable, and even necessary, that our geological friends understand, to a certain extent, the background of our work. Several chapters have been written with this in mind.

It will be observed that each chapter begins with a statement regarding the individuals for whom the chapter is written. It should be understood that a book such as this had to be written for a very diverse group. Certain chapters are written especially for people with particular interests and training backgrounds, and this specification is made at the beginning of each chapter for the convenience of the reader.

Since the aim of the book is primarily educational, perhaps a statement should be included to the effect that this does *not* mean that the book is written simply as a classroom textbook.

While it is hoped that the book will prove useful for teaching students in colleges and universities, it was written primarily with actual workers in mind.

I have spent most of my time as a geophysicist doing field work, and I feel the need of offering something useful to those who do the most important part of the exploration work, the *interpreters.* After all, the *purpose of all the field work* is to secure data for the interpreters. It is this vital interpretation link which I feel needs greatest emphasis now and in the future. Sometimes there has been a tendency in the other direction as the processes of securing data for the interpreter have often been of considerable difficulty, requiring the skills and energies of many people together with the expenditure of large sums of money. Under such circumstances, emphasis is likely to be placed where skill and energy and money are spent in greatest quantities. If this is carried too far, so that interpretive work is neglected, the job of exploration is not even half finished.

Since my aim is educational, it may be well to express a small part of my background philosophy in this matter. This has been received from many sources, but one of the outstanding ones was Alfred North Whitehead, the late mathematician and philosopher at Harvard University. Whitehead published several years ago a little book called *The Aims of Education.* This has now been published by Mentor Books in a low-cost paper-bound edition. Whitehead, in his kindly and incisive manner, analyzes the process of learning into a repetition of three essential stages. The first stage he calls the *stage of romance.* This is the stage in which the learner first becomes vitally interested in a certain part of the world. The reasons for his interest may vary widely from one individual to another, but they have this in common—they are vital and broad. The next stage Whitehead terms the *stage of precision.* At this stage the learner asks himself a series of critical questions and investigates his interests in greater detail. He seeks to achieve greater precision of view. In the third stage the learner consolidates his gains in precision and returns to the broader view. This Whitehead calls the *stage of synthesis.* Although the return to the first stage is made in the last stage, it is made with a new viewpoint of greater precision so that the learner now looks at the overall problem in a sharper, clearer way. It may be noted that in the intermediate stage the tools of *analysis* are likely to be useful. The learner makes *measurements,* seeks greater precision; but analysis is a process of taking apart. It is a process of studying little pieces, in a certain degree; it is a process of *specialization.* With due regard to the limitations which all of us possess with respect to ability and energy and time to do our various tasks, it is an unfortunate thing when the investigator becomes so concentrated in the work of analysis that no time is left for putting together the picture as a whole. It is as if no time were left for what was sought in the first place!

Whitehead has referred to the *rhythms* of education. Perhaps we can liken the process to that of ascending a spiral staircase and, as we ascend, occasionally coming to an open window from which we can see a broader panorama of our world. But as we go up this staircase we gather into our kit of tools ever more useful and precise instruments of observation, so that as we come to each successive open window we are better prepared to view the panorama before us as a whole with greater precision.

I like this view of Whitehead's because it sets forth the relationship between the various parts in the learning process. It appears to me that each of these stages is equally important, and that if any one of them is left out, a serious deficiency is likely to result. Because of the fact that the middle stage, the stage of precision, is often so difficult, this stage usually receives a maximum of emphasis. But it ought to be remembered that this stage of precision, though it requires much of our energy, is not, after all, the primary purpose of scientific investigation. Neither is it the primary purpose of seismic exploration. The primary purpose is rather a temporary synthesis of all the information. I say temporary because of the evident fact that as time goes on, new circuits in the spiral will add to the synthesized picture.

In planning the present book, the five parts have been separated from one another. The first part includes what might be regarded as a panoramic view. The second consists of a series of chapters dealing with the routine tools and operations used almost every day by the seismic field party. The third part is a chapter on the interpretation of reflection seismic data. A whole book could be written on this subject alone. The fourth part is a brief review of refraction seismic work suitable for workers acquainted with reflection seismic work. The fifth part consists of three background chapters, two of which are concerned with details in the physical picture which ought to be part of the background of the interpreter. The last chapter discusses what might be termed the interpreter's view of the instrument situation. No instrument details are given; the instrument problem is looked at from the point of view of the person primarily interested in the interpretation of currently available seismic records.

The illustrations in this book were drawn first and the book written—or rather dictated—afterward, on the basis of them. This procedure has made necessary the inclusion of all formulas on the illustrations themselves, and this has turned out to be a very excellent thing, since the quantities in mathematical equations always have specific reference to the diagram. Mathematics has been used to a limited extent simply as a *language*, as a descriptive means in which relations are emphasized. It is hoped that the traditional attitude of the mathematician has not inadvertently crept into this work. I refer to the attitude in which the particular meaning of a symbol is regarded as of no importance. For the mathematician, his symbols are empty pockets which have only a very few properties, the most important of which is that they may be distinguished from one another where they are different. On the other hand, the person who wishes to apply mathematics must assume an entirely different attitude; his interest resides primarily in the meanings to be attached to the symbols. It is hoped that by placing all the mathematical formulas on the diagrams to which they apply, the intimate connection between the symbol and its meaning may be maintained in each case.

Every book possesses limitations, and this is no exception. Part of its limitation is intentional. I thought that the book should not be very large. Another part of its limitation arises from the limitation of my background. My interests have been primarily interpretive, so primary emphasis has been placed on the interpretive problem. Another limitation is the reader's background with respect to mathematics and physics. It has been assumed that the reader will have had two years of college physics and two years of college mathematics. Thus it is assumed that he will know at least the descriptive terminology of differential and integral calculus. At each stage, however, an effort has been made to present results using mathematical means at as elementary a level as possible consistent with reasonable brevity. There is another limitation which always applies to workers in this field, namely, the commercial limitation, according to which processes and results of a confidential nature may not be disclosed. Several important aspects of the general problem of applied seismology have had to be minimized for this reason. It is to be hoped that this latter limitation will not prove to be a serious one. I do not believe that this will be the case for the field worker, because many matters that are now held confidential are really very well known throughout a wide segment of the industry.

Acknowledgment must be made to Dr. Raymond A. Peterson of the United Geophysical Company, who has kindly looked over the entire manuscript and the illustrations and has checked certain items which might be considered the confidential property of that company. Permission to publish several such items has been received and is gratefully acknowledged. I am also indebted to Mr. Henry Salvatori of Western Geophysical Company for permission to publish the maps of the Wilmington oil field. Acknowledgment, further, should be made to the authors of many papers published in *Geophysics*, and in the *Case Histories*, Volume 1, published by the Society of Exploration Geophysicists. This latter volume has made the author's task much lighter than it otherwise would have been; it also supplies many examples which may be used in connection with the material in the present book. The very excellent books by Heiland, Nettleton, and Jakosky on exploration geophysics have been much used in a general way. I would also like to acknowledge my indebtedness to many friends who have discussed geophysical problems with me. Special mention should be given M. King Hubbert, whose many hours of discussion have served to broaden my viewpoint. I should also say that it was Dr. Hubbert who first suggested that I write this kind of book. Acknowledgment should also be expressed to my students, who have often presented fresh viewpoints of considerable value. The help of Mrs. Berniece Tomczak and Mrs. Florence Wiltse in typing and retyping the manuscript is gratefully acknowledged. Miss Dorothy Thompson of Harper & Brothers made many improvements in the text for which the author is grateful. It is a pleasure to thank my wife, Inger, who made the final drawings for the book. I regard this collection of figures as the most important part of the book, partly because this portion of the book can be read in any language.

C. H. D.

July 1, 1952

PREFACE TO THE SECOND EDITION

The original preface is still applicable. This book is intended to help the people that do the various jobs understand why they do what they do. An effort has been made throughout to avoid a dependence on authority. Too much dependence on authority tends to promote a kind of blindness rather than comprehension and understanding.

The term understanding applies to technical work but it is recognized that some operations need team work with carefully organized routines. Thus, people have to be understood too.

I am grateful to the Division of Geological and Planetary Sciences at the California Institute of Technology for providing typing and drawing corrected figures in the rewritten parts of the book. I further thank Chevron U.S.A. Inc., for the record section comprising Figure 8a.

As always, our hope for the future depends on the young people.

October 1980 C.H.D.

Part I

INTRODUCTION AND GENERAL ORIENTATION

CHAPTER 1

Summary of Exploration Methods Commonly Used

This chapter is presented for all interested in exploration for oil, geophysicists and geologists alike.

The principal problem is to find oil. The finding process has to be an indirect one in almost all cases. The reason is that oil is usually located at considerable depth under the ground. At the present time no direct means of detecting the presence of this buried oil is known (except drilling; see Section 1.9). All the methods used depend for their success upon the discovery of geological structure in some form or other. The reader may refer to Figure 1.1 for a collection of traps. He will see there an anticlinal trap, a fault trap, a stratigraphic trap, and a reef trap. These traps may or may not be findable by known procedures. Our problem is to find such traps where possible.

We need only mention, in passing, the principle upon which the exploration is necessarily based. It will be observed that all the traps illustrated show gas on top, oil, and then water below, in a porous sand section or in a limestone reef. The arrangement is based upon the fundamental principle of hydrostatics according to which the less dense fluid floats on

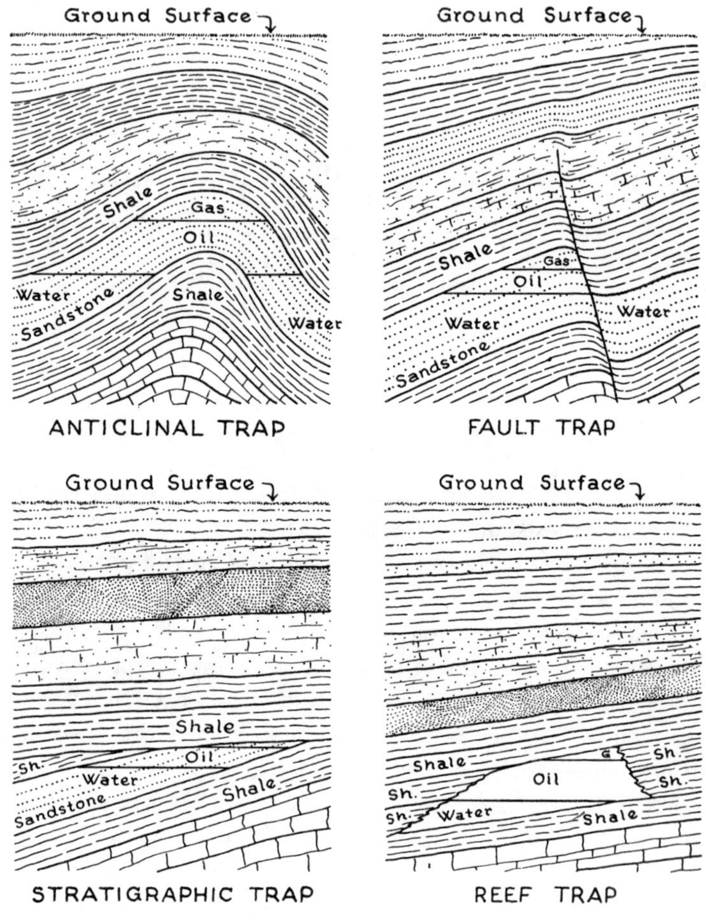

Fig. 1.1. Examples of oil traps.

top of the more dense fluid. This principle applies, not only in a pure fluid, but also in a fluid which is contained in a porous medium such as a sandstone or a limestone.

Although the problem of finding oil is always a structural one, it is not always a *simple* structural one in the sense of *structural geology*. The "structure" involved may not be simply an anticline or a fault trap. The "structure" may be due, instead, to the fine-grained characteristics of the section, or, rather, to the variations of this fine grain. For example, a porous section may grade into a less porous section, thus presenting a lithological trap. Such a trap, although it is not ordinarily called a structural trap, is nonetheless due to structure in the wider sense.

1.1. LIBRARY STUDIES

One of the first methods of exploration is to go to a good library and look up all the published material in an area of interest. This material usually consists of geological reports covering the surface geology, structure obtained from drill holes, if any occur in the area, data on seepages, soil studies, and possibly geophysical reports.

The first step in the library work is to accumulate a good bibliography covering a given problem.

But making the bibliography and reading the reports are not sufficient. It is necessary also that this information be somehow applied to the problem at hand. Usually the reports found in the library are not directed toward the particular exploration problem. So it is necessary for the investigator to *apply* the knowledge gained from his library studies to the exploration problem.

1.2. SURFACE GEOLOGY

In making a field survey, the geologist maps dips and strikes as obtained from the exposed outcrops. Occasionally he may supplement this information by digging pits and getting dips and strikes from the walls of the pits. These dips and strikes are indicated on his map. This map usually gives a certain

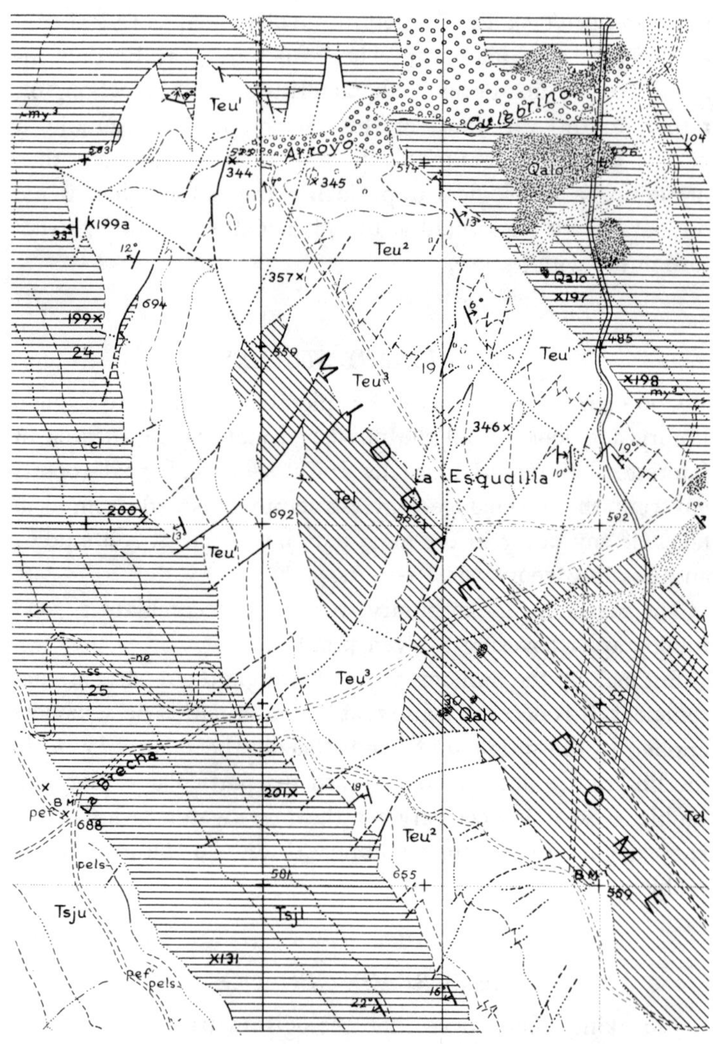

FIG. 1.2. Example of surface geology results. (From map in U.S. Geological Survey, *Professional Paper 195*.)

amount of the topography of the area. In addition, where possible, the map shows the contacts separating sections of different geological ages. Naturally, any wells or seepages in the area, as well as the drainage, are shown.

This information may be considered as *raw data*. From these data the geologist makes an *interpretation* which he includes with his report, usually in the form of a cross section or perhaps several cross sections through the area. His interpretations are based upon his data and upon *reasonable geological assumptions.*

The amount of oil that has been found with surface geological exploration as the primary key to the finding is a very impressive amount indeed. Wherever surface geological studies can be used, they furnish a valuable exploration tool.

The process of exploration by surface geological investigations is illustrated by Figure 1.2. This figure shows a small part of the surface geological map of the Kettleman Hills region in California, and gives the strike and dip symbols and other geological information. The map was made long after the Kettleman Hills field was discovered, but it illustrates an exceptionally good piece of surface geological work which could have been used for the discovery of this field.

Here, as in almost every other case of exploration, much depends upon the *interpretation* of the data that are accumulated. An interpretation that brings too much prejudice and not enough imagination to the problem may often be of little value. In some cases an interpretation can even have negative value. This may be the case if one is led to undertake an expensive subsequent exploration program, or to abandon an area which really has a good supply of oil. These constitute serious mistakes on the part of the interpreter. The important point is that responsibility for a major step in the exploration work rests with the interpreter.

1.3. CORE DRILLING

A commonly used method of exploration is to drill holes, say to a depth of 500 to 1500 feet, at carefully planned locations over an area. From this drilling two kinds of information are obtained. As the drilling progresses, samples are taken of the cuttings; they can be used, in some cases, to make correlations from core hole to core hole. It is usual, however, to rely largely upon the correlation of electrical logs taken in the core holes.

Under favorable circumstances this core drilling method of exploration offers a very fine supplement to the surface geological studies. It often has to be used in place of surface geological studies where no surface structure of interest is available. The core drilling exploration is usually somewhat more expensive than surface geology investigation, because special truck-mounted drills must be used and the drilling fluid must be supplied by means of water trucks.

An important advantage of core drilling exploration is that the data supplied the geologist consist of information that he is accustomed to handling. Thus he feels very much at home with this kind of information. Since, after all, he is the man who is to make the interpretation of the data, and since the data, by themselves, are valueless without adequate interpretation, it is obviously important that they be of such a character that he feels he can use them to the best advantage.

1.4. AERIAL MAPPING

Photographing the land from an airplane which is flown carefully at the proper height may yield information of considerable value. This is especially true in areas that are somewhat inaccessible. Also the aerial photographs supply very fine maps for planning subsequent exploration. Often there can be seen the trace of a geological contact, where the soil conditions vary across the contact in such a way as to affect the vegetation. Of course where outcrops occur regularly along the surface, these may also be seen on an aerial map. In some cases such a map has brought out otherwise hidden surface evidence of faulting by showing linear arrangements of ponds.

1.5. MAGNETOMETER SURVEYS

A great deal of the world has been mapped by magnetometer. In the early days the vertical component magnetometer was used on the ground. More recently the magnetometer has been used extensively in airplanes. This aerial magnetometer at the present time measures the total magnetic field of the earth rather than any component such as the vertical or the horizontal.

In earlier days the magnetometer was given a good deal of credit for finding a few oil fields, the most outstanding of which was probably the Hobbs field in eastern New Mexico. Magnetometer surveys are regarded as useful for finding basement highs, for uncovering the presence of igneous plugs, and in some cases for showing a little indication of shallow sedimentary structure.

The interpretation of magnetometer data is, of necessity, a very difficult technical task. It is at this interpretation stage that magnetometric exploration meets its principal difficulties. In spite of these difficulties, extensive magnetometer surveys have been conducted by most of the oil companies as a *reconnaissance* exploration method.

The value of aerial magnetometer surveys of large inaccessible areas can hardly be overestimated. The data furnished by these surveys, with careful and competent interpretation, may be used in planning a more extensive exploration program that does not need to be scattered over a vast area, but can be concentrated in certain smaller regions indicated by the interpretation of the magnetometer data.

1.6. GRAVITY SURVEYS

The second most extensively used geophysical method of petroleum exploration is the gravity survey method. In earlier days gravity surveys were made by means of the so-called Eötvös torsion balance. This torsion balance measured the horizontal gradient of vertical gravity, and in addition a quantity related to the curvature of the earth's gravity field. The most useful information was usually obtained by an interpretation of the gradient data to give vertical gravity at a series of points on the map.

The torsion balance has now been replaced by the gravity meter, which measures relative gravity at various station locations directly. With this procedure one lays out a large number of gravity stations in an area, surveys them very carefully, especially with respect to vertical control, and then takes the gravity meter to them and measures the difference in gravity between adjacent stations. When these differences are obtained, one then obtains a set of station values for gravity; these values may be corrected to a level datum. The corrected values may then be marked on a map and a gravity contour map drawn.

A structural high usually shows as a gravity high directly over the structural high. This is not always the case, for the high may consist of an uplift of less dense material which yields a gravity low, instead of a gravity high.

Another type of structural feature that may be found by means of a gravity map is a fault. The fault generally shows partly because of its semistraight-line character and partly because of a small but more or less definite change in the gravity contours as one crosses the fault.

Gravity surveys were originally used quite successfully to find salt domes. The general idea was that the salt, with a much lower density than that of the surrounding material, would lead to a gravity low. It was found, however, that this gravity low was often replaced by a gravity high or by a halo-shaped gravity contour map. It turned out that this reversal of gravity was due to the presence, on top of the salt plug, of a layer of high-density material which was pushed up with the salt. In many cases this high-density layer gave an effect which greatly altered the expected picture.

Although the gravity data may be treated by merely making a corrected gravity map, it is customary to make other maps as well. For example, it is usual to introduce a *regional correction* into the map which removes the effect of deep-seated and unimportant mass distributions. Although the removal of the regional effect is an interpretive procedure that suffers from personal bias, nevertheless it is a valuable procedure.

More recently, in a great many cases, another method of interpreting gravity data has been found very useful, especially for finding faults. Although we cannot go into this method in this book, we shall mention it briefly. It is called the *second vertical derivative* interpretation. This interpretation requires a considerable amount of scientific and technical skill for it to be valuable as an oil-finding tool. However, once a competent second derivative map has been obtained, any linear distribution of *zero second derivative contours* should be looked upon as possible, or even probable, fault trace locations.

Gravity data are capable of yielding very valuable exploration information. Here, as always, the information itself must be used by a competent interpreter. This is probably the greatest bottleneck in the exploration business at present, for gravity interpretation procedures are extremely difficult and require a high degree of technical skill and understanding.

1.7. SEISMIC REFLECTION EXPLORATION

We touch here very briefly upon the process of geological exploration by means of the reflection seismograph. This is the main topic of this book and will be greatly expanded upon in subsequent chapters. For orientation purposes, however, we include a brief description here.

We may suppose that an area to be prospected has been selected by a less expensive method. We go into the area and

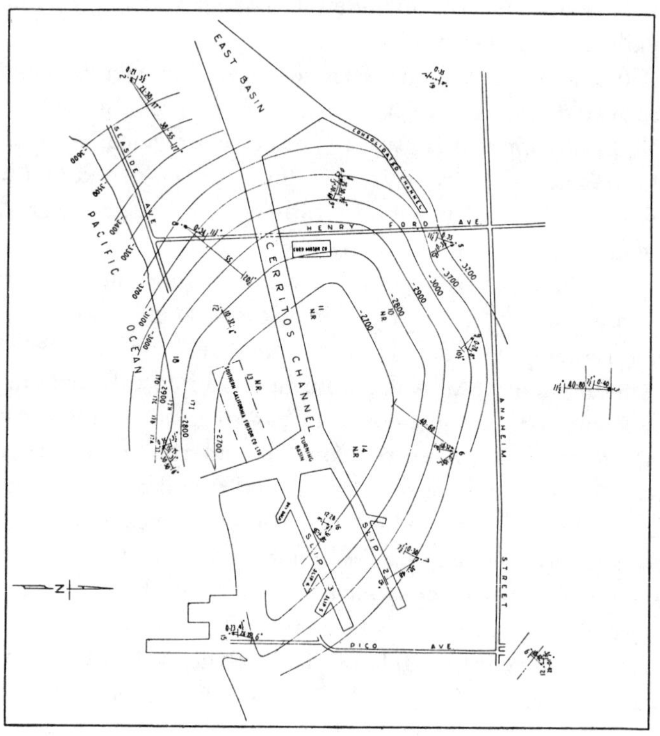

Fig. 1.3. Seismic discovery map for Wilmington oil field, California. (Map submitted by Western Geophysical Company to General Petroleum Corporation. This and Fig. 1.4 after Henry Salvatori, *Geophysical Case Histories* I, 1949.)

plan seismograph locations which are laid out by the surveyor. Then we drill shot holes from 20 to several hundred feet in depth and 4 to 6 inches in diameter. An explosive charge is loaded at the bottom of each hole and covered with a column of water for tamping purposes. This charge is capped with an electric blasting cap, the leads of which go up the hole and are ready for use at the surface. The shooter stands ready to

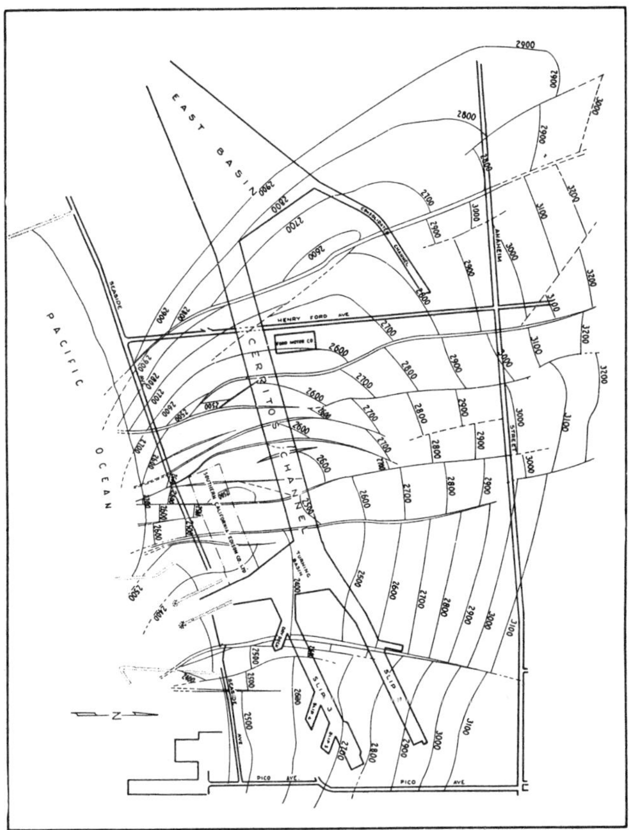

Fig. 1.4. Map showing subsurface structure obtained from drilling information on the Wilmington oil field.

fire the cap when ordered to by the seismic instrument operator, who has meanwhile placed his seismic detectors at suitable positions on the ground and is ready to take the seismic record.

When the operator orders the shooter to fire the charge, the operator starts his camera; it records the time at which the cap fires and also all subsequent times for a few seconds, usually about two.

The seismic record, thus taken, shows the so-called first breaks or first arrivals of energy from the shot. Also, in addition to a great deal of noise (on the average record), a few or sometimes many seismic reflections usually appear. These reflections come from contacts—for example, between shale and sandstone or shale and limestone or limestone and sandstone, and so on. The deeper the contact, the longer the time taken for the artificial seismic pulse to travel from the shot down to the contact, be reflected, and return to the surface, where it is detected and timed by the instruments.

As it is almost always possible to obtain information regarding the velocity with which the seismic pulses travel, it is also possible to transform the measured times into distances. These distances or depths are the raw material out of which a geological interpretation has to be made.

If we run a profile that crosses a "high," as we pass over the "high" the reflection time may be expected to decrease a little. This in fact is the usual method by which seismic "highs" are found.

It is possible to determine the *dip* and the *strike* by the seismic method. We shall go into this in greater detail in later chapters. For the moment, this will be illustrated by one of the outstanding examples of the use of the seismic method in the discovery of a major oil field. The particular case referred to here is the Wilmington oil field in southern California. Figures 1.3 and 1.4 are contour maps, one giving the earliest reflection seismic map of the area based upon a small set of strike-dip measurements. The other map shows approximately the same horizon, based on drilling data. The approximate coincidence is quite remarkable, especially since this major field was found by means of this very small number of shots in a highly industrialized area. The Wilmington case is exceptional, for very few oil fields are found with so small an amount of exploration information. However, a large number of fields have been found primarily by the reflection seismograph method, which accounts for its great popularity at the present time.

1.8. SEISMIC REFRACTION EXPLORATION

In the early days of seismic exploration the refraction method was used exclusively. The seismic refraction method was very successful in finding the shallow salt domes in the gulf coastal region of the United States. It has also been successful in finding basement faults and, to some extent, in finding structural highs.

Usually the refraction method is a little less accurate in its results than the reflection method. So it is resorted to only after the reflection method has failed in a given area.

However, there are circumstances when the refraction method is the most suitable one to use. One instance is when structure is controlled by relatively shallow basement faults. Faults in the basement complex can usually be found with greater certainty and ease by means of the refraction method than the reflection method. Therefore the refraction method may be used in preference to the reflection method, although this is rarely the case.

In the early days there was a tendency for the size of the explosive charges in refraction work to be excessive. This was due partly to the fact that the charges were not buried in deep shot holes and partly to the fact that an effort was made to shoot at longer distances than were economically feasible. Nowadays, with carefully designed instruments and proper attention to tamping the charge, it is possible to operate a refraction crew at a cost quite comparable to that of a reflection crew.

1.9. WILDCAT DRILLING

The drilling of wildcat wells may be classified in various ways, but for our purpose we shall refer to a "wild" wildcat as a well that is drilled on practically no information at all. Usually the ordinary wildcat, which is the other type, is drilled on the basis of geological and geophysical information which serves very well to locate the drilling site. In this case the drilling is very likely to find an oil field.

Random drilling of "wild" wildcat wells does furnish an exploration method, but it appears to be excessively expensive when one considers the cost of drilling a wildcat well. It is sometimes remarked, however, that this is the only method that can be used in certain areas, and this may be true in rare circumstances. Such remarks as "Oil is where you find it" appear in the literature and in conversation quite often. But it must be remembered that as a regular exploration procedure the cost of random drilling can mount up very rapidly.

Drilling must be regarded as the final step in the exploration procedure, since, obviously, oil cannot be found until a hole is drilled into it. There is no other way of finding oil directly. No other means is likely to appear in the near future, although some claims in this direction have been made.

1.10. MISCELLANEOUS METHODS

We list under miscellaneous a large number of methods, some of which are rarely used and some of which are regularly used. For example, electrical prospecting, which is rarely used for finding oil, may be divided into two procedures. One is the direct-current resistivity procedure and the other is the low-frequency alternating-current procedure. The latter has had some empirical success, though its actual operation is not well understood at the present time.

Another procedure that has been used extensively is the soil analysis method, which consists of taking soil samples and analyzing them for the various hydrocarbons and other gases and waxes they contain. This method is difficult to evaluate and for this reason has probably not been given as much credit as it deserves.

A method that is used very extensively is the logging method. When a well is drilled, resistivity logs may be taken, as well as self-potential logs. Other logs that may be used are radioactivity logs and dielectric constant logs. From these logs, correlations from well to well may be obtained. Information about the lithology may also be secured by interpreting the log. Since nowadays almost every well is logged by some process, this subsurface information is a valuable source of data which the geologist and geophysicist may tap in the finding process.

Another technique consists of micropaleontological correlation on the basis of the various foraminifera found in samples and in cores. This process is well known in the industry and is a standard procedure at the present time.

Other means of correlation include mineralogical studies, size, composition, orientation of grains.

Less exacting procedures, which nevertheless require a good deal of imagination and interpretive ability, involve the study of drainage patterns, topography, and regional tectonophysics. Model studies may be made in which geological structures are formed by applying controlled stresses to a properly designed model.

In the past, a certain amount of attention was given to methods which are classified, usually by the technical men, as black magic methods. These usually have to do with divining rods of one sort or other. They go back to an ancient time in our history when the wish was thoroughly mistaken for the fact. Sometimes it is difficult for a technical man to convince the management that the individual with the highly technical-looking divining rod can't honestly find oil structures with his gadget. So the gadget has to be tested. It seems only fair, however, that the test be conducted along objective lines similar to those that would be used for any bona-fide technically competent method of exploration. It is at this stage that the divining rod operator usually cancels the deal.

Finally there is another source of information for the exploration business. This is scouting and what might be termed "barroom gossip." Scouting is an attempt to get information from other people. Barroom gossip, on the other hand—it doesn't necessarily have to take place in a barroom—consists of general discussions of problems, maybe not at all in a serious vein, in which some individual makes a remark that is useful to a listener. The person who uses this information has to be a rather keen fellow who is vitally interested in the exploration problem, or is more aware of the possibilities than most of his associates and hence perhaps more imaginative in his approach to the problem than the average man.

In this chapter we have given a hasty look at the main methods used for the exploration for oil. Our purpose was to show how seismic methods fit in with respect to the whole exploration field. It may be objected that we have not done this very well because all we have done is to list the various methods in common use today. Such a pile of methods does not necessarily tell how the seismic method fits in.

1.11. COMMENTS TO THE SECOND EDITION

There have been great developments in general geology and geophysics. The sea floor is better known, leading to plate tectonics. There has been much off-shore exploration and discovery of reservoirs. There is a great accumulation of magnetometer data. A bore-hole gravity meter is being used to measure rock densities, undisturbed by drilling muds and hole cavings.

The reflection seismograph still gives the most valuable information on the chances of finding oil in most areas of the world.

CHAPTER 2

Summary of Seismic Methods Commonly Used

This chapter is presented primarily to geologists interested in seismic prospecting as a geological tool. It is not meant to be a technical discussion, although some results are presented diagrammatically. It may be used as an introductory chapter for geophysicists for orientation purposes prior to more detailed study.

2.1. SEISMIC DIP WORK

The field information, in this case, is usually obtained by seismic records shot with the so-called "split-dip" setup. In this setup the shot point is located and drilled. Approximately on a line through this shot point are located a series of geophones, the same number and the same spacing away from the shot point in each direction. The situation is illustrated in Figure 2.1, in which S represents the shot point, on either side of which are four geophones; g_E indicates the most easterly geophone and g_W the most westerly. The shot is fired and the record is taken. Let us suppose that at some point, *R*, there is a reflector which will send back the artificially created seismic pulse. Thus the pulse travels from S to R_E back to g_E. During this same time the pulse also travels from S down to R_W and then to the point g'. Thus it is clear that the reflected pulse comes back to the point g_E before it arrives at the point g_W. Hence, when the dip is like that indicated in the figure the easterly geophone g_E has a smaller reflection time than the westerly geophone g_W.

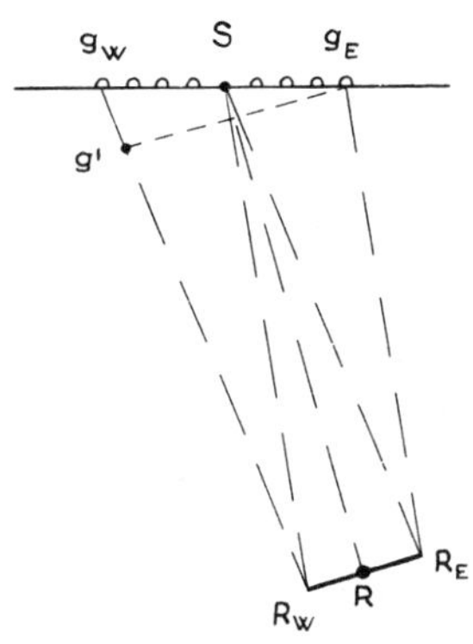

FIG. 2.1. Vertical section showing arrangement of shot and geophones used for dip determination.

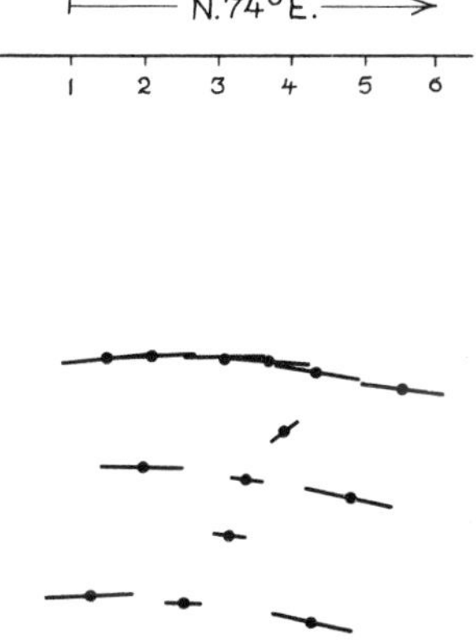

FIG. 2.2. Cross section showing plotted reflections.

The *difference in the reflection times* can be measured. It may be interpreted to give a measure of the amount of dip of the reflector at point R. Evidently the larger the difference in reflection time between g_E and g_W, the larger the dip of the reflector at R. Thus, it is easy to determine the direction of the dip from the way the reflection appears on the seismic record taken with this particular field setup.

If, further, we have a certain amount of information about the *velocities* with which the seismic waves travel, we can make a computation not only of the dip but also of the depth, and so we can compute point R and the little segment through R, namely, $R_W R_E$. Details of the methods of making this computation are given in Chapter 8.

A dip survey can be carried out along a line of several shot points, as illustrated in Figure 2.2, in which shot points are numbered 1, 2, 3, 4, 5, and 6. The little dip segments are plotted in their proper places in the cross section below the shot points, as indicated.

We should note, in passing, that Figure 2.2 shows plotted segments of different lengths. Sometimes the length of the dip segment plotted on the cross section is just half the surface length of the spread. The reason for this may be easily seen from Figure 2.1, in which the length of the segment $R_W R_E$ is approximately one-half the length of the geophone spread $g_W g_E$. This way of plotting has been used partially in Figure 2.2. However, for the very short segments we have deviated from this practice. *Very short segments* represent dips based upon *very poor reflections.* Our object is to give a rough indication of the quality of the data. We see that there is one discordant dip. It is helpful to indicate that this discordant dip is not very well established. This method of grading has a certain appeal, in that it makes the dips on the cross section *easy to see* in case they are *reliable* on the records, whereas they are somewhat more *difficult to see* in case they are *not reliable* on the records.

The cross section in Figure 2.2 illustrates a dip cross section taken across a little anticline.

On the credit side it is plain that an approximate geological interpretation of this cross section is not very difficult. On the debit side, however, it should be recalled that this cross section represents information obtained only along the line of the profile; we have completely ignored, or perhaps failed to take, any data across the line of profile. We should therefore be on our guard not to interpret this cross section as a vertical cross section beneath the line of shot points and geophones. It is approximately a vertical cross section only when all the dips in all directions are fairly small.

We may summarize by stating that the depth and dip may be obtained with the proper field setup from the reflection time and the reflection time variation as we have indicated.

2.2. CONTINUOUS PROFILING

Reference to Figure 2.3 will illustrate the principle involved in *continuous profiling*. Imagine that a shot is first fired at shot point 1 and recorded on the geophones located between shot points 1 and 2. This gives what is called the *subsurface coverage AB*. Then the shot is moved from 1 to shot point 2 and recorded over the same set of geophones located between shot points 1 and 2. This gives the subsurface coverage *BC*. The two cases just described have one path in common, namely, the path from shot point 1 to reflection point *B* to shot point 2 and the reverse of this path. Now it is assumed, very probably correctly, that the direction along such a path makes no difference to travel time along the path. Thus the travel time from shot point 1 to point *B* to shot point 2 is the same as the travel time from shot point 2 to point *B* to shot point 1. This coincidence of travel time gives a means of tying shot point 1 to shot point 2 by what we call a *time tie* on the records shot at these two shot points. In a similar manner shot point 2 may be tied to shot point 3.

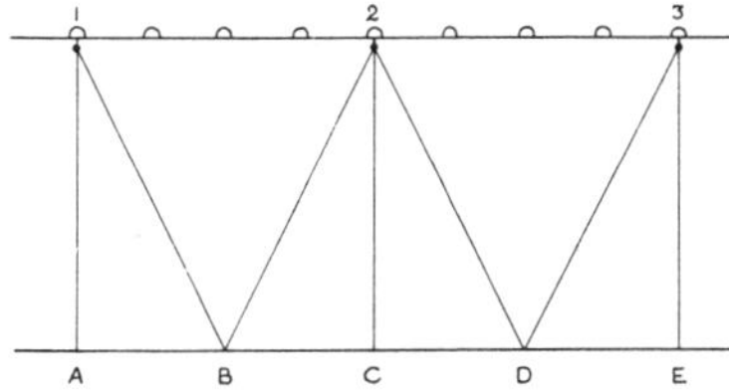

FIG. 2.3. Vertical section showing arrangement of geophones and shots in continuous profiling.

As long as the subsurface geology is sufficiently simple, with low dips and very small changes in dip, the method of continuous profiling yields very good results. Such a cross section is illustrated in Figure 2.4. Part of the profile is shot N. 43° 16′ W. The other part is shot N. 18° W. We show the break in the direction of the profile by making a break in the cross section, as indicated. In this cross section we have plotted depth points directly under the shots. We have assumed that this is a sufficiently good approximation for our purpose

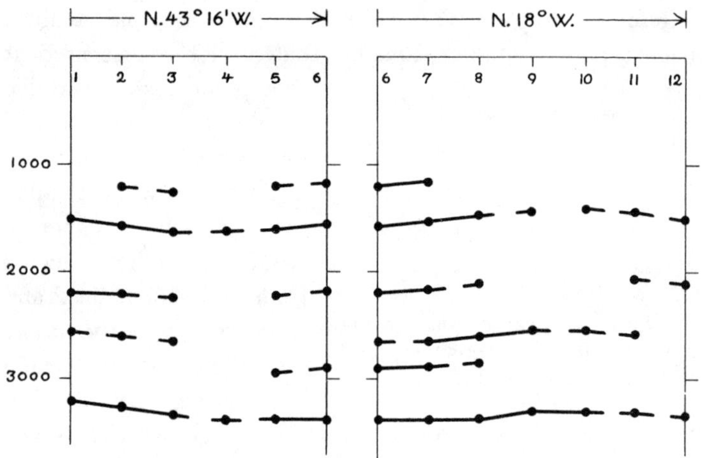

FIG. 2.4. Vertical section showing simplified type of plotting sometimes used for small dips.

because the dips are all small. This practice is very widely used. It will be noted that some of the points are joined by solid lines and others are not joined. The solid line connection is meant to indicate that the *tie* across from shot point to shot point is considered thoroughly reliable. A break is meant to indicate that there appears to be such a tie but it is not considered reliable. When no tie is indicated at all, we need to show that we see no evidence of tie between the two points. This is illustrated in one of the reflections between shot point 9 and shot point 10. We can see from the cross section that the reflector probably does tie across this gap but for some reason the reflection tie cannot be seen from the records themselves.

The method of continuous profiling has had very widespread use in the petroleum industry since about 1936. It proved so useful and so valuable in many of the low-dip areas that attempts have been made to use it in more complicated regions. Like any other exploration tool, however, the method is capable of misuse. In cases where the geology is very complicated and folding fairly sharp, the method breaks down. This breakdown will be described in some detail in Section 8.2.4. and in Chapter 11. It should not be supposed, however, that misuse of a method necessarily condemns it. The method of continuous profiling has been, in many instances, an extremely valuable exploration tool.

2.3. THE REFLECTION CORRELATION METHOD

The reflection correlation method was used before continuous profiling was invented. In the correlation method a number of reflection records were shot in an area and compared. This comparison was made by placing the records side by side and attempting to correlate reflections on one record with those on neighboring records. For correlation purposes some basis is of course necessary.

The bases for correlation were usually what were called *character* and *interval.* Character correlation depended upon the particular shape of the reflection on the record. This is very difficult to measure accurately because of the presence of a high level of background noise. It was therefore necessary for the interpreter to look at the reflection and mentally average out the effects of random noise, thus trying to imagine what the reflection would look like if the random noise were absent. Such a process is exceedingly difficult and requires a rather high degree of skill on the part of the interpreter. In spite of this, a great deal of very useful correlation exploration work was done in the early days of seismic reflection prospecting.

It so happened that some of the earlier areas to be explored had reflections which could be correlated. One example is the Viola lime in Oklahoma; another is the Yates sand in west Texas. In many areas where these two reflectors occurred, the reflections appeared to have a peculiar shape, quite different from the shape of other reflections. This was probably due, in the case of the Viola, to the presence of another good reflector a few tens of feet above the Viola. In the case of the Yates, this may have been due to the fact that the Yates was not a very thick section, so that reflections from top and bottom occurred. In either case, the reflection appeared as a *composite reflection,* in contrast with some of the more *simple reflections.* In all cases, however, except the most favorable, correlation by character is a very difficult interpretive task. It is nearly always subject to a great deal of uncertainty, especially when the reflections are a little weak. Hence other bases have to be used.

A second base for correlation is the series of *interval relationships* that occur between successive reflections on adjacent records. Often a sequence of reflections may be seen on a record with a sequence of corresponding intervals between them. If this sequence of intervals can be compared with a similar sequence on adjacent records, a correlation can be made. The process is much like the procedure used in correlating electrical logs, except that in the case of reflection records we often have to work under somewhat unfavorable circumstances because of the high level of background noise.

When both interval and character correlation can be used, we have two checks on our correlation, and thus the degree of reliability of our work increases.

In some cases a third base can be used, namely, the *strike* of the reflection.

For the most part, where the Viola reflection is a good correlation reflection, it is also a very strong reflection, standing out well above its neighbors. This *energy* of the reflection itself can be used as a reliable correlation crutch, provided that the character of the reflector does not change from one location to another.

A very important influence upon the reflection character comes from the curvature of the reflector. If a reflector is concave upward, the reflected energy tends to be a little more *concentrated*; hence the reflection may be expected to be a little stronger than it would if the reflector were a plane. On the other hand, if a reflector is convex upward, the energy is *spread out* and this makes the reflection smaller. Thus, as we pass over an anticline, reflections at the crest generally are very poor. Therefore correlation based upon the energy of the reflection is always subject to considerable reservation.

Cases exist in which the correlation interpretation has proved to be more reliable than the continuous profiling interpretation. In such cases we generally have to deal with fairly sudden structural changes. These usually occur between areas where the changes are much more gradual. Such a case arises when there is a certain amount of faulting. For example, there might be just the amount of faulting that would shift the reflections by one whole cycle. In this situation we might run a continuous profile across the fault without seeing it at all on the profile. Such an ever-present possibility emphasizes the need for correlation wherever it is possible.

It sometimes happens that correlation gives a wrong result and continuous profiling gives a correct result, no matter how detailed the shooting may be. For example, if a reflection is coming from the top of a limestone section which contains numerous stringers, over only one part of the area an upper stringer may become thick enough to indicate the beginning of the limestone reflection, whereas over another part of the area this upper stringer may thin out or entirely disappear. A correlation from the place where the limestone upper stringer is strong to the place where it is weak or absent gives a stratigraphically misleading result. It is sometimes possible in this situation to make a continuous profiling survey which will give approximately the correct result.

The correlation procedure itself is of such a nature that it can hardly be adequately described in a book. It is necessary for the beginning correlator to watch someone who has had experience in this particular art. During this time he will have to ask a lot of questions about why such and such is regarded as a good correlation. Where the correlation of one reflection record with another is very easy, little needs to be said. Almost anyone can understand such a correlation. On the other hand, this is a rare occurrence. The usual thing is for the correlation to be so difficult as to be almost impossible. It is for this reason that correlation procedure can hardly be described in words. A great deal of experience is required, experience that involves correlation which is easy at first and becomes more and more difficult.

The *threshold of impossibility* is reached by different interpreters at different levels. The important thing is for each interpreter to understand his limitations. Obviously it is foolish to go ahead and correlate when no correlation is possible. This involves giving a definite interpretation that is almost sure to be misleading and therefore very expensive.

A little of what may be regarded as the scientific background of correlation work will be given in Chapter 11.

2.4. REFLECTION STRIKE-DIP SURVEYS

In a region where the geological structure tends to be a little complicated, the *strike* of the reflector, as well as the *dip*, becomes important. The strike may be determined by a proper setup of seismic detectors.

Figure 2.5 shows a plan view in which S is the shot point. Geophone lines are laid out perpendicular to one another, each passing through S. The line *AB* and the line *A′B′* are two such mutually perpendicular geophone lines. Let us now suppose that a shot is fired at S, that the pulse travels downward, that it is reflected by the interface, and that it travels upward to the surface. This reflected pulse will have a small amount of

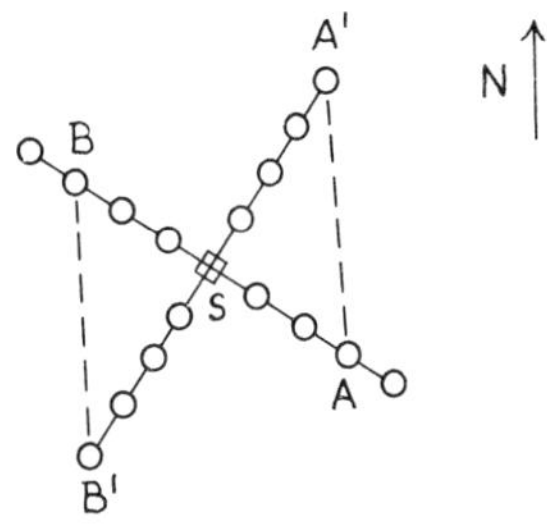

FIG. 2.5. Map view of geophone and shot arrangement when both strike and resolved dip are to be determined.

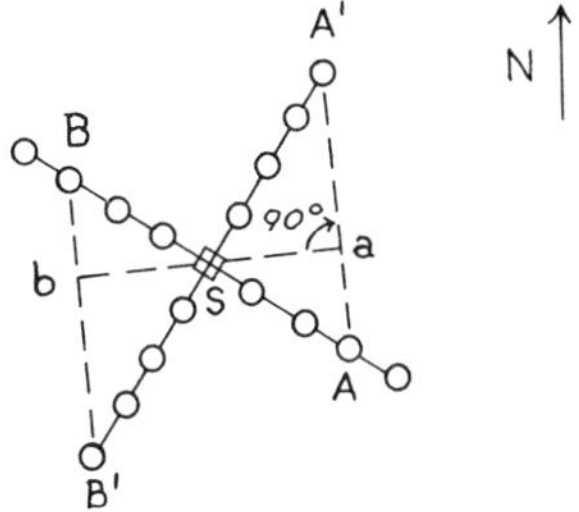

FIG. 2.6. Map view of cross spread showing strike direction and dip direction.

curvature, but for purposes of this explanation let us suppose that the reflected pulse is plane. The pulse will arrive at the surface, say at point *A* on the figure, and also at point *A′* at the same time. A dashed line is drawn between A and A′. This dashed line has the property, that, if detectors were placed on it, the approximately plane reflected pulse would arrive at all these detectors at approximately the same time. Suppose that at a somewhat later time the same reflected pulse arrives at point *B*. Then, at the same time as it arrives at *B*, it will arrive at *B′*, and at all the intermediate points on the dashed line joining *B* to *B′*. Now, under the simplest circumstances, which are those we use mostly in interpretive work, the velocity will depend only upon the depth; that is, it will not vary laterally to a significant extent. If this is the case, the line AA', which is parallel to the line BB', will also be *parallel to the strike of the reflector.* No proof is given for this, it being supposed that the reader will imagine the three-dimensional situation with sufficient clarity so that he will see directly that AA' is indeed parallel to the strike of the reflector.

Having determined the strike of the reflector, as described in the foregoing paragraph, we are now ready to consider the determination of the dip of the reflector. Refer to Figure 2.6, which is like Figure 2.5 except that we have added the perpendicular to the line AA' that passes through the shot point. This is line ab. Suppose now that we place a pair of geophones at a and b and reshoot the shot at S. We can then take a record of the reflected pulse, measuring the time at which it arrives at a and also the time at which it arrives at b. According to our approximate assumption, the time must be the same at a as it is at A and A'. Also the time the reflection arrives at b must be the same as the time at B and B'. Therefore, we did not really need to take this latter shot. We could have obtained the same result by computation, without actually making the shot, and this is what is done in practice. We measure the time difference between the arrival of the reflection at point A and its arrival at point B. We use this time difference in conjunction with the distance from a to b for the dip computation. In this way we obtain the dip measured perpendicular to the strike, that is, the maximum or total or resolved dip.

We have described the principles involved in making strike-dip determinations by the seismic method. The actual procedure will be described in some detail in Section 9.3. The important thing here is to realize that determining the strike and the dip by the seismic method is very simple, in principle.

Once we have determined the strike and dip, we may plot a

strike-dip symbol, as illustrated in Figure 2.7. This shows the symbol pointing directly toward the shot point. The offset has to be computed and the dip has to be written beside the symbol.

It is customary to conduct strike-dip surveys whenever the geological structure is sufficiently complicated. In such cases it is usually difficult to correlate reflections from one strike-dip shot point to another strike-dip shot point. But where such a correlation is possible, very valuable information is obtained. Generally, when a set of strike-dip determinations has been made, a strike-dip map is constructed. Unfortunately, such a map has a tendency to collapse the data from a large number of horizons on a single surface. This often leads to a great deal of confusion. An alternative is to project the strike-dip information on vertical cross sections. This is described in Section 10.1.3.

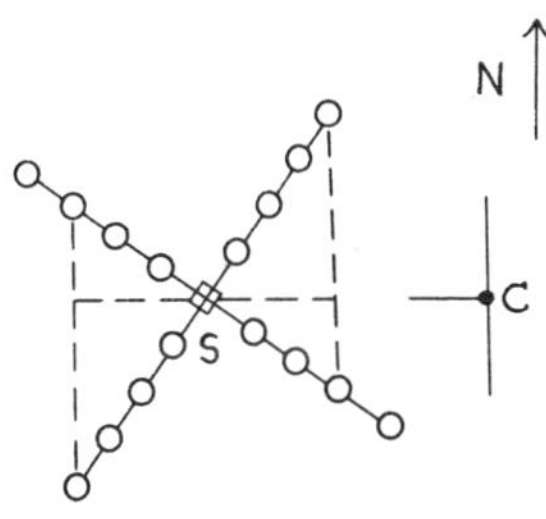

FIG. 2.7. Map view of cross spread together with resulting strike-dip symbol.

When it is difficult to move equipment from one location to another and a reconnaissance reflection seismic survey is required, a strike-dip survey is recommended. The amount of additional work involved in measuring the strike is usually small, compared to the work of getting the equipment to the location. Furthermore, the additional information given by the strike for the various reflections is of considerable value in filling in between shot locations.

2.5. REFRACTION SURVEYS

Refraction surveys may be divided into two main classifications. One is the *fan* shot, the other is the *line* shot.

Fan shooting was used in the early days of geophysical prospecting in the Gulf Coast for salt domes, especially shallow domes.

We may illustrate the technique by Figure 2.8, which shows two fan shot points S_1 and S_2. The energy from S_1 was recorded at points *a, b, c,* and the other points located along that straight line. The pulse from shot point S_2 was recorded on the other line at point *d*, etc. For this particular area, it was possible to calculate the *normal travel time* from the shot to the various distances where the detectors were located. This normal travel time was compared with the *actual* travel time. When part of the energy went through a large salt dome, the actual travel time was considerably shorter than the normal time. This results from the very high velocity with which seismic pulses travel through salt. The difference between normal and actual times was plotted on the ray between the shot and the detector location, as indicated. The difference between the normal time indicated by the circular arc and the reduced time indicated by the polygonal line is shaded. The figure shows two shaded regions from the two shot points pointing toward a roughly outlined dome.

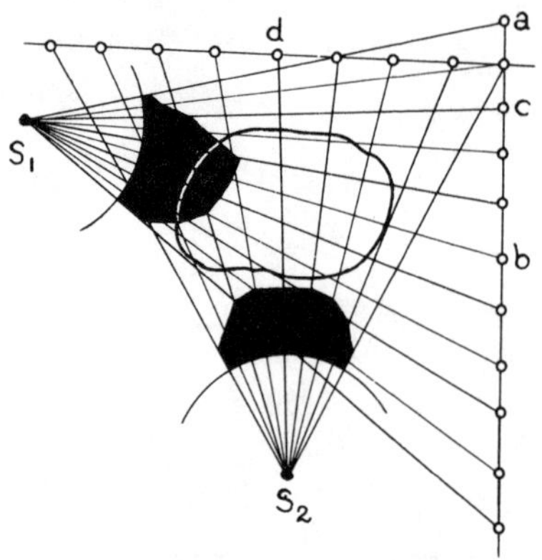

FIG. 2.8. Map view of refraction fan shots showing detection of outlined salt dome.

Fan shooting was very successful in finding shallow salt domes. However, like any other instrument of exploration, it had its limitations. It was not too successful in locating very deep domes or structures of small relief. Sometimes anomalous results were obtained, especially when a shot was located on top of a dome. Nowadays we understand this anomaly and do not give it this name. We merely recognize that we have placed one of our shots approximately on top of the dome. In connection with fan refraction shooting, there are numerous articles in the *Geophysical Case Histories* volume.

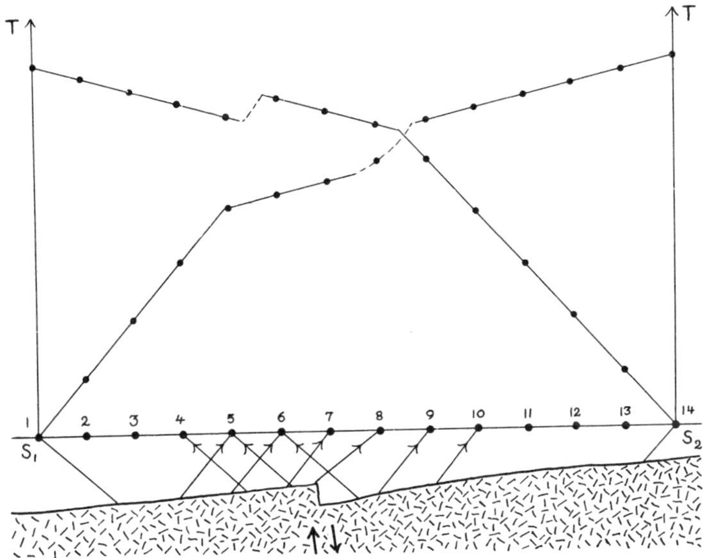

FIG. 2.9. Vertical cross section showing buried fault and time-distance plots of corresponding refraction data.

We now pass to a consideration of *line* shooting. Suppose, for example, that geophones are laid out as indicated in Figure 2.9. Shot points are S_1 and S_2. Geophones are 1, 2, 3, 4, 5, 6, 7, 8, 9, 10, 11, 12, 13, and 14, placed on the line between the two shots. The shot is fired at shot point S_1. Part of the energy travels in the shallower, lower-velocity material, arriving at geophones 1, 2, 3, and 4 by this path. By the time we get out to geophone 5, however, we find that the travel time by the direct path from S_1 to this geophone is a very little bit longer than the travel time from shot point S_1 down to the high-velocity material, along that material, and then up to geophone 5, as indicated. This path, which we call the *refraction* path, takes advantage of the very high velocity of the material and carries energy that arrives earlier at geophone locations 6, 7, 8, 9, 10, 11, 12, 13, and 14 than it would have if it had traveled along the upper, more direct, and shorter path. The time-distance data are plotted in the figure, the ordinate representing the travel time T. Travel through the shallower, lower-velocity material is indicated by the straight line corresponding to geophone locations 1, 2, 3, and 4. Travel in the higher-velocity material is indicated by the approximately straight graph corresponding to the travel time to geophones 5, 6, and 7.

Let us now consider, by reference to Figure 2.9, how we would determine the depths, at least qualitatively. It is fairly evident that if the high-velocity interface were deeper than it is in the figure, the times corresponding to travel along this interface would be longer. In fact, if we moved the high-velocity interface deeper down, geophone 5 would probably receive its first impulse by means of the direct path, rather than the path taking advantage of the high-velocity material. On the other hand, if the high-velocity interface were moved up, say almost to the surface in the present case, geophones 4, 3, and possibly 2 might take advantage of the travel path which includes the high-velocity material. It should be observed, however, that if we move the interface up or down without changing the dip we merely move the graph corresponding to the high-velocity part down or up. Details of the actual computation will be discussed in Chapter 12.

Figure 2.9 illustrates a very important problem, namely, the problem of finding a fault. The effect of a fault is shown in the time-distance graph. As we shoot off the edge of a fault, we increase the travel time, as indicated. This increase in travel time occurs between geophones 7 and 9. There is a small increase in travel time between geophones 7 and 8, but the main increase occurs between 8 and 9. Now, when we shoot in the other direction from shot point S_2, we find that there is a more or less sudden decrease in travel time. This decrease occurs between geophones 6 and 5. Thus the effect of the fault shows at different geophone locations when shot in different directions. This represents a great advantage of the refraction procedure over the reflection procedure. In refraction shooting we *undershoot* a fault. That is to say, the occurrence of the fault is shown at the surface at different places. In contrast, when we detect a fault by reflection shooting, the fault must be detected more or less above the fault location itself. This is often a difficult procedure, because a fault is likely to extend through the section, thus making the section somewhat complicated relative to the propagation of seismic pulses, and entirely obscuring the reflections in its neighborhood. Thus, as we can see from the figure, the larger the fault the larger the change in time on the time-distance graph corresponding to the refraction line.

As we have tried to indicate in the figure, the location of the fault is usually more accurate on the part of the refraction line shot *up* the fault.

The depth and dip aspects of refraction line work are illustrated in Figures 2.10 and 2.11. Figure 2.10 shows the reversed refraction line, in one case with a relatively shallow high-velocity interface in which the travel time, say from shot point 1 to shot point 2, is indicated by the point R_1. Since the travel time in reverse is the same, R_2 is the same ordinate as R_1. Now suppose that the high-velocity interface is made deeper (to the dashed horizon). Then R_1 will move up to R'_1 and R_2 will move up to R'_2. The new time-distance graph is part of the old graph, plus the dashed new parts. This indicates, again approximately, the way timing is used to compute depth. Details of this computation will be given later.

Figure 2.11 shows the way the dip is indicated by refraction shooting. When we shoot *up* the dip the graph is as indicated from S_1 to B_1 to R_1. When we shoot *down* the dip the graph

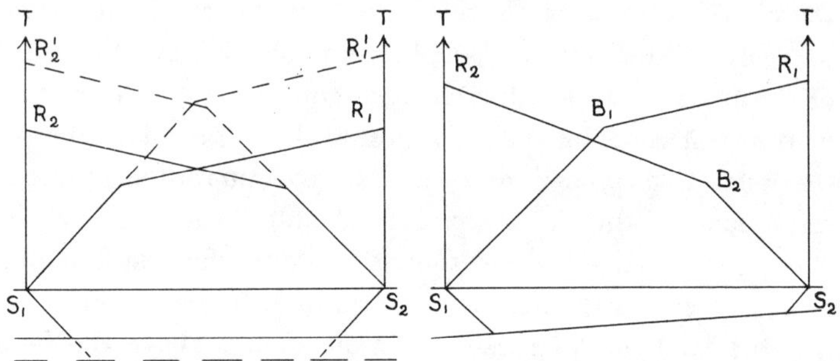

FIG. 2.10. Influence of depth upon refraction time-distance plot.

FIG. 2.11. Influence of dip upon refraction time-distance plot.

is as indicated from S_2 to B_2 to R_2. The slope of the line B_1R_1 is smaller than the slope of B_2R_2. In other words, when we shoot *down* dip, the time increases more rapidly than when we shoot *up* dip. In fact, the dip might have been steep enough so that when we shot up dip the time at B_1 would be the same as the time at R_1.

Shallow salt domes can be found by line refraction shooting as well as by fan shooting. By making a series of determinations of dips and depths, a structural anticline can be found by the refraction method. However, the principal usefulness of the refraction method, especially line shots, is in the determination of faults by the process illustrated in Figure 2.9.

2.6. RECONNAISSANCE SURVEYS

Reconnaissance seismic surveys can be carried out so as to be very valuable. What we mean by a reconnaissance seismic survey usually is the same thing as a detailed survey, with much more thinly scattered observations. Thinly scattered observations will detect a very large structure just as well as closely spaced observations. But thinly scattered observations may very well miss important small structures. When we make a reconnaissance survey, we are taking a gambler's chance on not losing a valuable structure because of the coarseness of our sieve. Also, it should be remembered that a reconnaissance survey can be designed so it can be followed by a more detailed method. Thus, any indications or leads obtained by a reconnaissance survey should be followed either by a more detailed seismic survey or by further investigation of some sort. If we do not proceed with this "follow-up," it must be assumed that we were interested only in very large structures such as would surely be found by the reconnaissance survey itself.

2.7. DETAILED SURVEYS

In contrast with the reconnaissance survey, the detailed survey is one in which the data are very close together. In this case we are trying to verify the existence of a particular structure that we have already found. We are trying to get more details regarding its shape. It is valuable to obtain such details by the seismic method, so that when a well is drilled the information obtained from it will have more exploration value. It quite often happens that the first well does not open up the field. However, the first well, in conjunction with detailed seismic information, may point to a more favorable location which will indeed open up the field. If, on the other hand, the structural information obtained is very fragmentary, the information given by the well may not be very valuable.

There is yet another aspect of this situation that is well known to prospectors, and that is the relationship between detailed seismic prospecting and *scouting*. For example, if a reconnaissance survey has been carried out over a broad area and the crew makes a detailed survey of a particular small portion of this area, this will indicate to the scouts of other companies that something very interesting has been found in this particular section. Hence, in order to protect the original company, it is necessary that the seismic crew avoid showing its hand by such a procedure. Before it makes a very detailed survey of this kind, the oil company should take time to take up leases on the land in the area or at least get ready to take them up.

In the foregoing sections we have described the main seismic procedures used at present. Since seismic operations cost so much, a good deal of attention has to be given to their efficiency. This is the subject of the next chapter.

2.8. COMMENTS TO THE SECOND EDITION

There are many differences in reflection seismology techniques that have been developed since 1950. First is the great progress in digital computer technology. Second is the digital recording of seismic records which gives data that can be fed to the computer. Third is the method of gathering and stacking field data invented by W. Harry Mayne in 1962. Fourth is the development of sources such as Vibroseis,[1] Dinoseis,[2] air guns, and so on. For the immediate future, perhaps a fifth development might be mentioned, that is, the one-bit recording that permits very good recording from 1,024 independent seismic channels on land work without using over ten 96-channel recording trucks, cables, and personnel. I shall outline the third development, as it is the main noninstrumental development and is very much used.

[1] Trade name of Continental Oil Company.

[2] Trade name of Sinclair Oil Company.

To describe briefly the "roll along" procedure and "common depth point" (CDP) stacking, we refer to Figure 2.12, which

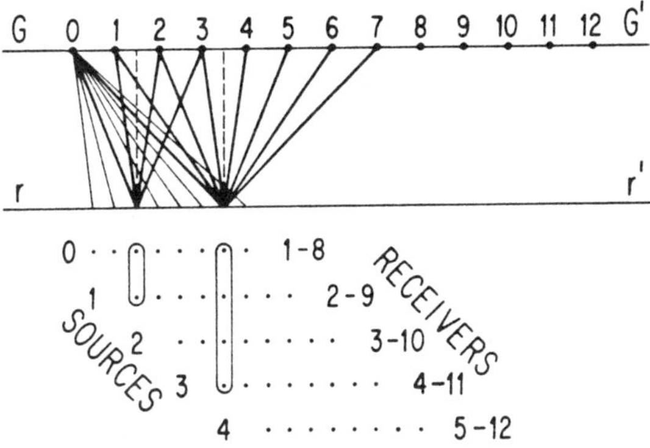

FIG. 2.12. "Roll Along" on Ground, G to G′ showing 2-fold and 4-fold Gathers.

shows a straight line *GG′* of equally spaced seismic detectors, numbered for reference, over a level plane reflector, *rr′*. We suppose for each shot there are eight receivers. The first shot, 0, is received at stations 1, 2, . . . , 8. Then the second shot, 1, is received at stations 2, 3, . . . , 9, and so on. Thus we "roll along."

We have indicated the "reflection points" for shots 0, 1, 2, . . . as dots below the reflector, eight for each shot, exactly below the reflection point in each instance.

We have "gathered" 0 to 3 and 1 to 2 depth points. This is a two-fold gather. We also show "gathered" 0 to 7, 1 to 6, 2 to 5, and 3 to 4. This is a four-fold gather. Right next to this we have another four-fold gather, namely, 0 to 8, 1 to 7, 2 to 6, and 3 to 5.

Look again at the two-fold gather, 0 to 3 and 1 to 2. The reflection path 0 to 3 is *longer* than that from 1 to 2. We can subtract the extra time in 0 to 3 and reduce it to the time

1 to 2. Actually, we make a "normal-move-out" (NMO) correction to station 1.5 for both 0 to 3 and 1 to 2 (see Figure 8.4 right side). When we have made these corrections, we can add (stack) the two records. The reflected signal adds coherently (is doubled). The noise adds in random phases (shows an average increase by a $\sqrt{2}$ factor). The two-fold stack may increase the signal-to-noise ratio by a factor $\sqrt{2} = 1.4$. When we go to our four-fold gathers, apply the NMO, and stack, the reflection signal adds coherently to four times the individual amplitude, and the random noise background adds to $\sqrt{4} = 2$ times the average individual noise background. Thus this signal-to-noise ratio is, on the average, up to twice as good (100 percent better) for the four-fold stack, than for the individual unstacked channels. If we want to double again, we have to go to sixteen-fold (and if another doubling is sought, we must go to 256-fold, which is easily possible with 512 independent recording channels).

We must keep in mind the question of the perfection of the coherence of the NMO corrected signal. The improvements in the signal-to-noise ratios just given are upper bounds to ideal averages, supposing exact NMO corrections known, exact source repeatability.

Returning to our 8-channel illustration (Figure 2.12), observe that the two four-fold gathers have "depth points" only half a detector spacing apart. If 1024 channels are used instead of 8, it may be advantageous to have close detector spacing, such as 10 or 15 meters, corresponding to 5 or 7.5 meters between "depth points." For such close spacing, the two four-fold gathers might be added to give eight-fold stacking. The greatest improvement obtainable would be about 40 percent for the signal-to-noise ratio, if dips are low and detector separations are small. This is discussed again at the end of Chapter 8.

CHAPTER 3

Routine Seismic Crew Operation. Cost Reduction

This chapter is written for people who are interested in, and somewhat familiar with, the practical operation of a seismic crew.

3.1. EQUIPMENT AND PERSONNEL. OVERHEAD

A seismic crew, for minimum operations, must have a rather large quantity of equipment. The office must have the usual furniture, tables, chairs, filing equipment, and machines required for drafting and computing. Field work requires an instrument truck and seismic instruments, including amplifiers, recording camera, cables, cable reels, geophones, and communication equipment. Also necessary is a shooting truck with its loading poles and, for deep holes, loading winch, cable, and spoon. The shooting truck must be capable of carrying a certain amount of dynamite with caps and also a certain amount of water to provide for tamping the charge. This truck also carries the necessary blaster and communication equipment. In almost all cases the shooting truck carries an up-hole geophone together with the cable required for attaching it to the recording truck. The drilling equipment, besides the truck-mounted drill, must include one or more water trucks to provide the drilling fluid. The surveyor needs a truck to carry his equipment and take him from place to place in the field. A separate car is often required for the "permit man." A car should be provided for the party chief.

The above list calls for many thousands of dollars' worth of equipment which must be maintained in good order, an expensive operation.

In addition to this large amount of equipment, which must be maintained and replaced when necessary, there is a fair-sized crew consisting of the party chief, a chief computer and his assistants, a surveyor and his rodman, a driller and his helper, and a driver for the water truck. Also very essential are the instrument operator, sometimes called the observer, and his assistants. He may have only one assistant if there are a small number of geophones and a small cable problem. However, if he has a large number of geophones and hence a major problem with their layout, he may need as many as three assistants for efficient operation. In some cases a "permit man" is required to secure permission to trespass on private property.

This rather large collection of personnel and equipment involves a rather heavy *overhead* cost, that is, the cost of the crew whether it operates or not. Of course, every effort is made to keep the overhead to a minimum, but this minimum has to be consistent with maximum operating efficiency; after all the crew is not set up not to operate.

3.2. GEARED AND UNGEARED OPERATIONS

In this section we wish to make a distinction between two kinds of operations. *Geared operations* include all the field procedure, the office routine, and the maintenance of the field equipment. This part of the operation has to be carried out on a teamwork basis; otherwise the efficiency will be very poor indeed. We refer to it as a geared operation for several reasons. For example, the rate at which the observer can operate may depend on the rate at which the drilling equipment can supply holes for the shooter to load. This rate, in turn, may be affected by the quality of the holes; if the holes are of very poor quality and difficult to load, the shooter may be the determining factor. If, on the other hand, the holes are relatively easy to load and to drill, but records are very difficult to obtain, a fair amount of experimental work may have to be done at each hole by the instrument operator, using different filter settings, charge sizes, and shot depths. When the equipment is in very poor condition, the people in charge of keeping it in shape may be the bottleneck of the operation. The main fact we wish to emphasize by this classification into geared and ungeared operations is that there is an extensive semiroutine operation which consists primarily of producing records ready for interpretation. This semiroutine operation is a team job. It depends upon the work of all the team members.

On the other hand, there is an *ungeared* operation—the interpretive work on the records. It is a type of operation which often cannot be geared to the crew's other operations. Usually an effort is made to have the interpretive work kept up with the rest of the operations. However, when this work is very difficult, this often cannot be done. Usually, in such a case, the interpretive work is only partially done. This results, we suspect, partly from misunderstanding the importance and value of this work.

3.3. FUNCTIONAL ANALYSIS OF GEARED OPERATIONS

We now analyze the geared operations of the crew. Instead of starting at the beginning, we start with the end product. We choose this order because the end product is our primary interest. Naturally, a stoppage at any point along the way, from the beginning to the end, will affect the production of the end product. Since one person's failure to produce may stop production, occasionally one of the men concludes that his particular operation is the most important one. This is indeed a mistake. The most important thing is for the *crew* to produce records of good quality, ready for the interpreter.

Proceeding with our analysis, we examine a day's records. These consist of several groups of records. Each group is shot from the same shot point and with the same spread of geophones. The records are arranged in order in stacks so that the best record is on the top of each stack. The time lines are marked and counted. The datum corrections are made, for both depth and dip computations. The face of the record shows the amount of charge used, the hole depth, the filter settings used, the shot point number, and a minimum of surveying data showing the location of the shot and of the geophones. The directions of the geophones from the shot point must be shown. These records have previously been dried and rolled backward to prevent or reduce curling. It is assumed that they have been thoroughly fixed and washed and that the instrument operator has taken the proper number of them to obtain usable information without excessive cost. It is also assumed that he has conducted his part of the operation—that is, operating his truck and equipment and managing the shooting—as efficiently as possible. In order to do this, the instrument operator needs the coöperation of his own equipment people in the laboratory and shop, as well as the full coöperation of the surveyor. For example, if the operator has to keep repairing his equipment out in the field, he will be continually slowed down, and this slows down the whole operation.

Another operation that may be of importance from an efficiency standpoint is the simple one of turning the truck around. Suppose the operator is making his survey on a narrow country road with deep drainage ditches on each side, and his truck is arranged so he must turn it around for each geophone set up. Each time he will have to run it to a driveway where turning around is possible. This, of course, takes considerable extra time. It would be much better if he could just back up such a road, instead of having to turn his truck.

Another matter of coöperation involves the directions given by the surveyor. If the surveyor, each time, plans the path to be used by the shooter and the instrument operator so it will get them into the location by the shortest and easiest route, he will save both these men a lot of time and energy.

Often it is possible for the drillers to load the first charge in the bottom of the shot hole immediately after they have finished drilling it. This means that when the shooter comes to the shot point he is practically ready to shoot the first shot as soon as the observer is ready. This has other advantages in many cases. For example, the first shot may be used to clean out the hole, so that subsequent charges can be loaded more easily. Of course, the surveyor has the same obligation to the driller that he has to the observer and the shooter. If it is necessary for the driller to get heavier equipment into his shot location, it is even more important that he know the best way to drive in. Usually the surveyor can give him this information. But since the driller usually follows immediately after the surveyor, the coöperation between them is more direct and easier to maintain than that between the surveyor and the instrument operator, who comes in some time later.

Of course, the surveyor cannot carry on his work efficiently unless he is given a program by his party chief. The party chief cannot give him this unless he himself receives his instructions well in advance from the people in the oil company who are in charge of directing field activities.

It is thus clear that the responsibility for the efficient operation of a well-knit crew is distributed all along the line.

Each man may think himself the most responsible man in the crew if he regards his job as the most important job. This is not quite the way that one would look at the situation if one saw it from an overall viewpoint. After all, what is sought is a set of records ready for the interpreter. The object is to get as many seismic profiles per dollar as possible, consistent with usable and valuable information.

3.4. MAXIMUM RECORDING EFFICIENCY

We have already referred briefly to the work of the recording operator and the shooter. The operator's work is usually somewhat more complicated in nature than the shooter's work. So it is usually considered desirable to gear the work of this group to that of the instrument operator. Maximum efficiency in this operation is a matter of great importance because it is likely to determine the efficiency of the crew's entire geared operation.

For example, if we use round numbers, we may take the cost of the crew as $10,000 per month of twenty days. This corresponds to $500 a day. Suppose that, under these circumstances, it is possible to average about ten seismic profiles per day in a particular region. This corresponds to $50 per profile. Now suppose that, by adding one extra helper, at say $200 a month, we increase the average number of profiles per day to eleven. A short calculation will show that we have reduced the cost of the operation from $50 per profile to $46.36 per profile, a saving of $3.64 per profile or $36.40 per ten profiles. Thus the man's wages of $10 induce a net saving of $26.40 for the day's operations.

Another very important consideration in the instrument operator's work, which we have already referred to, is the *quality* of the instruments themselves. Only the best parts should be used, because instrument breakdowns are extremely costly. Nowadays, it is customary to use plug-in amplifiers so that if anything goes wrong with one of them it can be pulled out and replaced by a spare. The damaged one can be repaired either in the field office or in the headquarters shop of the company. The main thing is to have equipment that doesn't go bad, because any loss of time due to bad equipment is exceedingly expensive. This, of course, applies to truck equipment as well as electronic equipment.

Another means of increasing the operator's efficiency is to use so-called *double-ended cables*. In double-ended cables the conductors go all the way from one end of the cable to the other. Thus the instruments in the truck can be plugged into the cables at either end, leaving the geophones in place. This requires that more weight be carried in the cable, but usually this added weight greatly increases the efficiency of the operation. Another way of increasing operating efficiency, in some cases, is to use a separate reel truck so that cable can be laid in advance. However, an extra set of cables and geophones is required to secure the greatest efficiency of operation.

It is the observer's responsibility to give the shooter as much time as possible to get to the new location and begin the hole-loading procedure. This is especially true where the holes cannot be preloaded. In some cases it is desirable to drill several shot holes, preload each one of them, and not plan to reshoot any of them.

Although the writer's experience is somewhat limited in this field, the experience he does have indicates that a reasonable degree of comfort should be provided for the observer so he can operate at greatest efficiency. This is especially true in very cold and very hot weather. In hot weather the instruments give out a good deal of heat themselves. Thus, unless special provision is made, the interior of the instrument cab is likely to be much hotter than the outside air. One means of taking care of this problem is to install a cooling mechanism in the ceiling of the cab.

3.5. MISCELLANEOUS FIELD EFFICIENCY CONSIDERATIONS

We have already considered the simple problem of adding a helper to the recording crew. There are other considerations of this sort, large and small. One of the large ones involves the use of multiple drills. In some cases, where drilling is very difficult, multiple drills increase the efficiency of the crew operation. Another consideration involves preloading shot holes; this was mentioned above. Obviously, if this can be done, it will contribute to increased efficiency. Another idea which sometimes helps, especially in cases when the water truck has to make long hauls for drilling water, is to carry a water trailer along with the drill. When this is done, the water trailer serves as a storage unit for the drill, so the water truck can unload its water into the water trailer to be used as needed by the driller. The water truck is thus free to travel more of the time, since it does not have to stand by to supply water for the driller.

Another aid in some cases is a portable mud pit which is carried with the drill. This allows the driller to set up for his drilling operation without having to dig a pit on the location. This is especially valuable when digging such a pit is very difficult.

The author recalls one case in the tropics where a tractor-mounted drill made the crew operation possible. Without this, the crew could not have operated.

Sometimes it is desirable to use a bulldozer to make roads. Such an operation is very expensive, but the alternative of not using a bulldozer may be even more expensive in terms of dollars per profile.

Equipment maintenance in general is a serious problem. There is such a thing as maintaining equipment too well. The decision as to how well to maintain it requires a good deal of judgment. If no repairs are made until the equipment breaks down, there will be a lot of breakdowns and a very inefficient operation. The author recalls one company that maintained a very efficient inspection service. Each piece of equipment on a big job was given a systematic thorough inspection at least once every two weeks. Any weaknesses that were uncovered were taken care of at this time.

Of course, there is no substitute for careful operation of equipment by the personnel. If the personnel is so inclined, it can always abuse the equipment badly. When the manpower situation is very difficult, as it was during the war years, this creates a serious problem, because some of the people simply are not capable of taking care of equipment. For efficient operation, there is no substitute for capable men who coöperate with one another in an effort to do good work and care for the equipment.

3.6. OFFICE ROUTINE

There is a large amount of routine work to be done in an office where seismic records are prepared for interpretation. We may suppose that the records have been dried and rolled backward to minimize curling, and are ready for the office staff at the beginning of the morning's work. The records are then distributed for counting and labeling. When they are counted and labeled, they are arranged according to profiles, with the so called "specimen" or "best record" placed on top of each profile stack. The various profiles are then distributed among the office staff for computing the datum corrections. This computation is usually done most efficiently on the record label itself.

The direct information, such as charge size, hole depth, filter setting, and so on, should be transferred from the data sheets to the record label. The computed quantities are inserted in the proper places on the record label. When all the datum corrections are made, the records are ready for the interpreter to make a *first interpretation*. This generally consists of picking all the good reflections, those that are of good quality and easy to pick. It is best not to try to pick all the reflections at this stage, for this may hold up the office's routine operations.

When the reflections of good quality have been picked and marked on the records, these records are turned over to the computing staff so that the necessary depth and dip computations can be made for purposes of plotting. The plotting operation is then carried out. The result of the plotting is used by the interpreter as a background for his further study of the records.

One of the men in the office should be an *office manager*. He will fit into the day's schedule the work of office bookkeeping, and keeping track of the time records, expendable supply records, maintenance cost records, and the multitude of other items which modern seismic crew operations entail.

3.7. DANGER OF LOSING SIGHT OF THE PURPOSE OF THE GEARED WORK

The routine geared work of a seismic crew is very difficult to do. So far as the volume of work is concerned, it is the major work of the crew. It is the work that involves the most serious breakdowns, if any breakdowns occur. It is likely to occupy the major portion of the party chief's attention. This is probably as it should be.

However, one should never lose sight of the fact that the whole purpose of the geared work is to obtain records for the purpose of interpretation. Therefore the importance of the interpretive part of the work should never be forgotten. The information furnished by the interpretive part is, after all, the exploration information that is sought in the first place. Without this interpretive part, there would be no point in getting records.

The prime consideration of the remainder of this book will be the problem of interpretation. However, before we proceed to the discussion of interpretive procedures and problems, we will give a short description of planning exploration programs in general.

3.8. *COMMENTS TO THE SECOND EDITION*

Section 3.1 would be very different after thirty years. The degree of technical complication has grown, forcing greatly increased specialization. There have been many little details to keep in mind through the years, and this fact has made routine a prime necessity for the operation. The routine should be developed in an evolutionary way, thoughtfully and slowly. Large sudden management changes are hard for a team to bear.

Today's instruments are quite wonderful compared to thirty years ago. They are rugged and stable, and they store data in very reliable reproducible fashion, accessible for all kinds of treatment by computers. Operations are bigger and faster.

Specialization and organizational routine tend to narrow individuals. Seismic recording and source equipment are so good that they can almost run alone. The routines are mainly in the computer programs. The programers give little tweaks to programs to reduce cost. The field people gathering data seldom do more than just gather it fast. They always have gathered it fast (as fast as possible). It is difficult for a fast gatherer to be always alert to questions of data quality, especially when, as now, not much can be done to really improve the quality without severe risks for the efficiency. Since the field people tend not to see the results and since they work from routine orders that are not always optimum for the actual field conditions, losses can still occur, especially if the data processing is sent to a distant place and perhaps takes its turn. This latter consideration may be avoided in some cases with modern equipment, for example, equipment containing its own processing in the recording truck. In this case, there is a real challenge to interest a competent geophysicist-geologist in the local field problem (unless the problem is too simple). Clearly, the problem of choosing personnel for the field work depends on the equipment.

By and large, this chapter is still applicable today. Costs are up by a rather large factor. Office routine varies from just facilities to ship off unprocessed data to facilities that plan and interpret the field data, usually in a much better way than could be done thirty years ago.

Section 3.7 is good today. I have always felt that it is better to be led than to be driven. It helps to have as many employees as possible see and take pride in the successful outcome of the work.

CHAPTER 4

Efficient Exploration Programs

This chapter is written primarily for people concerned with the management of exploration programs.

4.1. SYNTHESIS, ORGANIZATION OF EXPLORATION PROGRAMS

In the first chapter we outlined a number of commonly used exploration procedures. The list is probably not complete, but it is sufficient to show that a large number of operations are possible and that many kinds of information can be used.

It is a matter of some importance to consider the *organization* of exploration. After all, merely piling up raw data in a corner isn't going to find oil fields. It is necessary that this information be organized and interpreted.

In planning an exploration program, it is desirable that the program be laid out in such a manner that the information is accumulated, and used, in a really *favorable order*. Otherwise, there will be a great deal of unnecessary duplication of work and also greatly increased costs for the exploration operation. This increased expense cuts down the efficiency of the exploration. Hence, in the sections that follow, we shall consider the general problem of organizing an exploration program so that it may be efficiently carried out.

4.2. WHAT IS EXPLORATION INFORMATION?

Exploration information may consist of a great variety of data with accompanying interpretations. The fact that the data must be interpreted to be useful emphasizes the prime importance of the users of the data. Although, of course, these people always have certain limitations, they must be of exceptional capability in order to make unusually valuable use of this information.

Information may in some cases have definite positive value. For example, the magnetometer information which helped to find the Hobbs oil field in eastern New Mexico had a great deal of value. This may be taken as an illustration of information that has positive value. But when magnetometer information was used in the neighborhood of Hobbs shortly afterward, several dry holes were drilled. Naturally these dry holes cost a good deal of money. We could therefore consider that some of the magnetometer information in this region had negative value, since it led to an outlay of money that was not directly productive.

4.3. PROBABILITY OF FINDING OIL

After oil is found, estimates can be made of the productive capacity of the field. Also, on the basis of probable market conditions, the dollar value of the oil that has been found can be estimated. But before the oil is actually found, no direct accounting procedure can be used. Instead, a much vaguer term has to be employed, namely, the *probability of finding oil.* This is very difficult to evaluate with any degree of precision. However, we may state roughly that if the probability of finding is zero, the chances are very poor indeed. On the other hand, if the probability of finding is very close to 1, we may consider ourselves practically on the verge of finding oil. This latter case might occur, for example, if we were working in a region of known productivity and obtained a seismograph picture that was considered very reliable and showed a very good closed high structure. In such a case, a miss is unlikely. Hence, in such a case we may consider that the probability of finding is very close to 1, just before the first discovery well is drilled.

In the earlier stages of exploration the probability of finding is always fairly low. In those stages, too, the procedure is not so well organized; it consists partly of a sort of stumbling procedure. We try this and that, preferably methods that involve minimum expense. Then we may get a small lead and our probability of finding, based upon this lead, increases materially. We follow up this lead, perhaps by means of a reflection seismograph. If this does not prove the lead worthless, we may say that the probability of finding is still further increased. If the reflection seismograph definitely improves the picture, our probability of finding is increased even further. Thus the probability of finding can be roughly estimated.

Once the probability of finding has been estimated, we can make a second estimation, but this is necessarily pretty much of a guess. This is the estimation of the *probable amount of oil.* In a favorable producing area, such as the Gulf Coast, we may be justified in making a favorable estimation of the probable amount of oil. In an entirely new region where no oil has been discovered within a radius of several hundred miles, our estimate would not be so favorable.

When we have made an estimate of the probability of finding and the probable amount of oil to be recovered, these two figures can be multiplied to get an *estimated expectation.* This estimated expectation gives an estimate of the probable amount of oil that is reduced by the factor of the probability of finding.

Probabilities and *expectations* have been used extensively in games of chance. One of the best examples is a poker game.

The finding of oil is, in some respects, analogous to a poker game. This is especially true when several companies are engaged competitively in prospecting a given area.

It may be of value to consider the way this analogy breaks down. In a poker game, the players know the size of the pot, so they know what risk they wish to take with respect to that particular item. With regard to an oil field, the players—in other words, the oil companies—do not know the size of the pot; they can only guess at it. It is this guess which corresponds to the estimate of the probable amount of oil. Usually, no such estimate is made, because any individual knows that in making such an estimate his chances of being wrong are enormous. He hesitates to commit himself in this way. On the other hand, a guess is a valuable thing in this connection, for it helps to bring out the nature of the exploration problem. Without this guess, the nature of the problem, at each stage of the exploration, must remain unknown, not explicitly stated. Since it helps to make an explicit statement and since such a guess has a certain amount of utility, if an exploration is to be properly organized such a guess should be made with the full understanding that it is a guess.

4.4. EXPECTED EFFICIENCY

If, as was said in the last section, we guess at the probable amount of oil, estimate the probability of finding, and multiply these two quantities together, we obtain the estimated expectation for this particular problem. Now, if we convert this estimated expectation into dollars, we can divide it by the actual cost of the exploration to date. This gives what we might call the *expected efficiency*. The actual cost, of course, is a matter of accounting. The estimated expectation is an estimated quantity.

We are dealing in quantities which were formerly regarded as lying in the domain of what we call mature judgment. Here we are trying to make an analysis of just what goes into the use of *mature judgment* in the case of an exploration program, so that such judgment may be—in part, at any rate—a social asset, rather than a purely private one.

At each stage in an exploration program, we may look back at what has been done and make an estimate of the expected efficiency. At each stage it is further possible to consider several alternative procedures. For example, at a certain stage we may decide either for or against doing any reflection seismograph work. Our decision ought to depend on our estimate of the expected efficiency for an interval of time in the near future. We should choose our procedure so as to secure the *maximum expected efficiency*. In other words, we should keep the cost at a minimum and the estimated expectation at a maximum.

Planning that considers only the factor of minimizing the cost of exploration can never be thoroughly successful because it leaves out of account the major question of finding oil. Usually the procedure is somewhat as follows: When a lead is obtained in an area, a relatively inexpensive method of exploration is used first, as for example surface geological exploration, magnetometer work, or gravity work. When this work has been interpreted and considered in relation to the particular area, a decision is made as to whether to follow up some of the leads by making a detailed seismograph survey. If the seismograph work is carried out, it may indicate the presence of a structure. Then it is necessary to come to some decision as to whether or not the structure is large enough to warrant drilling. If it is decided to drill, a wildcat well is located and drilled. During the drilling operations, proper attention has to be given to logging and testing, to make sure that oil, if present, will be found by the drilling.

At each stage in the process it is possible to make a rough estimate of the expected efficiency. One of the determining factors is the *competitive factor*. Usually we have sitting in on this game a number of competitors who are also interested in finding this oil first. Whereas in other circumstances it would be better to go about the exploration more slowly, taking time to digest each piece of information as it comes along, in practice this cannot always be done. It sometimes becomes necessary to use several methods at once and just hope that the operator's particular company will be the one that reaches the goal first. A racing element is involved in such a case.

A matter of major consequence is the fact that at each stage several possible alternative procedures are open to the operator. The course actually chosen depends to a large extent upon his judgment, which, of course, depends upon his experience and general ability.

4.5. AN ANALOGY

The following is an account of an experience which occurred a few years ago when the writer was doing some seismic exploration in South America. The crew had been operating in a very difficult area, just prior to the rainy season, which was known to be very severe. We had driven out to this area over a dirt road which was quite passable in dry weather but became an impassable mess during the rainy season. The last four days of our operations in this outpost area, heavy rains fell. This was very discouraging, but we finished our work. Our next problem was to get ourselves and the equipment back to our headquarters, about 30 miles away. We started early in the morning of the fifth day, which fortunately was a dry day, with the sun shining. We had a long procession of cars and trucks. The front car, an ordinary passenger car, was followed by three pickup trucks. Then came the trucks with the heavy equipment, water trucks, drills, instrument truck, shooting truck, surveyors' trucks, and so on.

Before we started out, we realized that we had two choices:

either not to start or to start. If we waited a day, the roads would dry out somewhat. If, however, we delayed a day, we had a very good chance of running into more rain, in which case the roads would be worse than ever. Hence we decided to go on.

We went some distance before we came to a place where the going looked very difficult. The car in front stopped and signaled to the remainder of the train to stop. We then investigated the possible alternative paths. We found that, a short distance farther on, the road had been completely washed out and was cut across by a deep gully. There was therefore no possibility of using this road. Other paths, off the road, appeared to boil down to about three possibilities. We weren't very long in deciding which of these possibilities seemed most favorable. Therefore we took what appeared to be the most favorable alternative path and traveled it successfully with our entire train.

We went on until another similar situation developed. We again stopped and considered the alternatives. This time there were only two. One consisted of going along the road, which was exceedingly muddy and went over a rather steep hill. The other consisted of driving very close to the edge of the road but along the edge of a very high cliff, which constituted one side of the road cut. In this case, it would be necessary for the trucks to go within two or three feet of the edge of the cliff, which was about 50 feet above the road. Both alternatives seemed equally bad. One man in our group had a pointed stick; he measured the depth of the mud. Another man was especially interested in the steepness of the slope. Both of these men were *specialists*. We held a conference, but no decision was possible. Thereupon the writer pulled out a coin and announced that he was going to flip it to decide between the two alternatives. This gave the necessary impetus to the group, who then decided to take the more conservative way on the road. In this way we proceeded until finally we got back to headquarters. We brought all the equipment back safely.

The procedure we used was analogous in detail to the *exploration procedure.* At certain stages it was necessary for us to stop and consider the various *alternative possibilities.* It was helpful to employ the services of *specialists,* experts in their particular lines. But we knew that it was necessary to synthesize all the information. Our specialists came to us with *fragments* of information. They had made a detailed *analysis* of certain aspects of the problem. We had to be careful that these isolated bits were properly put together into a synthesized, organized whole. This is always the case when dealing with fragmentary information of the type that is supplied by specialists.

It is not the writer's intention to make derogatory remarks about specialists. Specialization is a necessity, primarily because of the simple fact that each human being is limited in his energies. This limitation, which applies to all of us, makes all of us specialists. We are specialists in the sense that we take account of only a small part of the world around us and solve only some of the smaller problems.

The exploration problem is, in itself, a kind of specialty. It involves the use of fragments of information taken from a great many different specialized lines of investigation. A certain degree of broadening is necessary in organizing and synthesizing information, which may perhaps be felt as a lack of specialization. It should be remembered, however, that the exploration process itself is an extremely difficult one and requires the full energy of many individuals. It is by gathering together the leading people engaged in a given bit of exploration and discussing as much of the problem as can be discussed in such a group, that better solutions of the exploration problem can be arrived at. Such a process makes use of some of the *latent information* stored away in the imagination of the individuals in the group. This latter information is brought together with the information that has been put in reports and on maps, cross sections, and the like.

4.6. THE PLACE OF SEISMIC PROSPECTING IN THE OVERALL PICTURE

We are now ready to assemble the ideas that we have previously considered piecemeal. We do this by reference to Figure 4.1, which shows a three-dimensional coördinate system with its origin at the point *O*. Vertically upward is plotted the expected efficiency. To the right is plotted the cost of exploration. Into the depth is plotted the probability of finding. We plot the relation, shown at the bottom of the figure, between the expected efficiency, the probability of finding, and the cost of exploration, assuming a fixed probable amount of oil. This relation is represented by a surface which contains the cost axis and also the hyperbola, *JGUL*. This hyperbola is obtained by setting the probability of finding equal to 1 and plotting the resulting relation between cost and expected efficiency. We can picture the surface graph of this general relationship better if we start at point *G* and, without changing the cost, allow the probability of finding to decrease from 1 at *G* to 0 at C_G. The path between *G* and C_G is the straight line, GC_G. Other similar straight lines in the graph surface are *JK*, UC_c, LC_L. Thus the surface which is the graph could be formed if we held a straightedge against the hyperbola *JGUL* and also against the cost axis and moved the straightedge so that it always remains perpendicular to the cost axis. Imagine that such a surface has been thus constructed from concrete and pictured in Figure 4.1.

Low cost, low probability of finding, and low expected efficiency correspond to very early stages of exploration and are near the origin, *O*, in the figure.

We shall suppose that the actual exploration can be represented by a path on the concrete graph surface. This path we illustrate with an early part, S, and an intermediate part, *H*.

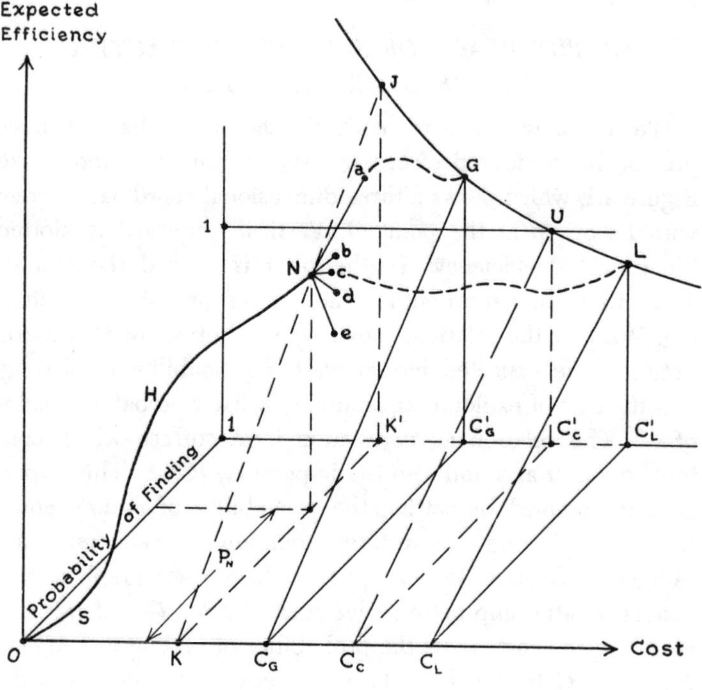

$$\text{Expected Efficiency} = \frac{\text{(Probability of Finding)(Probable Amount of Oil)}}{\text{Cost}}$$

FIG. 4.1. Three-dimensional graph showing the relation of expected efficiency, probability of finding, and cost of exploration for a past and future exploration program.

On this path is the point *N*, which represents "now." "Now" the probability of finding is P_N, and the cost of exploration to date is *K*. Notice that the actual exploration up to "now" is represented by the single curve, *SH*. Observe also that now we consider alternatives *Na, Nb, Nc, Nd, Ne,* so our single path branches as we look into the future. As an example, let us suppose that gravity work has been done and an interesting fault possibility has been mapped. Suppose that *Na* represents a seismic survey to check the gravity results. Suppose *Nc* represents another alternative such as checking the gravity lead by a soil analysis survey. *Ne* might represent ending the gravity work in an unused and not understood but filed report and then holding the land. Suppose, if *Na* is the alternative selected, that we finally arrive at the discovery point *G* on our exploration path, perhaps because the seismic work verifies the gravity lead. Had we taken alternative *Nc* we would have arrived at another discovery point, *L*.

At discovery point *G* our expected efficiency is greater than 1, so we expect to make a profit. At discovery point *L* our expected efficiency is less than 1, so we expect to have a loss, but still to recover much that we put into the exploration. *U* is the division point between profit and loss when the expected efficiency is 1. *Cc* is the cost of exploration for discovery without profit or loss. No matter how clever the staff, no matter how excellent the equipment, a profit cannot be made after the cost has exceeded *Cc*, unless our estimate of the probable amount of oil proves to be too small.

The path *OSHNaG* represents a successful exploration program carried from the stumbling stage, *S*, through an intermediate stage, *H*, to the discovery point *G*. What was actually done is represented by a single path on the concrete surface. On this path the cost almost always increases. However, unless we are capable—and sometimes lucky—the expected efficiency will not always increase. Unrejected information of negative value may lower it seriously.

Thus an exploration program is like the trip over the difficult road described in Section 4.5. The actual path traveled is unique—it has no alternatives. But often during our trip over it we have to make a choice.

The seismograph is just one of the possible *exploration tools*. Like all other good tools it is useless unless it is in the hands of competent people. In some areas it is capable of contributing all the additional information needed for successful drilling locations. Usually, but not always, it is the most efficient tool to use in such cases. Sometimes it cannot be used. Most cases in practice are between these two extremes.

Although there are cases in which the seismograph should not be used for greatest expected efficiency, especially in the early stages of exploration, the reflection seismograph is the best modern oil exploration tool for the average operation that is well out of the stumbling stage. In some cases using it has been like traveling on a broad paved highway. Even so, resourceful imagination is often required to avoid wrecks at washouts on the way. A resourceful imagination is one that is free to find and consider sane alternatives. Physics and geology help to avoid wasting too much time in considering utterly improbable alternatives.

Thus the seismograph generally has a place on an exploration operation as a tool that is very valuable when used by skillful and competent people.

The purpose of the remainder of this book is to help people become more skillful and competent in using the seismograph as an oil-finding tool.

We have discussed, in the past four chapters, aspects of oil exploration designed to show to some extent what the general exploration problem consists of and how seismic processes may be employed. We have also included a description of the practical operations, which is designed partly to help fill the gap occasioned by lack of experience in the field.

We now come to the second part of the book, which covers

routine computing done in the office. Each explanation is fairly complete if the reader understands that, in each case, a number of simplifying assumptions have been made. These assumptions are necessary so that we can proceed to a method of computation which is easy enough for a great mass of data to be handled.

4.7. COMMENTS TO THE SECOND EDITION

In 1980 I would omit this chapter. Now the geophysicist is simply asked to supply a broad reconnaissance magnetometer and gravity map, plus a set of seismic reflection cross sections, with reports. From here onwards, what is done depends on the people that use these data. Abilities of people, individuals, and organizations vary in many ways.

So what happens next? Is there any indication of a possible reservoir? If not, how sure do you have to be before abandonment? It is very embarrassing to abandon and later have a competitor prove the area productive, especially if it turns out to be very productive. Limits have to be set, and these have to be set by company policymakers.

If there is promise of a possible reservoir, what next? This again is a policy question. I would always question the reliability of the seismic picture and of the gravity picture right over the interesting feature. In some cases, I would want greater gravity station detail over the area of special interest, unless topography and surface conditions make the value of such work questionable. Sometimes a few seismic cross spreads, as in Figure 2.7 and in Chapter 9, may resolve ambiguities or reinforce an interpretation. Sometimes further regular seismic lines may fill local gaps.

It is necessary to make estimates of chances of success. These can hardly be real objective measures nor can they be wild guesses. My feeling has always been that the people with the most comprehensive knowledge of the data details should get together and give an assessment of chances based on that knowledge. No one else can make an assessment of any value. This assessment is usually partly quantitative and partly qualitative. Sometimes, resolving power is sacrificed for smoothness of the presentation on seismic sections. Many other possible deficiencies can occur, for example, those caused by poor statics, use of wrong gathering velocities for stacking, and so on.

Originally I used the terms "probability of finding oil" and "expected efficiency." At that time I had just read *Theory of Probability* written by the great geophysicist, Sir Harold Jeffreys, in 1939. His treatment sought to make this discipline more generally applicable by defining "probability of a proposition p given data q" as "a measure of reasonable degrees of belief"; not a bare guess but an estimate of comparative chances based on an analysis and a synthesis of the data and background knowledge of the time. This somewhat subjective theory of Jeffreys is not accepted by most workers in the mathematical theory of probability, but there is some subjectivity in every human choice.

If I were writing this chapter today, I would hesitate to use the terms "probability of finding oil" and "expected efficiency," mainly because we really cannot all get together for an assessment. In my innocence I thought the problem was purely technical.

Part II

ROUTINE SEISMOLOGY

CHAPTER 5

Basic Technical Tools

In the following short chapter we describe some of the basic technical instruments called for in Part II.

5.1. TERMINOLOGY AND USE OF FIGURES

In order to make a reasonably precise presentation in a reasonable number of pages, it is necessary to use a certain amount of mathematical terminology. This terminology has been placed almost entirely on the figures.

Some of the figures contain a certain number of mathematical formulas. These formulas play an essential part in deducing the results. So it may be well at this point to explain the order in which this material is always presented. Whenever the formulas are derived from the drawing or diagram, they are placed below it. Whenever they lead to the diagram, they are placed above it. Thus the reader should always start at the top of a figure and read down.

A result of this arrangement is that almost every formula is directly connected with a specific diagram. Thus all the mathematical work is directly related to some geometrical problem. The geometrical problem is directly related to a corresponding detail of the interpretive seismic problem.

In every case where we deal with a seismic pulse we have to deal with a source and a receiver. When we speak of *travel time,* we regard it as being the time from a source to a receiver. The *velocity* between a source and a receiver is the distance between these two points divided by the travel time from one point to the other. By rearranging this relationship we find that the *distance* between the two points is the velocity multiplied by the travel time between them, and that the travel time is the distance between the two points divided by the velocity from one point to the other.

5.2. PULSE FRONTS. RAYS

We must now consider two fundamental ideas that are much used in this book. In order to present the ideas, we shall use an analogy which is illustrated in Figure 5.1. At the right-hand side of this figure are the ideas we are trying to illustrate. At the left are ideas which are presumed to be already familiar to the reader. At the upper left is a series of elevation contours, a high point *P*, and a lower point *Q*, with a path of steepest descent joining *P* to *Q*. The series at the upper right is identical with that in the upper left, except that *P* represents the source of a seismic pulse and *Q* represents a receiver. The contours of equal elevation at the left correspond to the *equal travel-time curves* or *pulse fronts* at the right. The contour interval at the left—that is, the change in elevation, or ΔE_l—corresponds to the time interval, ΔT, between pulse fronts at the right. The path of steepest descent, *PQ,* in the left drawing corresponds to the path of least time or the ray, *PQ,* in the right drawing. The travel distance per unit time change in the right drawing we refer to as the velocity, whereas in the left drawing this same quantity corresponds to the horizontal displacement per unit vertical displacement. In each case the displacements are taken along the ray paths. The *path of steepest descent*—that is, the path joining *P* to *Q*—is *perpendicular* to each of the *elevation contours.* In the same way, the *rays* are *perpendicular* to the *pulse fronts.* This latter is a basic idea which is used many times in the succeeding pages. We give it without proof, as a fundamental, directly understandable idea. If the reader requires proof, he can easily find it in the more mathematical textbooks dealing with ray optics.

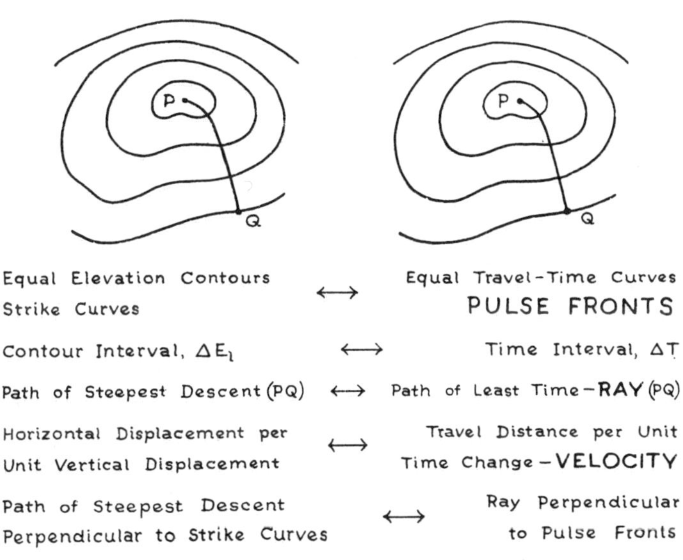

FIG. 5.1. Analogy between elevation contours and pulse fronts.

5.3. SNELL'S LAW

In order to present Snell's law in accordance with the analogy just given, we need to use a mathematical formula that is fundamental to the presentation of results of structural geology. The deduction of this formula is illustrated in Figure 5.2. In this figure, are a level plane LP and a dipping plane DP. Our problem is to relate the apparent dip, θ_C, along the direction CG, to the maximum or resolved dip, θ, perpendicular to the strike, CH, that is, along the direction HG. This is easily found by using the relationships below the diagram. The first and second equations are applications of the definition of the cotangent. The third one is obtained by using the definition of the sine and dividing both numerator and denominator of the sine ratio by the same quantity, GG'. When this is done, we evidently have, for the sine of angle i, the ratio of the two cotangents, which may be rewritten in the form of equation 4. It is this equation that we are directly concerned with in the present section.

Let us now consider Figure 5.3, in which the left-hand side gives the elevation contour analogy directly related to the pulse-front system shown at the right. In the case of the elevation contour system, using equation 4 of Figure 5.2 as applied to the triangle CHG, we have the first relation at the left.

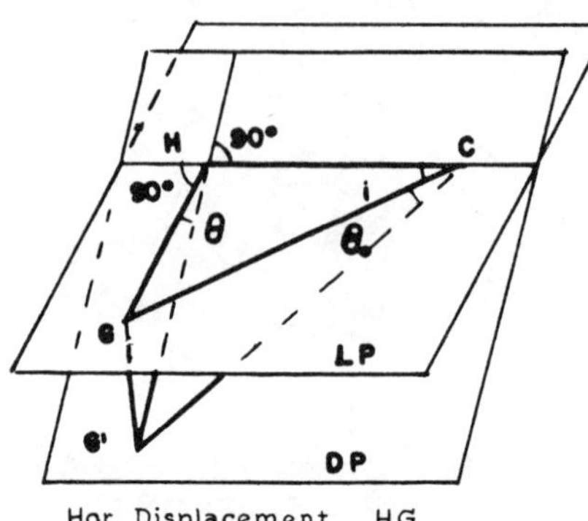

$$\frac{\text{Hor. Displacement}}{\text{Vert. Displacement}} = \frac{HG}{GG'} = \cot\theta \qquad (1)$$

$$\frac{\text{Comp. Hor. Displ.}}{\text{Vert. Displacement}} = \frac{CG}{GG'} = \cot\theta_c \qquad (2)$$

$$\sin i = \frac{HG}{CG} = \frac{HG/GG'}{CG/GG'} \qquad (3)$$

$$\frac{\cot\theta}{\sin i} = \cot\theta_c \qquad (4)$$

FIG. 5.2. Relation between total dip and dip component in vertical plane.

It should be noted that the elevation contours, such as $C'CC''$, make a sharp bend along the line BB'. This means that on the upper side of BB' there is one dipping plane, and, on the lower side, a plane with another dip. In the present case, the dip below the line BB' is less steep than the dip above this line. The second formula on the left gives equation 4 of Figure 5.2 applied to the lower set of contours, with the smaller dip. It is to be noted that the third relationship—that is, $FC = CG$—

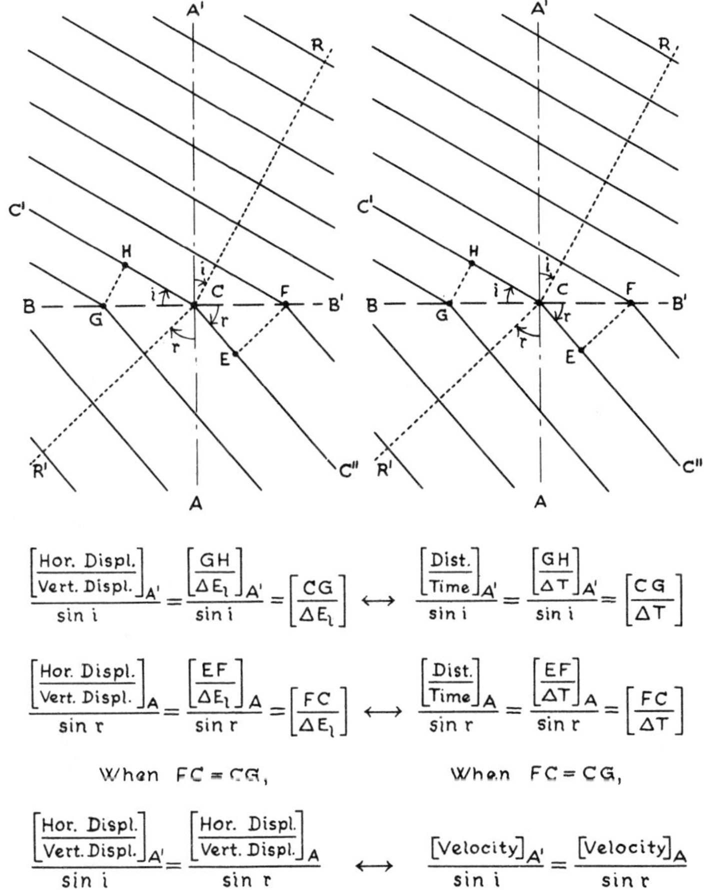

FIG. 5.3. Snell's law analogy for elevation contours and for pulse fronts.

is merely a statement of the fact that we have used equal contour spacing. Since by construction this relationship is taken to be true, the last relationship on the left follows.

This completes the deduction of what might be called Snell's law as applied to a set of elevation contours.

Let us now consider the analogy that represents our present principal interest. This is shown in the right-hand side of Figure 5.3. The correspondence is precise and detailed, just as it is indicated in the figure, and hence will not be described in further detail. However, the result is of considerable importance, namely, that the velocity in the upper medium divided by the sine of the *angle of incidence, i,* is equal to the velocity in the lower medium divided by the sine of the *angle of refraction, r.* This result is known as Snell's law and is frequently used in this book.

5.4. APPARENT VELOCITY

There is another idea that is quite different from the idea of velocity defined in Section 5.1. This is the idea of *apparent velocity,* which is illustrated in Figure 5.4. As before, the lines slanting from upper left to lower right represent pulse fronts. BB' represents an interface between two media with different seismic pulse velocities. If we suppose that the pulse sweeps across from F to C to G, we find that the pulse front travels along the interface, BB', with a certain velocity. We refer to this velocity as the *apparent velocity* of the pulse front along BB'. Thus we may measure the time from F to G and divide the distance from F to G by this time; the resulting ratio will define the apparent velocity of the pulse front from F to G.

It will be observed that the apparent velocity is equal to the actual velocity only when the *direction is taken along a ray.* When the direction for which the apparent velocity is computed is elsewhere than along a ray, the apparent velocity will be higher than the seismic pulse velocity.

Again referring to Figure 5.4, we note that the time from M to J to L is the same as that from F to C to G which, in turn, is the same as that from S to K to R. Since the corresponding distances are also equal, the *apparent horizontal velocity* in the upper medium is the same as the apparent horizontal velocity in the lower medium. That is to say, when we pass across the discontinuity BB', the apparent horizontal velocity is *unchanged.*

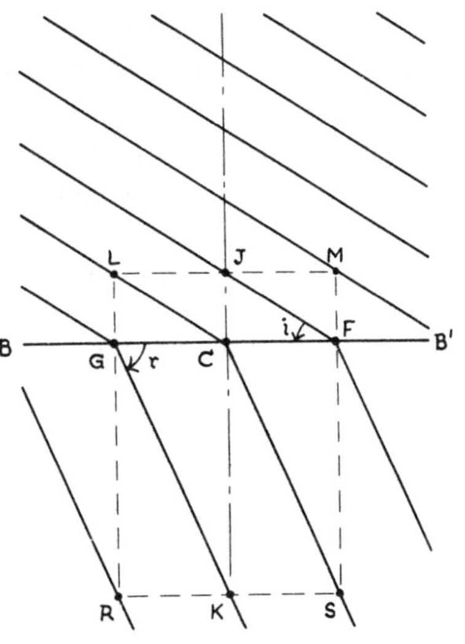

Fig. 5.4. Continuity of horizontal apparent velocity on crossing a level interface.

On the other hand, if we were to measure the apparent velocity from J to C and then from C to K we should find that the *apparent vertical velocity* undergoes a severe change in passing from one medium to the other.

We may summarize the results of our investigation of apparent velocity and Figure 5.4 by stating that the apparent velocity changes only when the velocity changes along the direction in which the apparent velocity is measured. When the velocity does not change in this direction, the apparent velocity does not change.

In most routine work, we avoid considering lateral changes of velocity; that is, we consider that the velocity depends only upon depth. This being so, the apparent horizontal velocity never changes for a given ray. Indeed the preceding is merely another way of stating Snell's law.

We now proceed to a consideration of the first steps in preparing seismic records for interpretation.

5.5. *COMMENTS TO THE SECOND EDITION*

If I were writing this today, I would introduce the concept, *wave slowness*. In Figure 5.5 there is an impulsive point source

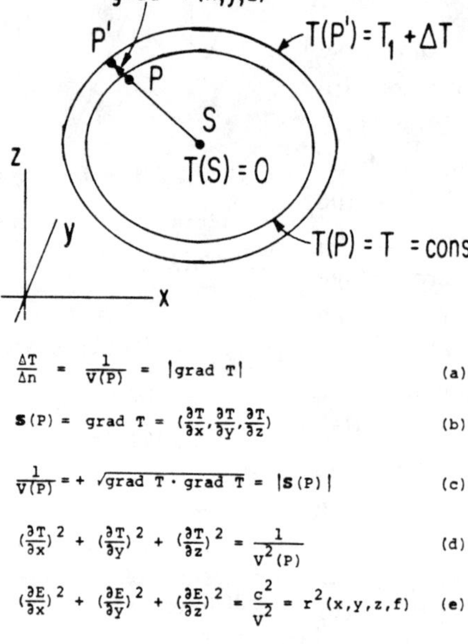

FIG. 5.5. Wave Slowness Definition.

at S that starts a pulse at time $T=0$. Later at time $T=T_1$ the pulse reaches the *surface*, $T(x,y,z)=T_1$. We pick any point, P, on this surface, draw the outward normal direction at P, move out along this normal to P'. $PP'=\Delta n$, which is arbitrarily small. $T(P')=T_1+\Delta T$. So in equation (a), we show the relation satisfied by ΔT, Δn, and V, the velocity of the outward motion of the pulse front at P. The *vector pulse slowness* at P is written $S(P)$ and is defined by relation (b). The numerical value of $S(P)$ is given by relation (c). Squaring this, gives equation (d), which is what we called the Eikonal Equation in Figure 7.8 for two dimensions. The Eikonal is used extensively in optics where the wave propagation velocity is usually dispersive, that is, depends not only on position P, but also on frequency, f. It is customary in optics to make (d) nondimensional by multiplying through by the square of the velocity of light in vacuum, c. Then (d) becomes (e), where r is the index of refraction and V is the velocity of light in the medium considered. In seismology, we have no universal reference velocity, so we use (d).

CHAPTER 6

Datum Corrections

The present chapter is written for seismologists, party chiefs, computers, and others interested in details of computing seismic records.

The general question regarding the *necessity of making a datum correction* may perhaps be answered by referring to Figure 6.1, which shows a rather exaggerated case. In it there are three shot points, three geophones, a rather irregular type of topography, and an irregular layer of weathered material. We suppose that there is a series of reflector points shown as *A, B, C*. The shot fired from shot point S_A generates a pulse which goes down to point *A* on the reflector and is then reflected back to geophone g_a. In doing so, it passes through the regular high-velocity materials down to *A* and up to the base of the weathering. The dashed lines show its path up through the weathering to g_a. At the next shot point the path goes from S_B down to the reflector, *B*, and then up to geophone g_b. It so happens that the path through the weathering is smaller for this pulse than for the preceding. This decrease in the amount of low-velocity material has the effect of decreasing the total travel time from S_B to *B* to g_b. If we plot the depths according to the observed times, we plot the depth point not at B but at B', a little above B, as indicated. The third shot point, S_C, gives a reflection from C up to g_c. In this case the path through the weathering material is longer than it was in either of the other two cases. This will result in a longer travel time from S_C to C to g_c. If this longer time is interpreted in terms of depth, we will plot the depth point, not at C, where it should be, but at C', somewhat below C, as indicated. It is evident that errors of this sort should be avoided. Hence it is necessary that we remove the effects of shallow irregularities.

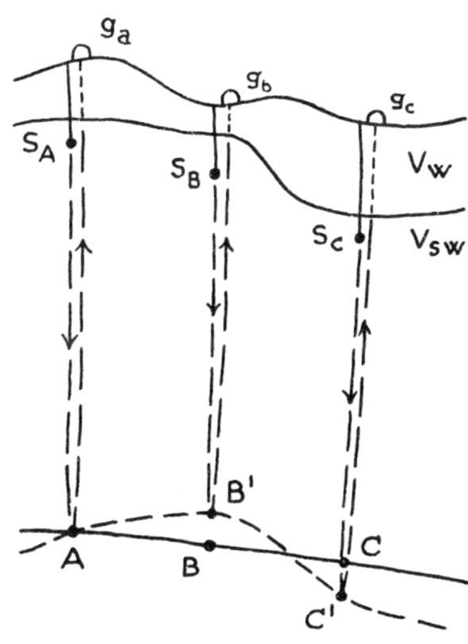

FIG. 6.1. The need for removing shallow irregularities.

The problem, then, is to make our calculation in such a way that the irregular low-velocity portion of the cross section is removed, as much as possible. This procedure is called a *correction to datum* in this book.

Reference to Figure 6.2 will show what is meant. In this figure is shown the surface, on which is located a geophone, g. The shot hole is shown. The explosive is assumed to be at S, the bottom of the shot hole. A little above the shot is the weathered layer base. Now we shall suppose that we wish to make a correction to a datum level, somewhat below the shot in this case. We wish to *reduce* the reflection time from the actual time (from S to the reflector and up to g) to the time we would have if the shot were located on the datum at S_D and the geophone were also located on the datum at g_D. It is this type of *reduction to datum* with which we shall be concerned in the remainder of this chapter.

6.1. DATUM CORRECTIONS BASED ON SHORT OFFSET GEOPHONE LOCATIONS

In the present section we shall discuss a case in which the geophone is located relatively close to the mouth of the shot hole, as illustrated in Figure 6.3. This figure gives most of the notation which we shall have occasion to use. We use *E* for elevation and *D* for depth. *W* indicates the weathered layer and *SW* the subweathered layer. *D* as a subscript indicates the datum and S the shot, in all cases. Elevations are always measured from sea level and we shall suppose that they are *positive* above sea level and *negative* below sea level. It will be observed that we have to consider several levels in the figure: the level of the shot point, *SP*, which has an elevation E_{SP}; the level of the geophone, *g*, which has an elevation E_g; the level of the shot, S, which we call E_S; the level of the datum, which we call E_D; and sea level. Below the figure are the times of primary importance with which we must deal. T_r is the time from the shot to the reflector to the geophone; this is the *total reflection time*

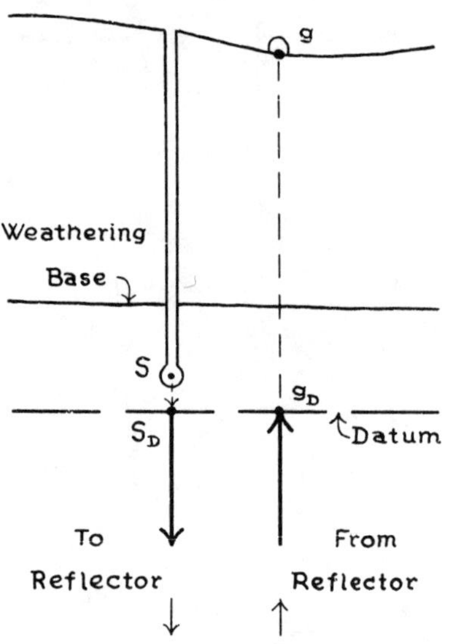

FIG. 6.2. Vertical section showing shot hole, shot, weathering base, geophone, and datum.

as found by measuring the time from the cap break to the reflection on the seismic record. T_{SD} is the time from the shot to the datum. T_g is the time from datum up to the geophone. T_C is the corrected time, that is, the time from point S_D, directly beneath the shot at the datum level, down to the reflector and back to point g_D, directly under the geophone but at the datum level. As the last relationship indicates, the corrected time is $T_r - T_{SD} - T_g$, a relationship which should be clearly understood.

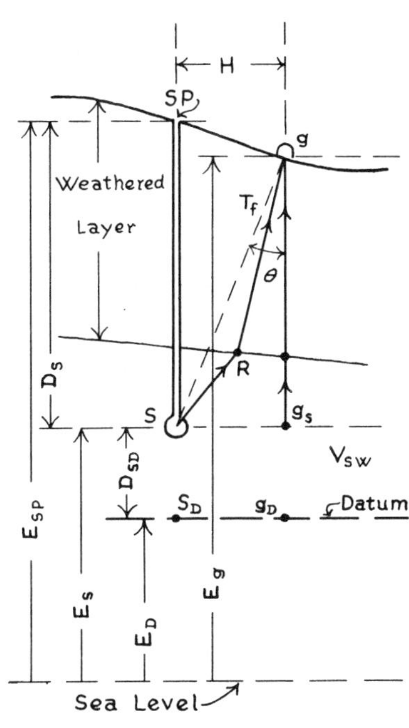

$$T_r = \text{time}\,(S \rightarrow \text{Refl.} \rightarrow g) \qquad (1)$$

$$T_{SD} = \text{time}\,(S \rightarrow S_D) \qquad (2)$$

$$T_g = \text{time}\,(g_D \rightarrow g) \qquad (3)$$

$$T_c = \text{time}\,(S_D \rightarrow \text{Refl.} \rightarrow g_D) \qquad (4)$$

$$T_c = T_r - T_{SD} - T_g \qquad (5)$$

FIG. 6.3. Correction to datum.

Now with the notations indicated in Figure 6.3 we consider the portion of a record shown in Figure 6.4. The upper trace shows the pulse which reaches the geophone first, the so-called *first break*. The second pulse toward the right indicates a reflection. The lower trace gives the cap-break record. As is indicated in the figure, the timing lines are one hundredth of a second apart. We have taken the zero point just to the right of the cap break, so that the cap break is 0.006 second to the left of our zero. Also, the depth of the shot is 115 feet and the horizontal offset 30 feet, as shown on the figure.

In Figure 6.5 we have tabulated the information that is necessary for the computation. First we get the elevation of the shot point E_{SP}, which, in our example, is 342 feet. Next we write the depth of the shot D_s, which is 115 feet. Then, to get the elevation of the shot, E_s, we subtract the depth of the shot from the elevation of the shot point, and get 227 feet. Next we write the elevation of the datum, E_D, which is 200 feet. Then we get the depth from the shot to the datum, D_{SD}, by subtracting the elevation of the datum from the elevation of the shot,

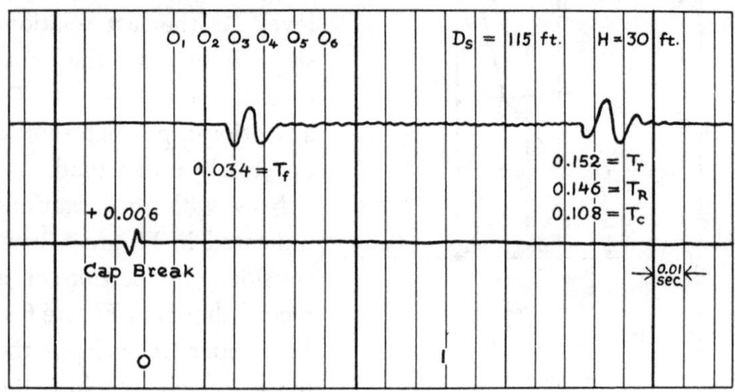

FIG. 6.4. Portion of seismic record showing cap break, first break, and shallow reflection.

as indicated; this is 27 feet. Then we divide the depth from the shot to the datum by the velocity of the subweathering material, V_{SW}, to get the time from shot to datum, T_{SD}. We have put this subweathering velocity at 5000 feet per second. This gives a time from shot to datum, T_{SD}, of 0.0054 second. Next we read the time of the first break, T_f, directly from the seismic record; it is 0.034 second. This we write in the next row. This is the *slant time*, which has to be corrected to vertical by multiplying by the cosine of the angle between the vertical and the direction from the shot up to the geophone, as viewed from the shot (see Figure 6.3). This gives a time,

T_{fc}, corrected to vertical, equal to 0.033 second. Our final step, so far as the calculation of the time from the datum to the geophone T_g is concerned, is to add the time from the shot to datum, which gives 0.0384 second.

The total correction, as we may recall from Figure 6.3, is the time from shot to datum plus the time from datum to geophone which in this case gives, very closely, 0.044 second. Thus referring to the record in Figure 6.4, we see that the time T_r from the cap break to the beginning of the reflection is 0.152 second. If, then, we make our correction by subtracting 0.044, we obtain 0.108 second as the *corrected reflection time*, that is, the time from the datum under the shot down to the reflector and back to the datum under the geophone.

		Correction Time
El. S.P. = E_{SP}	+ 342	
Depth Sh. = D_S	115	$T_{SD} + T_g =$
$E_{SP} - D_S = E_S$	+ 227	0.044
El. Datum = E_D	+ 200	
$E_S - E_D = D_{SD}$	+ 27	
$D_{SP}/V_{SW} = T_{SD}$	+ $0.005^{4}_{\pm}$	$T_s = 0.000^{6}_{-}$
T_f	0.034	$T_g = 0.038^{4}_{\pm}$
$T_f \cos\theta = T_{fc}$	0.033	$T_s - T_g =$
$T_{fc} + T_{SD} = T_g$	$0.038^{4}_{\pm}$	−0.038

FIG. 6.5. Table for calculation of weathering correction.

The above discussion has given the correction for an example as a means of explaining the principle involved. We are now interested in putting the calculation on a basis that will make it a little easier to use in practice.

Figure 6.6 illustrates the case. The time from datum to geophone g_1 is T_{g1}, and to geophone g_2 it is T_{g2}. These quantities are properties of the geophone location and of the elevation of the datum and have nothing to do with the location of the shot point. It is for this reason that this part of the correction, the geophone part, has been separated from the remainder.

We may regard the time from shot to datum as a part which is specifically related to the shot itself. It will be noted from the record in Figure 6.4 that the +0.006 second, from the zero timing line to the cap break, refers also to this specific shot. We could combine these two parts referring to the shot, as indicated by the first relationship in Figure 6.6, to give T_S, the *part of the correction at the shot,* which is T_{OC}, the time from the zero timing line to the cap break measured positively to the left from the zero timing line minus T_{SD}, the shot-to-datum time. If we introduce another bit of notation and call T_R the reflection time as read directly from the zero timing line, without including the $+0.006$ second correction, we can write T_r as T_R+T_{OC}. Then, using relation 5 in Figure 6.3, we get the second relation in Figure 6.6. The first and second relations in this figure lead to the third relation, which is in a form suitable for routine use. In other words, we have reduced the corrected time to a time read directly from the zero line plus a correction that is associated with the shot point minus another correction that is associated with the geophone location. The corresponding calculation for this example is given in the lower right part of Figure 6.5, in which T_S is found to be 0.0006 and T_g is 0.0384, so T_S minus T_g is very close to -0.038. If, then, we add to the read time, which is 0.146 on the record, this correction of -0.038, we obtain the corrected time, 0.108 second.

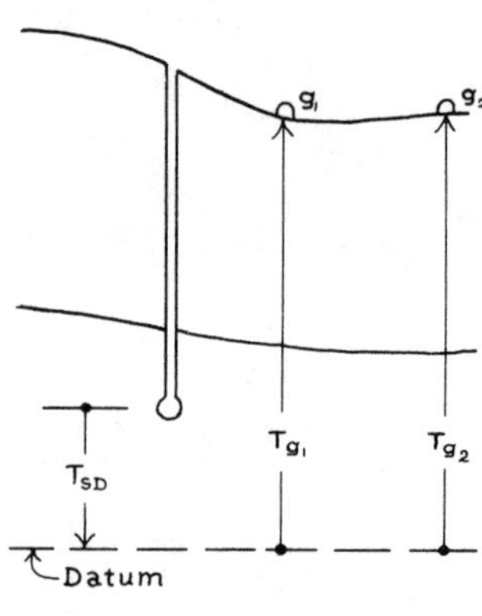

$$T_S = T_{OC} - T_{SD} \quad (1)$$

$$T_C = T_R + T_{OC} - T_{SD} - T_g \quad (2)$$

$$T_C = T_R + T_S - T_g \quad (3)$$

FIG. 6.6. Résumé of correction to datum.

The procedure that we have just described can be left as it is, or we can simplify it still further. If we regard the series of zeros at the top of Figure 6.4 as possible zero timing line markers in place of the one we have previously used, we will find an improvement of our procedure for correcting to the datum. For example, if we had taken the zero point at 0_1 instead of where we did, we would decrease the amount of T_R by 0.01 second, to the value 0.136. But it will be observed that at the same time that we decrease T_R, we increase T_{OC} (see formula 2 in Figure 6.6). Thus, the corrected time becomes $0.136+0.016-0.0054-0.0384$, which gives approximately 0.108 again. Also, it will be noted that instead of having the total correction equal to -0.038, as we had originally, we now have a total correction of -0.028. In a similar manner, if we had moved the zero line over to 0_2, the total correction would have been reduced in magnitude to -0.018, and if we had moved it over to 0_3, the total correction would have been reduced to -0.008. If we had moved it over to 0_4, the total correction would have been reduced to $+0.002$ second. If, on the other hand, we had made our move over to 0_4, our read reflection time, T_R, would have become 0.106, to which we add 0.002 to get 0.108.

Although shifting the time origin on the record may appear a little unnatural at first, it leads to quite a saving in labor because it reduces the correction that must be applied to each of the reflections on this particular trace to a very small value, in this instance $+0.002$ second. Almost everyone will find that it is much easier to add 0.002 second to the read time than to subtract 0.038 second from it. It is not only easier to do the smaller addition than the larger subtraction, but it can be done with much less chance of error. It is therefore probably advisable for anyone who makes datum corrections to consider the possibility of using this shifted time origin in order to make the arithmetic simpler and thus freer of errors.

The shift of time origin outlined above and recommended whole-heartedly is not the sort of thing that is simple to explain. It is suggested that the reader follow the procedure

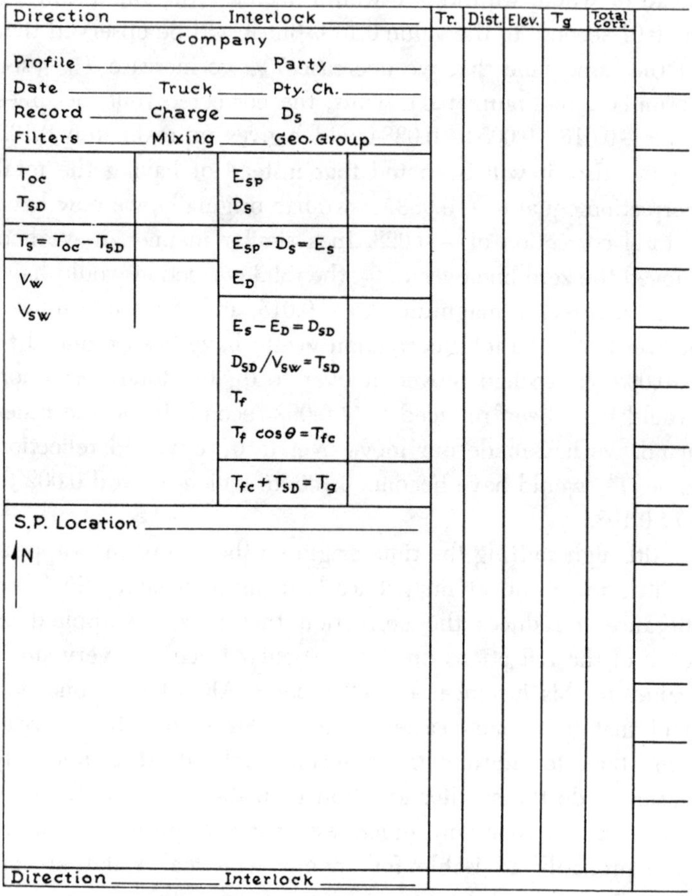

Direction ________ Interlock ________ | Tr. | Dist. | Elev. | T_g | Total Corr.

Company

Profile ________ Party ________

Date ________ Truck ________ Pty. Ch. ________

Record ________ Charge ________ D_s ________

Filters ________ Mixing ________ Geo. Group ________

T_{oc} T_{SD}		E_{SP} D_S	
$T_S = T_{oc} - T_{SD}$		$E_{SP} - D_S = E_S$	
V_W V_{SW}		E_D	
		$E_S - E_D = D_{SD}$ $D_{SD}/V_{SW} = T_{SD}$ T_f $T_f \cos\theta = T_{fc}$	
		$T_{fc} + T_{SD} = T_g$	

S.P. Location ________

N

Direction ________ Interlock ________

FIG. 6.7. Seismic record label arranged for presentation of data.

through with his own example, or perhaps several examples, to assure himself that no error is committed by making this shift. It is of course not advisable to carry out computations blindly; therefore, unless this process is thoroughly understood, it should not be used. On the other hand, the process is a labor saver and an accuracy promoter.

Figure 6.7 shows a record label that is suitable for use in case there is a geophone of small offset from the shot point.

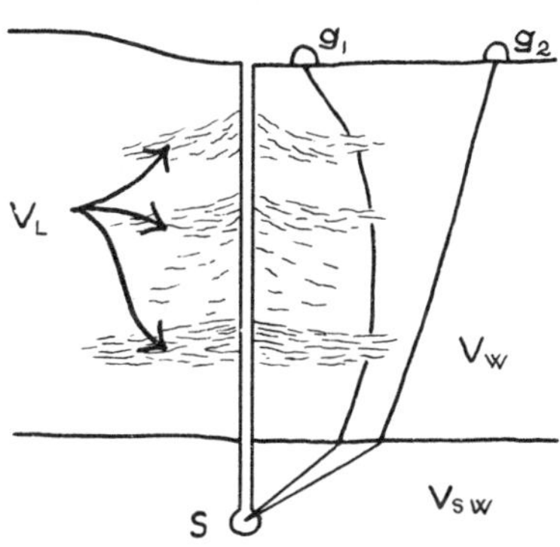

Fig. 6.8. Vertical section showing weathering disturbed by drilling fluids.

Before leaving this topic it may be well to consider some of the difficulties associated with locating a geophone too close to the mouth of the shot hole. Figure 6.8 shows diagrammatically a shot hole dug through the weathering and into the subweathering with two geophones g_1 and g_2. Observe the shading in the neighborhood of the shot hole that makes the hole look somewhat like a Christmas tree. The arrows pointing into the shading lead from the symbol, V_L, which means low velocity. During the drilling process the drill goes through a section that is porous, the pores being filled mostly with air. As the drilling proceeds, mud at high pressure is driven into the section. It may be supposed that in some cases this mud has a considerable effect upon the velocity in the weathered layer. This may be especially serious in case a large number of air bubbles become trapped in the muddy water of the section. Such a mixture of bubbles and loose material has an only slightly reduced density and a low incompressibility which gives rise to a very low seismic velocity. In extreme cases the situation may be like that illustrated in the figure, in which the path of the energy from shot point S to geophone g_1 is bent around the low-velocity material, whereas the path to geophone g_2 is much more normal. The situation illustrated in this figure can be verified by shooting with a close row of geophones across the shot hole. Perhaps the geophones should be only two or three feet apart. Often, the geophone nearest the shot hole mouth is not the one that shows the least time; instead, one located much as g_2 is located may be the one with the minimum time.

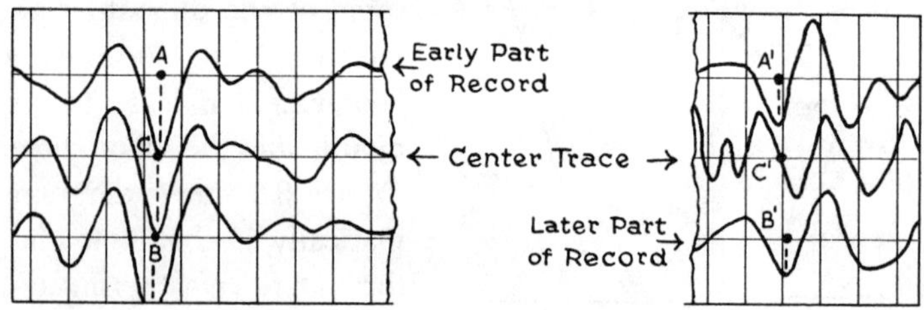

Fig. 6.9. Early and late portions of seismic record showing hole disturbance.

All we have just said is a warning against placing the geophone too close to the mouth of the shot hole. If we do this, we may have a weathering material which is very different from that a little farther away from the shot hole. This may be quite important, for we may wish to use the adjacent geophones for depth computations. In such a case we may be in error, perhaps even serious error, if we extend our notions about travel time in the weathering out to geophone g_2 from data collected from geophone g_1.

Another difficulty which is by no means serious, although it appears very much so at first, is illustrated in Figure 6.9, in which some of the reflections are greatly disturbed on the trace near the mouth of the shot hole. This usually happens with respect to the later reflections, the earlier reflections being quite clear and undisturbed. This figure shows the method of taking care of this situation. Points *A*, *B*, and *C* are directly above the trough bottoms for their particular traces and are located at equilibrium positions. The relationship of *C* to *A* and *B* remains more or less fixed throughout the record. This is not precisely so, because the curvature across the reflection changes, becoming smaller in the deeper reflections; but for adjacent traces the error introduced in this way is negligible. We therefore go out to a later reflection, as indicated on the right side of the figure, and interpolate C' between A' and B'. This gives the time C', even though the center trace is badly disturbed by the hole noise.

The above procedure may be objected to, on the grounds that it would be easier just to move the geophone nearest the shot a little farther away and thus avoid the hole noise. This is quite true. However, we have a method of computing that has a great advantage and that, in some cases, we would be foolish to give up without thought. The advantage is that we need not know the velocity with which the seismic pulse travels through the weathering. Our assumptions are not dependent on the value of this velocity. Since in regular field operations this velocity is seldom really measured, a method which does not require its use has a certain definite advantage.

Before ending this section it will be well to remind ourselves that, in all that has been done up to this point, we have supposed that the *datum is below the shot*. Suppose, however, that this is not the case, that the *datum is above the shot*. We may even suppose that it is up in the air, as indicated in Figure 6.10. This situation might occur in a region with considerable variations in topography, if we extended our survey into low-lying territory adjacent to the starting area and found that the original level datum was in the air. This has happened occasionally.

When the datum is above the shot level we first scrape off all the material from the shot level on up—in our imaginations, of course. This, we suppose, removes the irregular low-velocity material. Then we imagine a level slab of material inserted resting on the level plane through the shot with its level top at the datum level. This latter material we suppose has a uniform velocity V_{sw}.

Now let us calculate what the correction to datum will be in the case just described. We shall see that this gives precisely the result we would get if we use equation 3 in Figure 6.6 blindly, providing we use the correct algebraic sign for T_{SD}. We shall also see that this choice is made automatically if we use the first six rows at the left of the schedule in Figure 6.5.

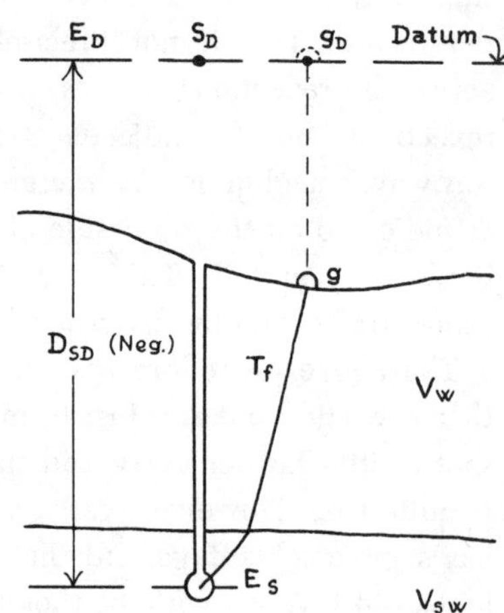

FIG. 6.10. Vertical section showing correction to datum above the shot level.

Referring to Figure 6.10 we first subtract T_{fc}, the time up to the near geophone corrected to vertical, from T_r, the time from shot to reflector to geophone. Then we add $-T_{SD}$ (which is positive since T_{SD} is negative in this case) twice, once at the shot to raise it to S_D and once at the geophone. Thus by reducing the reflection time down to the shot level (by subtracting T_{fc}) and then raising it up to the datum level (by adding $-2\ T_{SD}$) we get the whole datum correction. This can be expressed by $T_c = T_r - T_{fc} - T_{SD} - T_{SD}$, which is equivalent to relation 1 in Figure 6.10, if we take relations 2 and 3 into account. But relation 1 in this Figure is the same as relation 5 in Figure 6.3. So the same relation holds for both cases, the one with the datum above the shot and the other with the datum below the shot.

This description of how to handle the situation when the datum is up in the air applies equally well when the datum is in the weathering material, or when it is above the shot, but still in the subweathering material. In other words, the formula we have given is in such a form that it can be applied regardless of the location of the datum with respect to the shot.

6.2. DATUM CORRECTIONS BASED ON LONG OR MEDIUM OFFSET GEOPHONE DATA

We wish to consider the problem that is involved when there is a relatively larger offset distance from the mouth of the shot hole to the nearest geophone. We are going to consider a particular case and follow through all the computations with respect to it. A plan view of the layout is given in Figure 6.11, in which the shot point is shown to be 200 feet offset from the line of geophones opposite the center geophone. Laid out in either direction along the line are ten intervals, each of 65 feet, giving a total spread of 650 feet each way from the center geophone. The direction of the line of geophones in this case is N. 43° E. The top record trace is indicated at the top

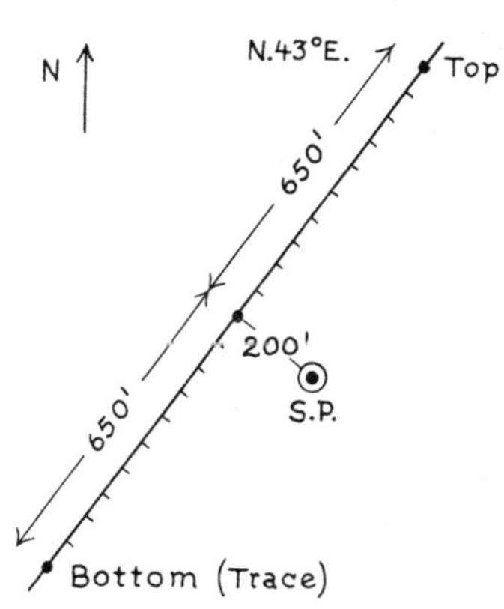

FIG. 6.11. Map view of shot and geophone layout.

of the diagram and the bottom record trace is at the base. These might have been reversed. However, it is important that the diagram indicate which is the top trace.

The record shot with this field layout is shown in part in Figure 6.12. There are 21 traces altogether, starting with the top trace and ending with the bottom one. On the third trace from the bottom is shown the up-hole time record. On the ninth trace from the bottom is shown the cap-break record. The timing lines are shown on the record and also the first-break impulses. The time is written at the top of each trace immediately to the left of the first breaks. Besides the time, we need the distances. Let us refer, for a moment, to Figure 6.13, which is a slight magnification of part of Figure 6.11. We have the offset of 200 feet and, for the fifth geophone from the center, a distance along the geo-

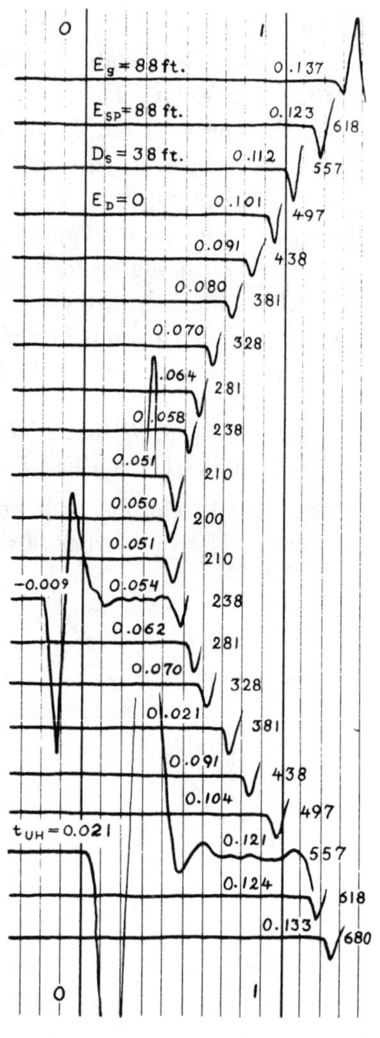

FIG. 6.12. Traced copy of early portion of seismic record with cap break, up-hole geophone trace, and corresponding first breaks.

phone line of 325 feet, which corresponds to a horizontal distance of 382 feet from the shot to this particular geophone. This is obtained by solving the right-angled triangle for the hypotenuse.

This problem—solving a right-angled triangle—appears very often in computing seismic records, and so it is desirable to have a quick and easy means of doing it. This we shall now describe. Figure 6.14 shows part of a slide rule. The upper scale is measured according to the square of the scaled distance, as is the identical scale on the slider. Although the *measures* are in terms of the square of the scaled distance, the *labeling* is in terms of the distance, so we have a means of adding the sum of two squares by putting the two scales together in the proper way.

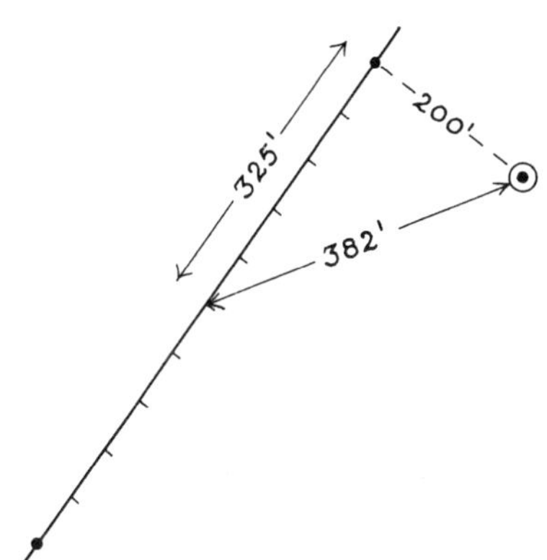

FIG. 6.13. Expanded detail of Figure 6.11.

For example, the square of the distance from zero to 200 plus the square of the distance from zero to 455 equals the square of the distance from zero to 497. These distances are indicated in the figure. This is a *square slide rule* that can be used for solving right-angle triangle problems. This slide rule is exceedingly useful for seismic computing. It doubtless would also be very useful for a great many geological computations. Each of the diagonal distances in Figure 6.11 was computed with a square slide rule, as illustrated in Figure 6.14. The values are written immediately to the right of the corresponding traces in Figure 6.12.

Our next step is to plot the time-distance data on the graph, as shown in Figure 6.15. Here *distance* is taken as the abscissa and *time* as the ordinate. Note that these points lie approximately on a straight line. Observe also that the two arrows

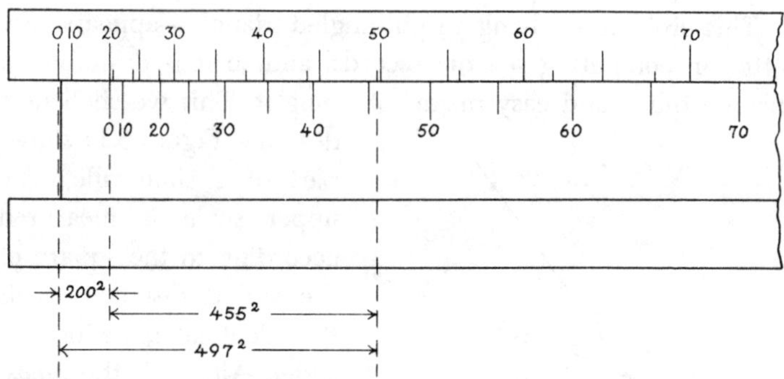

FIG. 6.14. Square slide rule for solving right-angled triangle numerical problems. (Courtesy United Geophysical Company.)

pointing to *A* and *B* correspond to a time difference of 0.1 second. Note also that this time difference of 0.1 second corresponds to 570 feet on the ground. Thus, in 0.1 second the pulse travels along the line of geophones for 570 feet. This corresponds to a velocity of 5700 feet per second. This is written on the graph, so that 5700 feet per second is the *subweathering velocity.* In arriving at this result we assume that the travel path (and the travel time) through the weathered layer is the same under each geophone and that the only difference is

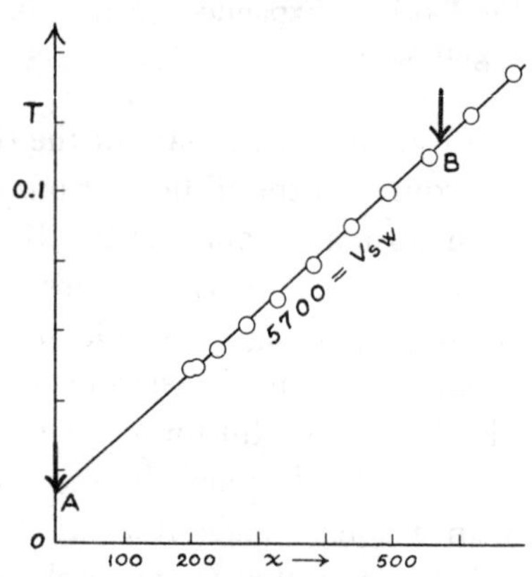

FIG. 6.15. Time-distance plot from first breaks.

due to a travel path in the subweathering layer. This assumption is approximately but not precisely correct, as will be seen by more detailed consideration of the question.

We do not have direct data by which we can measure the weathering velocity in this particular record, but from other records shot in the neighborhood we find that the weathering velocity is about 2000 feet per second. We shall use this value.

Now let us turn to Figure 6.16, which shows the situation

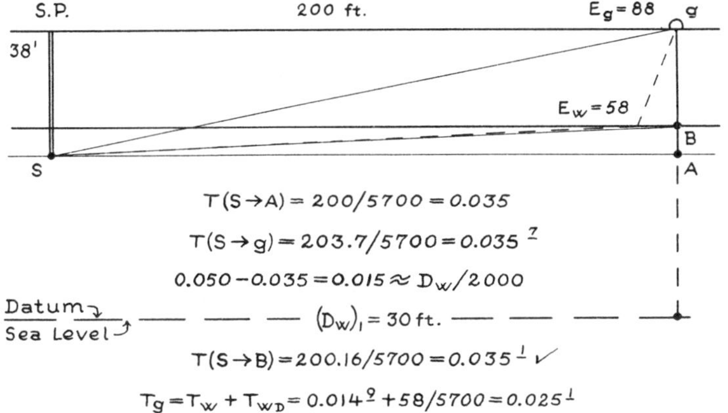

Fig. 6.16. Vertical section showing datum correction when geophone has long offset.

we wish to consider. We see the shot point, which in this case is 38 feet deep. We see the offset geophone 200 feet away from the shot hole mouth. It happens that the shot hole mouth and the geophone are located at the same elevation, namely, 88 feet. Let us consider two extremely different hypotheses. Let us suppose, in the first instance, that the energy travels exactly horizontally from S to A, A being directly beneath the geophone. The travel time for this path is easily found; it is 0.035 second. Let us suppose, on the other hand, that the energy travels directly from the shot S to the geophone g along a straight line at the subweathering velocity. This case corresponds to the case when there is no weathering material at all. By using our slide rule, we find that the slant distance is 203.7 feet, hence the travel time is 0.0357 second. In other words, the total difference in travel time corresponding to the two quite different hypotheses is only 0.0007 second. Thus the travel time from S to *B* ought to be very near 0.035 second. We suppose that the path is not the *true path*, but the path from S to *B* plus the path from *B* to *g*. We compute the weathering thickness on this basis. Since we have supposed that the time from S to *B* is practically the same as that from S to *A*, we may subtract 0.035 from 0.050, the observed time, which leaves 0.015. This is approximately the depth of the weathering divided by the weathering velocity. This gives a *first approximation* of 30 feet for the *weathering depth.*

We can check this value by calculating the time from S to *B* plus the time from *B* to *g*. This gives a time which differs from the first-break time, T_f, by only 0.0001 second. Hence this is a verification of the correctness of point *B*. If such accuracy had not been attained, it would have been necessary to make a slight adjustment. In some cases such an adjustment is necessary; in each case, however, it is easily seen what direction this adjustment must take and also how much it must be.

Although no adjustment is needed in the present case, we shall make it anyway just to illustrate the method. Suppose, for example, that we have to adjust for the 0.0001-second error. Since our time, calculated over the path *SBg*, is 0.0001 too much, we need a little less weathering. The amount less is very close to 0.0001 of 2000, or 0.2 foot. If we were to calculate the travel time from S to a point 0.2 foot above *B* and then from this point to *g*, we would obtain a much closer approximation to the 0.050 observed. Such a degree of closeness as is calculated in this example is of course entirely unnecessary (and illusory). The adjustment is carried through merely to illustrate the kind of procedure that must be used when such an adjustment is necessary.

As was said above, the depth of weathering is 30 feet. We can then compute the time from the weathering to the datum. Since we are taking our datum at sea level, the elevation of the weathering is 58 feet; this is 30 feet below the geophone, which has an elevation of 88 feet. We then have 58 feet to traverse at 5700 feet per second, which gives 0.0102 second. We have to add to this the weathering time, which is $0.050 - 0.0351$, or 0.0149 second. This gives a total T_g of 0.0251 second.

It may be objected that we have not used the proper travel path for our computation. We used a path from S to B to g. This is only an approximation. If we had wished to be very accurate, we should have used the dashed path in the figure. This would have given a somewhat different travel time. In fact, if we do the computation we find that we come out with about 0.0491 instead of 0.050 second. Thus we are in error by about 0.001 second. This kind of error is not regarded as serious for several reasons. One is that the error is always small; another is that it is always in the same direction; another is that it is a systematic error which does not vary much from one profile to another; and another is that usually we do not know the weathering velocity too accurately. It costs too much money to measure V_w at every shot point, so we make an approximation by guessing it. This error of 0.001 second corresponds to an error in the weathering velocity of less than 7 percent, or 140 feet per second in 2000 feet per second.

If we wish to refigure the computation on a somewhat more accurate basis, we may consider the formula in Figure 6.17.

In this figure the shot is supposed to be ideally at the base of weathering. If it is moved down a little, the move does not affect the accuracy of the computation very much. The path is along the interface to the critical point and then along the ray at the refraction angle up to the geophone *g*. The travel time is T_f. This travel time is the time which corresponds to every point on the pulse front perpendicular to the ray through *g*, as illustrated. Thus the travel time from *S* to *A* to *g* is the

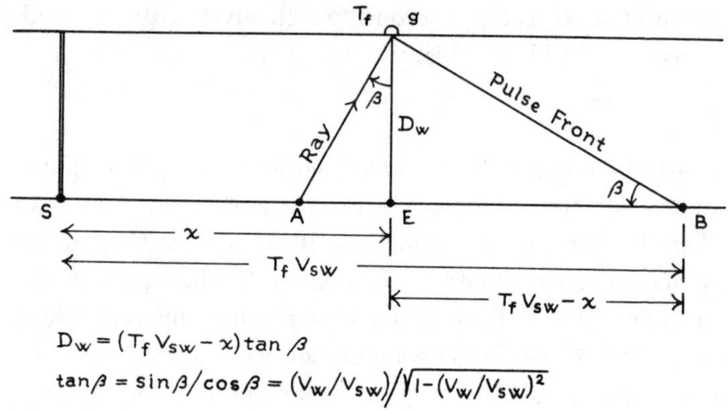

FIG. 6.17. Correction when geophone is at long offset distance from shot hole.

same as the travel time from *S* to *B*. If we multiply the travel time from *S* to *B* by the subweathering velocity, which we know from the plot in Figure 6.15, we obtain the distance from *S* to *B*. We subtract from this distance the distance x, from shot hole to geophone, and this gives us one of the legs of a right triangle, *BE*. We then know that the depth of the weathering is *BE* times the tangent of the angle β. The tangent of β can be easily computed, since the sine of β is the ratio of the weathering velocity to the subweathering velocity. Also the tangent of an angle is the ratio of the sine to the cosine. We have expressed the cosine as the square root of 1 minus the sine squared. This is the formula for the tangent of β, given below the diagram. This may be combined with the formula for the depth of the weathering.

This formula tends to become more accurate as we go farther away from the shot point. When we get close to the shot point, if the shot is appreciably deeper than the base of the weathering, the formula is very inaccurate. On the other hand, the approximate formula referred to in Figure 6.16 becomes more and more accurate as the offset distance approaches zero.

T_f	0.050
T_{SL}	0.035^{1}
T_W	0.014^{9}
T_{WD}	0.010^{2}
T_g	0.025^{1}

FIG. 6.18. Table for computation of datum correction for long offset case.

We thus have two kinds of computations for the depth of the weathering based on different approximations; one is very good for short distances and the other is very good for long distances. Probably, for most purposes, the simpler computation (Figure 6.16) is the better one to use, since simplicity reduces the chance of making numerical errors. Furthermore, the errors that are involved are systematic and contribute only small amounts to the difference in reflection times from one shot point to another.

The work may be organized simply, as illustrated in Figure 6.18. Here the top line gives the observed travel time. The next line gives the slant time in the subweathering material. The next line, the difference between these two times, is the weathering time. The next line shows the time from weathering to datum, which is the elevation of the weathering minus the elevation of the datum divided by the subweathering velocity. The last line gives the total T_g, which is the sum of the two figures immediately above it. The T_s part of the datum correction is computed as described in Section 6.1.

We have not solved the problem of finding the weathering velocity. This is considered in the next section.

6.3. UP-HOLE SHOOTING FOR WEATHERING VELOCITY

One of the commonest methods for finding the weathering and subweathering velocity is to shoot at several hole depths

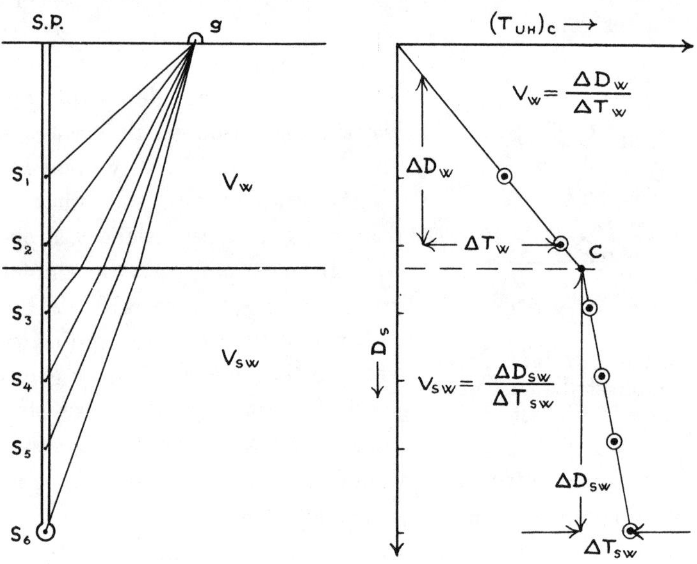

Fig. 6.19. Vertical section illustrating up-hole shooting to determine weathering and subweathering velocities.

Fig. 6.20. Time vs. depth plot from up-hole shooting.

in a deep shot hole, as in Figure 6.19. The geophone, *g*, records the travel times from the various shot locations as indicated. When these travel times have been recorded each one is corrected to vertical by multiplying the time by the cosine of the angle between the vertical and the direction from the shot to the geophone. These corrected times are then plotted on a graph, as in Figure 6.20, with corresponding depths as indicated. We then determine the weathering velocity by measuring the change in weathering depth divided by the change in weathering time.

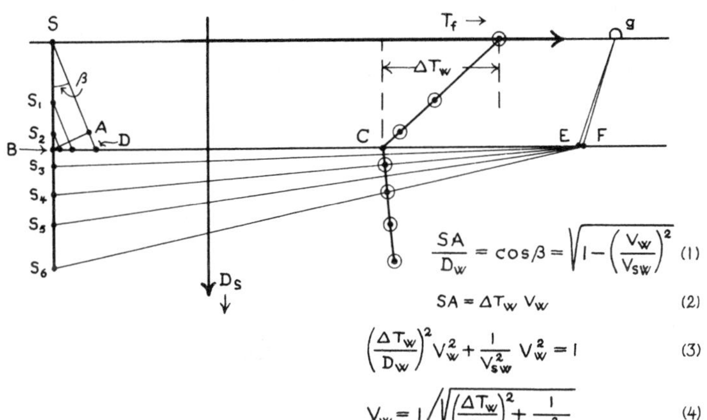

FIG. 6.21. Up-hole shooting when geophone has long offset from shot hole, illustrating plotting and calculation of velocities.

It is supposed that the break in the curve, located at point *C*, is at the base of the weathering. Thus, this measurement gives the depth of the weathering at the shot hole.

The subweathering velocity is found by taking an interval of distance in the subweathered layer, ΔD_{sw}, and dividing this by the corresponding time interval, ΔT_{sw}, as illustrated in the figure.

A second method of obtaining the weathering velocity is illustrated in Figure 6.21. The shot point is at the left of the figure and the geophone, with a relatively large offset, at the right. The shots are taken at various levels in the shot hole, as before. The travel times are plotted on the graph in the center of the figure. The base of the weathering is the point C, which again shows a sharp break if the weathering velocity changes sharply.

The interpretation, in this case, is not quite as simple as in the former case, but it can nevertheless be made. Let us imagine a shot fired at point E in the interface, as shown. The energy will travel, in part, to the left and will reach point D. It will then be partly refracted up to A and will reach this point at the same time that it reaches point B in the shot hole. Point B is at the interface level in the shot hole. Thus the travel time from E to D to A is the same as the travel time from E to D to B. So the difference in travel time from B to E to g and from S to A to D to E to g is entirely occupied by traveling from S to A at the low velocity. Now SA divided by the depth of the weathering is the cosine of the critical angle β, which is given by the square root of 1 minus the ratio of the squares of the weathering and subweathering velocities (see relation 1, Figure 6.21). Also SA is the difference ΔT_w, times the weathering velocity. We may substitute this value for SA in the first formula and, after squaring and transposing one of the terms, obtain the third formula in the figure. This formula can be solved for V_w^2 and from this we can obtain V_w by extracting the square root of the whole expression, as indicated in the fourth formula. This formula gives the value of the weathering velocity in terms of ΔT_w, the weathering depth, and the subweathering velocity.

Thus we see how we can determine the weathering velocity by shooting up the deep shot hole, either with a geophone of small offset, as shown in Figure 6.19, or with a geophone of large offset, as shown in Figure 6.21. Usually it is desirable to have two geophones for the up-hole shots so that both methods may be used for the computation.

Both methods are subject to considerable uncertainty. In the first place we have to measure rather small times, and even smaller differences in those times. Usually in such measurements we have only one significant figure for time differences. This can give us no more than about 10 percent accuracy, so the weathering velocity is likely to be subject to about this much uncertainty. We should remember also that the drilling has probably influenced the velocity of the low-velocity section in the neighborhood of the shot hole. This has not helped our determination of weathering velocities and it has rather seriously influenced the travel times measured by the procedures described above. Nevertheless, these methods are the ones usually used, because other methods are subject to uncertainties even more severe in nature.

6.4. CALCULATIONS BASED ON WEATHERING SHOTS

In the early days of reflection prospecting, special shallow refraction shots, called *weathering shots,* were taken. First a shallow hole was drilled, usually with a hand auger, at one or both ends of the seismograph spread, as illustrated in Figure 6.22, by points S_1 and S_2. Shots were then fired at these two points and records of travel time taken for the various geophone locations. The geophones nearest the shot points recorded the travel times over paths more or less straight from the shot to the geophone. These were the direct paths through the weathering and gave a measure of the weathering velocity, at least in the shallow part of the weathered layer. As we proceed outward away from either shot we find that the slope of the time-distance graph suddenly diminishes at a certain critical distance and we then begin to take advantage of the higher subweathering velocity. The energy travels from shot point S_1 down to the higher-velocity layer, then along that layer and again up to the geophone. By plotting the time-distance graph in both directions, we can obtain the two weathering velocities V_{w1} and V_{w2}, as well as the two higher velocities V'_{sw} and V''_{sw}, as indicated in the figure. The formulas for these four velocities are given immediately below the diagram. We then take our weathering velocity as the average of the

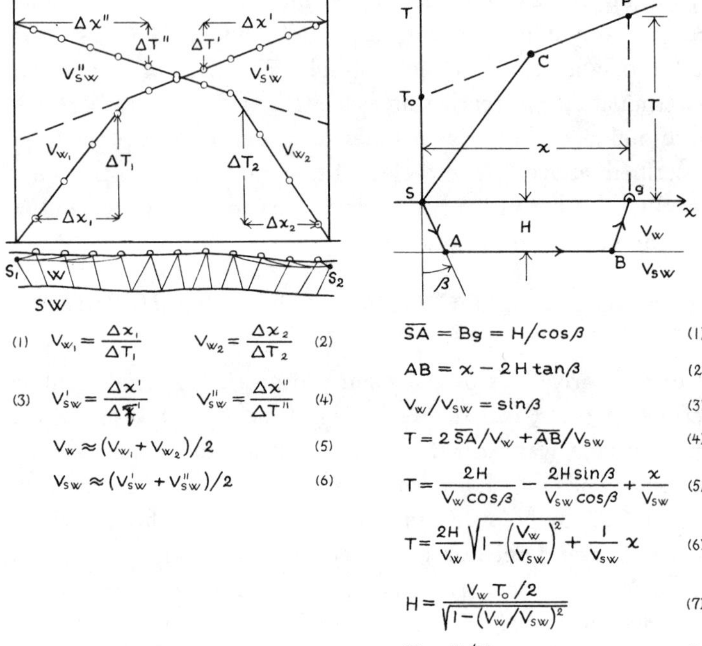

FIG. 6.22. Vertical section of weathering and time-distance graph for shallow refraction weathering shots.

FIG. 6.23. Calculation of weathering time from refraction data.

two weathering velocities thus obtained. The subweathering velocity is taken as the average of the two subweathering velocities. We thus have two velocities which we can use in further computations.

Refer now to Figure 6.23, which shows shot point S and

geophone g. The energy is supposed to travel from the shot down to point A, then along the base of the interface to B with subweathering velocity, and then up to g. In all this travel the path follows Snell's law; that is to say, the sine of β (Figure 6.23, equation 3) is equal to the ratio of the weathering velocity to the subweathering velocity. Since we have already determined both these quantities, we know the angle β. The distance SA is equal to the distance Bg, if there is no dip, which we assume in this calculation. These two distances are each equal to the depth H, divided by the cosine of β, as shown in relation 1. AB, the other part of the travel distance, is the spread distance x, minus $2H$ times the tangent of β, as shown in relation 2. The travel time is made up of two parts: one, the two paths in the weathering, which we may write as 2 times SA divided by the velocity of the weathering; and the other, the path AB with the subweathering velocity, which we may write as AB divided by V_{sw}. This result is shown as relation 4. Making use of relations 1 and 2, we can write relation 4 in the form of relation 5. Using Snell's law (relation 3) and a little algebra, we can rewrite relation 5 in the form shown in relation 6. This equation gives the time as a function of the distance, x. The slope of the graph of relation 6 is the reciprocal of the subweathering velocity. If we make x equal to zero in this formula, we obtain the intercept time, T_0, which can be used to calculate the weathering thickness H, as shown in formula 7. The weathering time, T_w, is the weathering thickness divided by the weathering velocity (relation 8). In this way we obtain both the weathering thickness and the weathering time.

We can carry out the above calculation for any geophone. We may also suppose that we know the elevation of the geophone. So, since we know the *depth* of the weathering, we know the *elevation* of the base of the weathering. We can then compute, as before, the travel time from the base of the weathering to the datum and add this time to the weathering time to get T_g, which always means the time from the datum up to the geophone.

The above computation has been described as if all the measured parts really lay on a straight line. In actual cases they do not lie on such a line. A small adjustment may have to be made in order to correct the weathering and the weathering time for this slight offset of the points from a straight line. This correction is illustrated in Figure 6.24, which shows an offset point, with offset time, ΔT. Here the actual time observed is longer than the time represented in the straight-line graph. Therefore, the amount of weathering under this geophone is somewhat greater than that computed for the straight line. The amount of increased weathering is the increase from B to E. The small triangle FBE has been redrawn much expanded, as indicated. ΔT is the change in time. This is made up of two parts. One is a subtracted part that arises because the path FB in the subweathering material does not have to be traversed. Thus the time corresponding to this latter path is subtracted. The other is the added travel time over EB, traversed at the weathering velocity. This gives the first formula under the diagram. The second formula shows the solution for the increased depth of weathering, FE, in terms of the weathering velocity V_w, the change in time ΔT from the straight-line position, and the cosine of the critical angle β, which we have previously calculated.

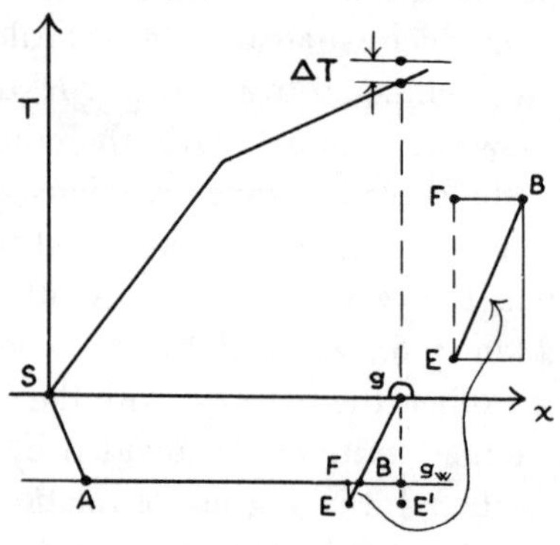

$$\Delta T = -\overline{FE}\tan\beta / V_{sw} + \overline{FE} / V_w \cos\beta \quad (1)$$

$$\overline{FE} = V_w \Delta T / \cos\beta \approx V_w \Delta T \quad (2)$$

FIG. 6.24. Adjustment for small irregularities in first breaks.

Thus the small adjustment in weathering depth, FE, is made on the basis of this formula. Since the adjustment is very small, we may approximate it by saying that FE is equal, or approximately equal, to the weathering velocity times ΔT. Since ΔT is of the order of only three or four thousandths of a second, the error involved in ignoring cos β is negligible in most cases.

By the procedure outlined above, a series of weathering base points, like E in Figure 6.24, can be constructed. These points can be joined by a smooth curve which will represent the base of the weathering. Thus the depth of the weathering, D_w, under each geophone, can be found. Using the elevation of the geophone E_g, we find the elevation of the weathering base to be $E_w = E_g - D_w$. The depth from weathering to datum will be $D_{wD} = E_w - E_D$, where as before E_D is the elevation of the datum. The time from datum to geophone is $T_g = D_W/V_W + D_{WD}/V_{SW}$. Thus we achieve our goal, namely, computing T_g from weathering shots.

Because of its added cost, the method just described is not much used at present. However, it has its uses, especially in velocity profiling (see Chapter 7), when a very careful weathering correction must be made. But it must be understood that the result achieved does not have much absolute reliability. The reliability is relative and is very good from geophone to geophone. Thus although the errors resulting from the method may be appreciable because of using too low a weathering velocity, they are not ordinarily serious, as they apply approximately equally to each geophone computation, so that the relative correction from geophone to geophone is very good.

6.5. *DIFFERENTIAL WEATHERING CORRECTION*

A differential weathering correction is the one that is made for computing dips when the shot point is symmetrically located with respect to the geophone spread. Such a setup is illustrated in Figure 6.25. The shot point, S, is in the middle. We assume that x is the distance from the center of the spread out to the end geophones.

The differential weathering correction—or, as it is abbreviated, D.W.C.—is defined as the difference between the T_g of the top trace and the T_g of the bottom trace on the record.

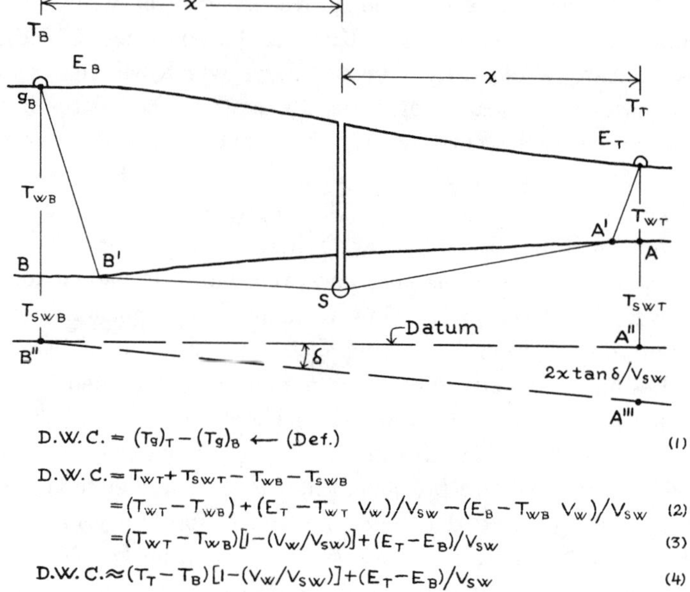

FIG. 6.25. Vertical section showing differential weathering correction to be used for split-dip computations.

It will be remembered that T_g is the time from the datum up to the geophone in each case. This equation appears directly under the diagram in Figure 6.25.

We compute the D.W.C. as indicated in the second and third relationships in the figure. Now we make the *approximation* that the first-break time for the top trace, T_T, minus the first-brake time for the bottom trace, T_B, is equal to the difference between the weathering time on the top trace and the weathering time on the bottom trace. This approximation is indicated by changing from the *equality* sign in relation 3 to the *approximate equality* sign in relation 4. We can see from the diagram that this approximation may not always be very good. However, if we compute the differential weathering correction for even a rather severe example, we shall find that it is at least in the proper direction. Figure 6.25 represents a case in which the weathering thickness and shot hole depth are unusually great compared with the spread from the top to the bottom geophones. For most cases, the present formula is sufficiently accurate; furthermore, it takes into account the dip of the base of the weathering very well.

If the base of the weathering does not dip, the formula can be simplified a great deal because we can replace formula 4 in Figure 6.25 by D.W.C.$=T_T-T_B$. We can readily see the reason for this if we look at relation 2, because the time in the subweathering for the top trace is the same as the time in the subweathering for the bottom trace.

We may also consider the problem of making a differential weathering correction when the datum dips, say with an angle δ, as indicated in the figure. We have merely to add to the differential weathering correction $2\ X \tan \delta / V_{sw}$, as indicated. A slanting datum is convenient to use sometimes when there is a rapid lateral variation of velocity.

6.6. CORRECTIONS FOR POINT PLOTTING

Some companies make a practice of plotting a depth point for every trace or perhaps every alternate trace on the record. When this is done, the depth point is usually plotted at a location on the cross section midway between the shot position and the geophone position. For this type of plotting it is necessary to make a datum correction at each geophone from which the depth is plotted. This may be done by means of any of the above methods except the first one. That method, of course, is applicable only for the nearest geophone.

Sometimes a quicker and, in some ways, a more satisfactory correction can be obtained by using one of the shallower good reflections. This reflection is plotted as if it corresponded to a plane reflector. The T_g values for the center trace and the end traces are then computed, and the intermediate corrections are forced so that they yield the plane plot. This method of point plotting requires a great deal of labor. In order to facilitate the work, a plotting machine has been devised by R. B. Hale; it is built by the Ruska Instrument Corporation.

6.7. CORRECTIONS TO A DEEPER DATUM

The primary object of corrections to datum is to remove the irregular effects of the highly irregular shallow material. It sometimes happens that this irregularity extends deeper than what we ordinarily regard as the weathered layer. When this is the case, it is necessary to make the correction to datum a little deeper. This may be done, in some cases, by making isopach maps based on the time from a shallow reflection to all the deeper reflections. When the shallow reflections are good and plot regularly, this procedure is highly recommended. It is especially recommended when it brings order out of chaos on the deeper part of the cross section, as it often does.

6.8. INTERRELATION OF METHODS. PREFERENCES

In the preceding sections of this chapter we have described almost every method of correcting to datum that has been used in the industry. We may not have emphasized each particular method as a particular case. For example, some companies use a different datum for every shot, namely, either the level of the shot itself, or the base of the weathering, assumed level. In each case, the principal recommendation for a method is that the people who work for the particular company are accustomed to using it. This is very important, because errors are not so frequent when a familiar method is being used. On the other hand, the author has certain preferences. For example, he definitely prefers the method described in Section 6.1 for the nearby geophone, because it does not require the determination of the weathering velocity; this velocity is very difficult to determine accurately. If maximum accuracy is wanted, the method discussed in that section should be coupled with the up-hole shooting described in Section 6.3 and the weathering shots described in Section 6.4. However, for most exploration purposes this degree of accuracy is not required; hence the method described in Section 6.2 is quite adequate. In each case, simplicity, with its accompanying decreased cost and decreased chance of arithmetical error, must be weighed against the more expensive but more accurate process involving the several different procedures. The particular job should dictate the particular method to be used.

We have described methods of preparing seismic records for calculation of depth and dip. This is a vitally important step in the routine. We might suppose that we are ready for the depth and dip computation, but this is not the case.

Before we can compute depths and dips, we have to be able to convert time and time variations into distances and distance variations. For this purpose we need to determine the velocities with which seismic pulses travel in the region below the datum level. This sometimes difficult but always important problem we consider in some detail in the next chapter.

6.9. *COMMENTS TO THE SECOND EDITION*

In this chapter the basic problems are not very different from those of 1950. The differences reside in the larger number of channels (traces) and the machine processing that goes with the large numbers.

Consider a land line rolling along (see the comments at the end of Chapter 2) with 100 receivers with small equal geophone-to-geophone intervals recording travel times from a succession of sources at geophone positions at one end of the line of receivers, as in Figure 6.26, where we follow Lincoln

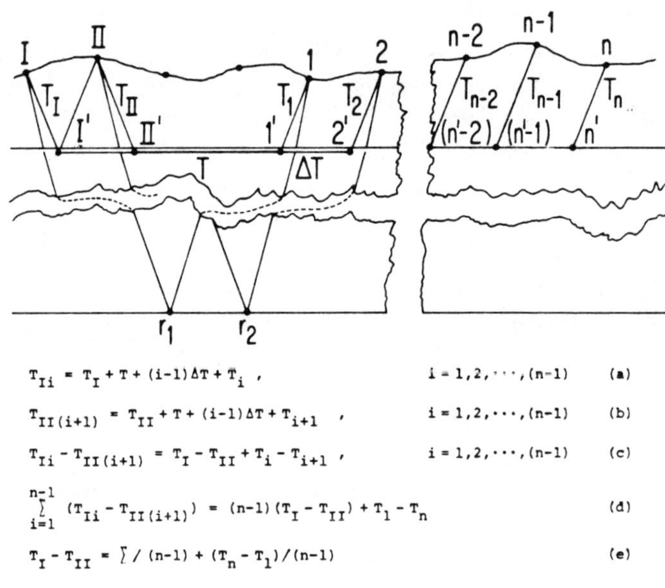

FIG. 6.26. Static Corrections.

A. Martin's patent description (4,101,867) of a suitable procedure.

Regard *I* as a pulse (or a Vibroseis®) source, where receivers are at *II*, . . . , 1,2, . . . , $n-1$, n, equally spaced, so that ΔT is the travel time from 1′ to 2′ (=2′ to 3′=. . .= ($n'-1$ to n'). Station 1 is taken far enough from the source, *I*, so that we have some assurance that the true refraction path, *I*,*I*′,1′,1 is being used, rather than a set of nearly critical reflection paths like I, I', II. Also, n is taken small enough so that we continue to have shallow refractions.

Let T_I be the travel time from I *to* I', T_I' from 1′ to 1, and so on. Then equation (a) gives a series of refraction times for source I, and equation (b) the same for source II. The differences of (a) and (b) are given by (c). Summing (c), we get equation (d). Solving for $T_I - T_{II}$, we get equation (e). The term $\Sigma/(n-1)$ is the mean value of the left-hand differences in equations (d). The term $(T_n - T_1)/(n-1)$ approaches zero as n increases.

Martin discusses the old "eye ball" approach and the modern cross-correlation techniques of picking $T_I - T_{II}$. The old direct approach is easily carried out but consumes much time. The computer cross correlation is quick. Note that these static corrections are not absolute, but consist only of station differences, which smooth irregularities without giving a correction to a real depth datum plane.

CHAPTER 7

Velocity Measurements

This chapter is designed for exploration seismologists and computers. However, the instrument operator who secures the records will find in it an explanation of what he is trying to obtain.

After the datum corrections are made on the regular seismograph records, a knowledge of the velocity of seismic pulses in the subsurface section is necessary in order to proceed with the interpretation. Usually, velocity measurements are made only at the beginning of a survey of an area. When the velocities have been determined, they are used throughout the area.

7.1. WELL SHOOTING

The most direct method of determining the velocity of seismic pulses in a subsurface section is to lower a geophone in a well and measure the travel time from near the surface down to various points in the well. This process is illustrated in Figure 7.1, which shows a well and a sequence of four geophone locations, G_1, G_2, G_3, G_4, in the well. The shot is located at the point S near a datum which is indicated by a dashed line and the letter D.

The only special equipment required for the well-shooting job is a special geophone that is arranged to operate properly under the high pressures encountered in a deep well. There are two ways of constructing such geophones. One method involves the use of a pressure equalizer so that the pressure inside the geophone is the same as the mud pressure. This may be accomplished by using a piston arrangement or by connecting the interior of the geophone to the exterior by a long oil-filled tube. The other and more satisfactory method, however, involves damping the well geophone by electromagnetic means, in which case the pressure inside this geophone must remain relatively small at all times. This requires a high-pressure sealing process.

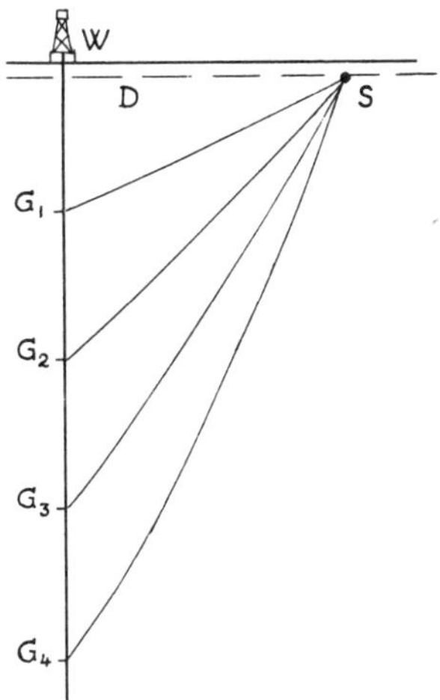

Fig. 7.1. Well shooting for determining seismic velocity.

The well geophone is attached electrically to the cable by which it is lowered into the well. This cable is usually the same one that was used in making an electrical log of the well. The geophone is lowered into the well to a certain measured depth relative to a reference elevation at the well. Usually the reference elevation is taken as the top of the derrick floor, but it may be the average ground elevation across the well. In any case, the depth has to be carefully measured.

Lowering a geophone into the well disturbs the equilibrium of the mud. Hence it is generally necessary to wait a few moments for the mud to become quiet again before shooting. The instrument operator can observe the noise on his camera traces created by bubbles and by temperature changes. When the geophone has reached a quiet stage, he orders the shot and makes the record.

The record should have the cap-break trace showing the time the shot was fired. The well geophone trace or traces must also show. Since it is difficult to arrange the amplitude of the well geophone trace in advance, it is common practice to use three traces for this part of the record. One is a low-level trace which is designed to get all the secondary parts of the well geophone pulse. Another is a very high-level trace which is designed to make as certain as possible that the very first of the pulse reaching the well geophone is picked up. The third is a medium-level trace which is designed as a sort of compromise, in case the high-level trace proves too noisy just before the pulse reaches the geophone. There should also be an up-hole geophone trace and a reference geophone trace. The reference geophone may be placed on one of the concrete piers of the well. It is very important that this reference trace always be included as a check for several reasons. One reason is that although the precise time of the firing of a cap is almost always known accurately, there is sometimes a little doubt. The reference geophone trace will remove this doubt. Another reason, which has nothing to do with the characteristics of the cap or of the charge, concerns a delay in the neighborhood of the shot, apparently due to firing several shots in the hole. By using the reference geophone trace any delays, or advances, introduced in this way can be corrected.

The record is developed and fixed and washed and inspected. If it is satisfactory, the well geophone is lowered to another level.

If the mud in the well is rather thick, so that some difficulty is experienced in lowering the geophone, it is customary to lower it clear to the bottom of the well, or almost to the bottom, so that successive locations for shooting can be obtained by pulling the geophone up instead of having to wait while the slow process of lowering it goes on. On the other hand, it is disappointing indeed to lower a geophone all the way down and then find that something is wrong. Therefore it is considered good practice to take at least one shot on the way down.

We will now suppose that we have a set of records shot in the well and are ready for the interpretation. For this purpose we refer to the graph in Figure 7.2. On this graph, depth is the ordinate and time is the abscissa. The curve passing through the origin shows vertical travel times. The other curve, indicated by square points, is obtained from the observed slant times.

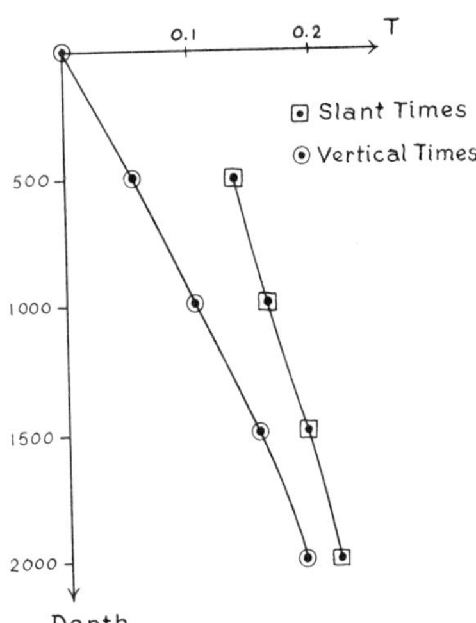

Fig. 7.2. Time-depth plot of well shot data.

We can describe the preparation of the records by reference to Figure 7.3, which shows a sample of a record label. This label may be pasted on the record, either before the cap break or at the other end of the record. The label has space for the well name at the top, for easy identification. The shot letter (*F*, in this case) means that this is the sixth shot of the survey. The weight of the charge is 4.5 pounds. The elevation of the shot point is 392 feet, and the depth of the shot is 85 feet. The elevation of the shot is the difference between the elevation of the mouth of the shot hole and the depth of the shot, or 307 feet. We have taken a datum at 292 feet. The depth from shot to datum is 15 feet. In this example we have taken a subweathering velocity of 5000 feet per second. The 15 feet divided by this velocity gives 0.003 second for the time from the shot to the datum. The directly read time on the record was 0.388 second. This is corrected to datum by subtracting from it the time from the shot down to datum, which gives 0.385 second. Thus, the last time on the record label is the *slant time* corrected to datum at the shot.

Well ____				Trace Meaning
Shot F Weight 4.5		El. Well Ref.	E_w +392′	Well G. High
El. Shot P.	E_{SP} +392′	Depth Well Geo.	D_G 3000′	" " Med.
Depth Shot	D_S 85′	El. Wl. Geo.	E_G −2608′	
El. Sh.	E_S +307′	Dpth. Dm. Geo.	Z 2900′	" " Low
El. Datum	E_D +292′	Hor. Offset	H 350′	
Dpth. Sh. Dm.	D_{SD} +15′	$H/Z = \tan\theta$	0.1243	Cap Break
Time Sh. Dm.	T_{SD} +0.003	$\cos\theta = C$	0.9924	Up-Hole Geo.
Read Time	T_R 0.388	Time Vert.	T_V 0.382	Ref. Geo.
T_R Corr. Dm.	T_{RC} 0.385	Date ____	Com. by ____	

FIG. 7.3. Record label for computing well shot data.

In the next column we write first the elevation of the well reference level, 392 feet. The depth to the well geophone below this reference level is 3000 feet, so the elevation of the well geophone is −2608 feet. The depth below the datum of the geophone is the elevation of the datum minus the elevation of the geophone, or 2900 feet. In this particular well job, the horizontal offset of the shot point was 350 feet. This horizontal offset and the depth are used to compute the tangent of the angle between the vertical and the direction from the geophone to the shot reduced to datum as viewed from the geophone. This gives the tangent of θ. The cosine of this angle can be found in a table of natural trigonometric functions. Since the tangent is 0.1243, the cosine is 0.9924. The vertical time is found by multiplying the slant time by this cosine, which gives 0.382. The date of the computation and the computer's initials ought to appear on the label for later reference.

The next column gives the meaning of each trace. This is important to know for each record, especially after the survey has grown cold in the oil company's files. The meanings of the six traces used in the present example are indicated. The top trace is the well geophone trace for high level, the second trace is that for medium level, the third that for low level. The fourth trace happened to be the cap-break trace and the fifth trace the up-hole geophone trace. The last trace is the reference geophone. Such a label can be made for each record of the survey.

When the records have been computed and labeled, as in Figure 7.3, a table of data for all the records can be prepared; an example of such a table is shown in Figure 7.4. This table provides space for the well name, location, elevation and reference level (which is described), elevation of the datum, the name of the company shooting the well, the date of shooting, the kind of amplifier used, and the cable and geophone used. The first column at the left shows which shot hole was used in the survey. An engineer's plot of the surveying data on the shot holes and well should also be shown. The depth of shot is given in the second column from the left. The grade of the "pick" is given in the third column; this has some importance for otherwise an odd survey point may be overvalued. The fourth column gives the depth from the datum to the geophone; this is taken from the record label in each case. The depth used for the label in our particular example is 2900 feet. The next column is the slant time corrected to datum. The 0.385 second on the last line at the left of the record label in Figure 7.3 is on the sixth line in this column. The vertical time is shown in the sixth column; our figure, 0.382, is on the sixth line of this column. The next column shows the average velocity, which is obtained by dividing the depth from datum to geophone by the corrected vertical time. In other words, the figure in the fourth column divided by the corresponding figure in the sixth column gives the figure in the seventh column. The interval velocity in the last column is

Well Name ______ Shot by ______
Location ______ Date ______
Elevation of ____=______ Amplifier ______ Cable ____
Elevation of Datum =______ Geophone ______

Shot Hole	Shot Depth	Grade	Depth Datum to Geo.	Slant Time (Corr.)	Vert. Time (Corr.)	Aver. Velocity	Interval Velocity
I	100'	A	400	0.080	0.060	6667	6667
I	100'	B	900	0.137	0.128	7031	7353
I	95'	B	1400	0.203	0.197	7107	7246
I	90'	B	1900	0.266	0.262	7252	7692
II	97'	B	2400	0.320	0.317	7571	9091
I	85'	B	2900	0.385	0.382	7592	7692
I	85'	C	3400	0.442	0.440	7727	8621
II	100'	B	3900	0.500	0.498	7831	8621
I	85'	B	4400	0.556	0.554	7942	8929
I	85'	B	4900	0.611	0.609	8046	9091
I	85'	C	5400	0.660	0.659	8194	10000

FIG. 7.4. Form for tabular presentation of results of well shooting.

obtained by dividing the difference in successive vertical depths by the difference in successive vertical times. Since the difference in vertical depths is 500 feet in each case, we divide 500 by the difference in travel time.

The results in the last five columns in Figure 7.4 are plotted graphically in Figure 7.5. This graph shows the time vs. distance graphs, both the vertical time (circles) and the slant time (squares). In addition are shown the average velocity vs. depth and the interval velocity vs. depth. It will be noted that the average velocity is a fairly regular curve, whereas the interval velocity has a tendency to scatter somewhat. This scatter is to be expected, because the interval velocity is computed from time differences which are not determined with great accuracy. For example, the interval velocity from 2400

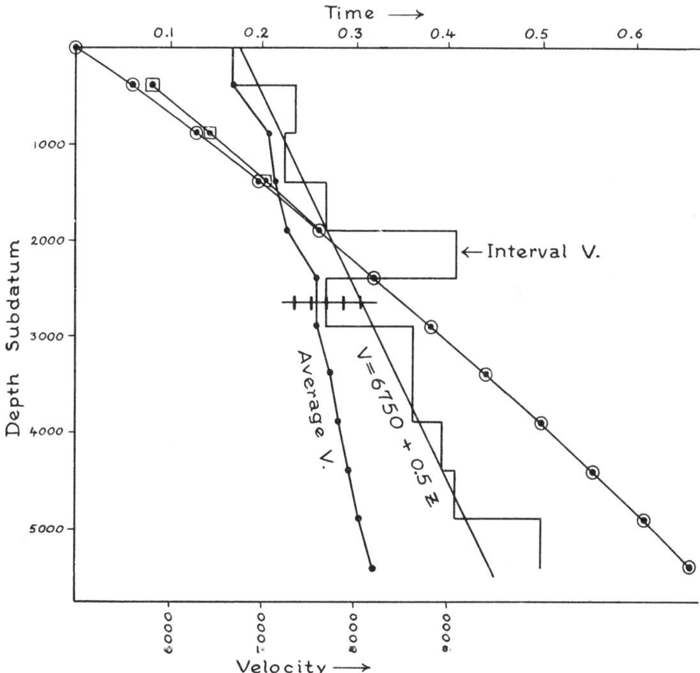

FIG. 7.5. Graphic presentation of results of well shooting.

feet to 2900 feet corresponds to a difference of depth of 500 feet, but a difference of time of 0.065 second. The two values from which this difference is derived can each easily be in error by ±0.001 second. Consequently, the difference has a corresponding standard error of ±175 feet per second. It is very unusual to have well shot records of this accuracy; ±0.002 second is more common, which gives a standard error of ±340 feet per second in the interval velocity. These two ranges have been indicated on the graph for the interval from 2400 to 2900 feet. An additional curve in this figure, the straight line, represents a velocity distribution which is 6750 feet per second at the datum and increases at the rate of 500 feet per second per 1000-foot increase in depth. It will be noted that the difference between the slant time and the vertical time for this small offset, 350 feet, is very difficult to see on the graph. If the offset were 1000 feet, this difference could be seen more easily.

The table (Figure 7.4) and the graph (Figure 7.5), together with a report describing the conditions of the work and giving details not included in the table and graph, complete the report on the well velocity survey.

The straight-line velocity vs. depth relationship shown in Figure 7.5 is obtained from the interval velocity graph by a direct eye fit. A more thorough method of making the fit has been described in detail by J. A. Legge and J. J. Rupnik.*

Another method has been devised by N. H. Miller but not previously published. We shall describe his method by reference to Figure 7.6, the upper portion of which is, in part, the same as Figure 7.5. The problem is stated directly below the drawing. We shall now carry out the various steps of Miller's procedure. The first step is to pick a deep reliable point on the graph, P_2, as shown. P_2 has time T_2 and depth z_2. The next step is to take point P_1 so that time T_1 is one-half T_2. T_1 must be interpolated between two of the observation points or else be one of these points. In the third step we use the relationship between the time and the depth, on the hypothesis that the interval velocity is linearly distributed with respect to the depth. In the last part of this step we use the fact that the difference between two squares is the product of the sum and difference. The fourth step makes use of the same formula for

* J. A. Legge, Jr., and J. J. Rupnik, *Geophysics*, 8:356 (1943).

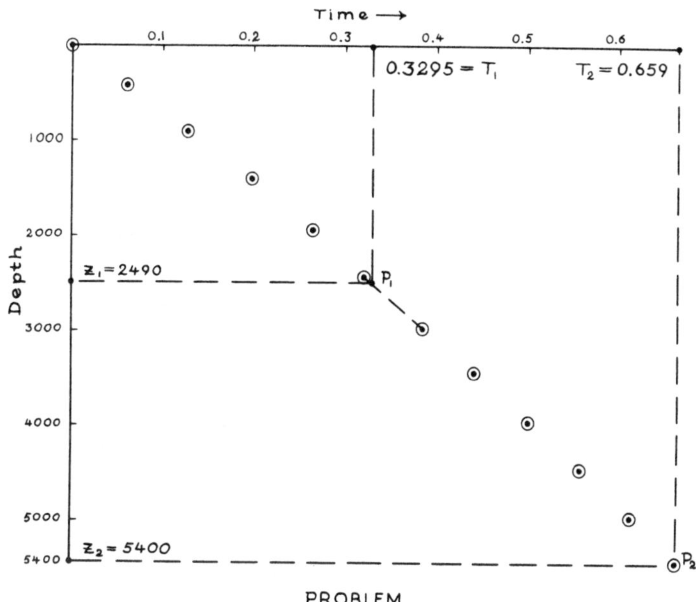

PROBLEM

Given $V = V_D + az$ and Vertical Time vs. Depth

Find V_D and a

MILLER'S PROCEDURE

(1) Pick a deep reliable point P_2 as shown

(2) Pick P_1 such that $T_1 = T_2/2$ and P_1 is on graph

(3) Note from formula Figure 8.20

$$z_2 = \frac{V_D}{a}\left(e^{aT_2} - 1\right) = \frac{V_D}{a}\left(e^{2aT_1} - 1\right) = \frac{V_D}{a}\left(e^{aT_1} - 1\right)\left(e^{aT_1} + 1\right)$$

(4) Since $z_1 = \frac{V_D}{a}\left(e^{aT_1} - 1\right)$ $\quad \frac{z_2}{z_1} = e^{aT_1} + 1$ or $e^{aT_1} = \frac{z_2}{z_1} - 1$ or

$$a = (2.3/T_1)\log_{10}\left[(z_2 - z_1)/z_1\right] \qquad \text{(I)}$$

Also $$V_D = az_1/\left(e^{aT_1} - 1\right) \qquad \text{(II)}$$

FIG. 7.6. Miller's method of finding velocity distribution from well data.

the depth of point P_1 in terms of the time T_1. This gives a relationship which we may solve for a, as indicated in formula I. The last step consists of solving for the datum velocity, as indicated in formula II. Using these formulas for a and for the datum velocity, V_D, we have our result. Let us now apply this to the values given. If we substitute the values in formula I, we find that a is 0.472. This value is then substituted in formula II to find V_D; this is 7000 feet per second. Thus we have for our velocity distribution $V=7000+0.472z$, which is in fairly good agreement with $V=6750+0.5z$, which was obtained by an eye fit. Miller's method of fitting is easy to use and gives a result which is probably sufficiently accurate, considering the nature of the data upon which it is based. His method could be extended to 3 points or even more. If 3 points are used, a quadratic equation will have to be solved. This is easily done. More points give equations of higher degree, which can likewise be solved.

Still another method has been described by Slotnick, Brooks, and Redding.* They select two or three points on a graph, as in Figure 7.6. For each of these points they insert the values of the depth, z, and the time, T, in $z=(V_D/a)[\exp(aT)-1]$. This gives, for each such point, a relationship between a and V_D which they graph. These graphs tend to intersect in a small region of the a, V_D graph plane. The center of gravity of the intersection points they select as the point which determines the a value and the V_D value they use.

An even simpler method of fitting the data is based on the assumption that the average velocity is a linear function of the vertical reflection time, i.e.,

$$V_A=2z/T=kT+k'.$$

By graphing V_A against T both k and k' can be determined,

* M. M. Slotnick, J. A. Brooks, Jr., and V. L. Redding, *ibid.*, 15:663 (1950).

when a straight line can be a reasonably good approximation to the plotted points. In fact, k' is the value of V_A (on the straight line) when $T=0$ and k is the rate of increase (or decrease) of V_A with increasing T. By differentiation of the above relation with respect to T we find the velocity

$$V = dz/d(T/2) = 2kT + k' = \sqrt{8kz + k'^2}.$$

Although this method is widely used in many areas of gentle dip, an elaborate ray tracing process is required for the correct interpretation of larger dips. If due attention is given to proper determination of the dip and of the location of the point of reflection, this distribution of velocity may be used safely in many areas.

For most practical purposes when we have obtained the fit for the well velocity data we are ready to use this information.

7.2. MISCELLANEOUS PROBLEMS ASSOCIATED WITH WELL SHOOTING

Occasionally we need more accurate shallow data than we can get with the regular procedure outlined in the preceding section. In order to obtain such data, we have to make a more accurate correction to vertical. This correction requires shooting at more than one offset distance. Figure 7.7 shows pulse fronts and rays approaching a well. Also shown is an interface, II', between layers of different velocities. The information that we obtain directly is Δz and a corresponding ΔT. Δz is the change in depth and ΔT is the corresponding change in time. From this information alone we cannot compute the velocity. We have also to obtain Δx divided by ΔT.

We refer now to Figure 7.8, which in a small triangle illustrates Δz, Δx, and the normal to the pulse front, Δn. This is a right triangle with the right angle at A, the legs being AB and AC. AE is perpendicular to the hypotenuse, BC. One of the theorems of elementary geometry is that Δn is related to the legs of the right triangle according to the first equation under the figure. This equation can be multiplied through by $(\Delta T)^2$, giving the second relationship. This equation is known as the *eikonal equation.* It relates the velocity, V, to $\Delta T/\Delta x$ and $\Delta T/\Delta z$, the two components of the *wave slowness.* The wave slowness is the reciprocal of the wave velocity. ΔT is a

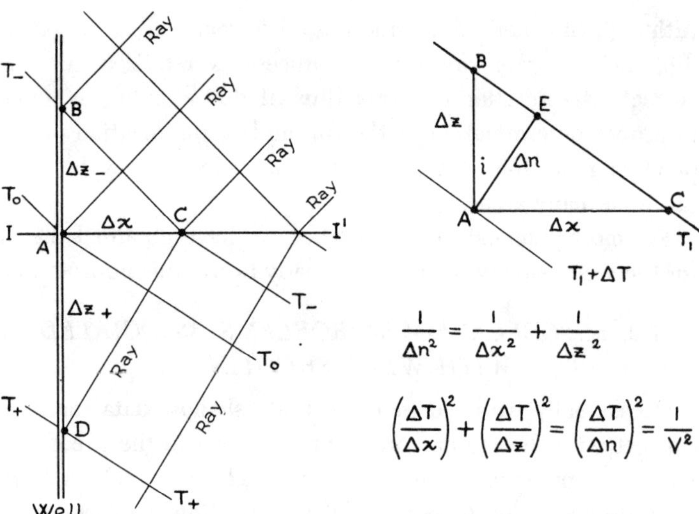

FIG. 7.7. Vertical cross section in well showing interface pulse front and rays.

FIG. 7.8. Deduction of the eikonal equation.

quantity which is measured and Δz can also be measured. Δx can be measured if we shoot at two different offset distances. By computing the quantities involved in the eikonal equation we can solve for V^2 or V.

The above method of finding V presents the same difficulty as any determination of interval velocity, namely, the inaccuracy of the determination of ΔT. However, the method allows us to face the problem squarely.

Another factor in the well velocity data is the effect of the dip of the section through which the well is drilled. If we shoot the well down dip from the offset shot point, we may expect decreased travel time because the path takes greater advantage of any high-velocity material in the section. This is based upon Fermat's principle that the first energy travels over the path which takes the least time. Snell's law is merely an expression of this principle. If, on the other hand, we shoot up a gentle dip, the assumptions made in correcting to vertical by using the cosine factor (described in Section 7.1) are more nearly correct than when the section has no dip at all. Unless the dip is very high, it is generally advisable to shoot the well up dip. On the other hand, it may be that the well will have to be shot along the strike. This is not serious if the dip is not strong. But if the dip is strong, the energy will travel over a curved path that does not lie in a plane. In such a case it may be very difficult to make the interpretation more precise than that in Section 7.1.

A very disturbing effect on the records, of course, is noise which occurs before the record of the pulse at the geophone. This noise may be electrical in nature, in which case steps may be taken to eliminate it. Sometimes noise on the records has been supposed to be due to the casing, the energy being presumed to travel to the casing and down it to the geophone. Direct measurements made by the author have shown that the casing is not very likely to transmit noise. For example, a cap attached to the casing created a very small amount of energy which traveled to the geophone. It is the author's opinion that the reason for this small amount of energy being transmitted by the casing is not the fact that the casing does not transmit energy, but primarily the fact that the coupling between the casing and the well geophone was not very good.

Another source of noise is the cable. If a disturbance gets into the cable, it travels down the cable to the geophone. The coupling between the cable and the geophone is excellent. When the pulse that is initiated at the shot point travels outward, the front of it may be supposed to be a compression. This compression may influence the cable in such a way as to create noise which travels down the cable and disturbs the record. We may understand this perhaps by looking at Figure

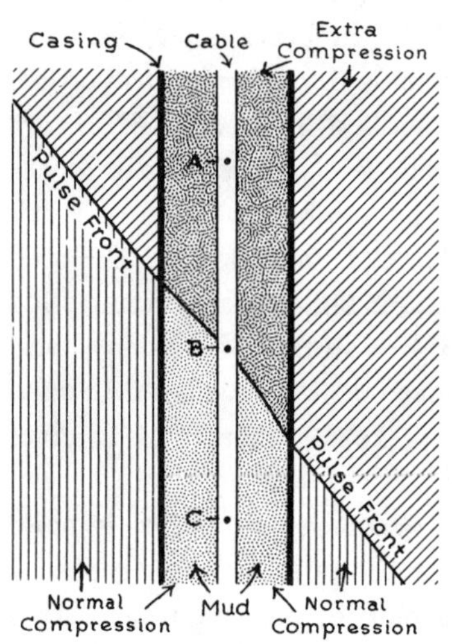

FIG. 7.9. Pulse front passing through well and squeezing cable.

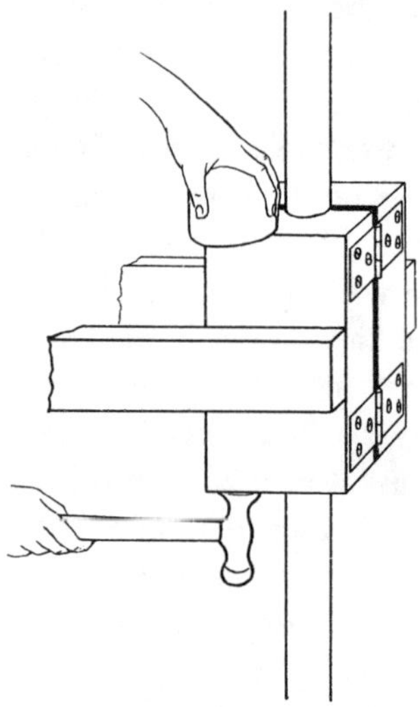

FIG. 7.10. Method for starting pulse at top of well geophone cable for determining cable velocity.

7.9, which shows the casing with the cable hanging in the middle, and a pulse front that passes more or less roughly through the interior of the well as shown. The compression part—or the *extra compression* part, as we shall call it, because the whole system is under high pressure—is the part above and to the right of the pulse front. The *normal compression* part is below and to the left of the pulse front. Three points in the cable are marked *A, B,* and *C.* We may suppose that, at the instant considered, the portion of the cable between B and C is under normal compression. The portion between A and B is under extra compression because of increased pressure in the mud, which, in turn, is due to the extra compression in the section. This extra compression from A to B will tend to lengthen the cable momentarily. This momentary lengthening will initiate a pulse that will travel down the cable and be picked up by the geophone. This pulse can be observed unless the mud has damped this energy out sufficiently.

A great many well velocity determinations have been seriously disturbed by cable noise generated in this manner. This may be verified by measuring directly the velocity with which pulses travel down the cable. For this purpose a wooden clamp may be constructed as illustrated in Figure 7.10. This clamp is held on the cable and a geophone is placed on one of the wooden blocks of the clamp and the lower part of the same block is tapped with a small hammer. The upper geophone gives the time at which the pulse starts down the cable. This time is recorded on a regular seismograph record. The time at which the pulse arrives at the well geophone is also recorded. In this way we determine a time interval, the time of travel from the clamp down to the well geophone. Since we can measure the distance from the clamp to the well geophone, we have a means of measuring the velocity with which the pulse travels down the cable. This velocity may be compared with velocities measured in the well velocity survey.

If we determine the cable velocity, which is likely to be around 9000 feet per second, and shoot at two different offset distances and plot the slant times on the same graph sheet, then, if the energy travels down the cable, the slant times should line up in parallel lines with slopes corresponding to the velocity of the cable. If, on the other hand, the paths are the desired geological paths, the slant times should converge with increasing depth.

One way to avoid trouble because of noise traveling down the cable is to suspend the geophone at the bottom of the cable by means of a spring system. Then most of the energy going down the cable will be reflected back up, thus reducing the coupling between the cable and the geophone. Another method is to increase the offset distance. In parts of the San Joaquin valley of California the velocity of the geological section between the surface and about 6000 feet is often such that the cable noise comes a little before the regular pulse which is desired. In such a case the information above 6000 feet is likely to be faulty. In regions where the velocity of the geological section is well above 9000 feet per second, naturally there are no difficulties from cable noise.

An interesting side line that the author has noted concerns the coupling of the mud to the cable. Cable velocities were measured in two different wells, the same cable being used. In one well a thin mud of low viscosity was used. In the other a heavy mud was used. The measured velocities were appreciably different, being much lower with the heavy mud. The author believes that this is due to an additional loading caused by the mud caking on the cable.

An electrical effect which may be very important is created by the distributed capacity in the cable. Ordinarily the distributed capacity in a cable is so small per foot that we don't think of it, but if the cable is very long, say 15,000 feet from beginning to end, the distributed capacity may become very large. This introduces a capacity in the line between the geophone and the input to the amplifier. This capacity may be so large that we introduce an oscillatory system with a natural period close to the natural period of greatest interest to us. In this case the record will be very poor in the sense that the energy will build up gradually and the record will then oscillate for a long time, perhaps even for two or three seconds. This gives a very poor recording for calculations. As we cannot do anything about the distributed capacity in the cable, we have to make the adjustment at the geophone end. This can be done by making the inductance of the geophone very small or very large depending upon its design. The problem, of course, is to move the natural period of the oscillating system, consisting of inductance and distributed capacity, well away from the period of greatest interest to us.

An effect that is difficult to measure but which may be suspected in many cases is a disturbance that is due to the penetration of drilling fluids into the section while the well is being drilled. Such an effect is often observed in connection with shot holes and may be expected to be of even greater consequence in connection with a drilled oil well.

7.3. SURFACE VELOCITY PROFILING

This section deals with a method of measuring the subsurface velocity when wells are not available.

The process may be described most simply by means of Figure 7.11, which illustrates the simplest possible case. We suppose that S is the shot location and that very near it is located a geophone, g_o. We also locate another geophone, g_x, at a distance x. We suppose that there is a constant-velocity layer of velocity V, with no dip at its base. Then the points of reflection are R_o and R_x for geophones g_o and g_x. The reflection times are T_o for O distance and T_x for x distance. S' is the image point of the shot, S, reflected in the reflector. That is to say, $SR_o = R_oS'$. In triangle $SS'g_x$ we have, according to the Pythagorean theorem, the first formula shown under the figure. This formula can be solved for V^2, as indicated by the second formula. Thus, if we have a distance x and the two reflection times T_o and T_x we can use this second formula to find the velocity, V.

Details on velocity profiling have been given in papers by

Green and Gardner.* Green points out that if an x^2, T_x^2 graph is made, the points will fall on an approximately straight line. Gardner takes account of dip effects and shows that in place of an x^2, T_x^2 plot an ($x^2 \cos^2 \theta$), T_x^2 plot (where θ is the dip angle) will correct for dip effects. This is illustrated in Figure 7.12.

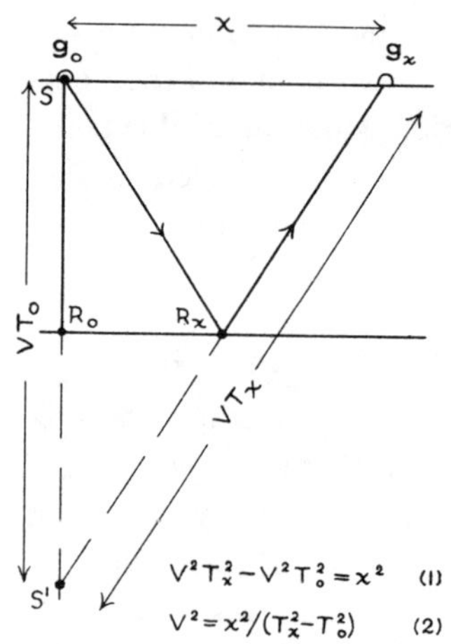

FIG. 7.11. Vertical section showing means of determining velocity from surface seismic shooting.

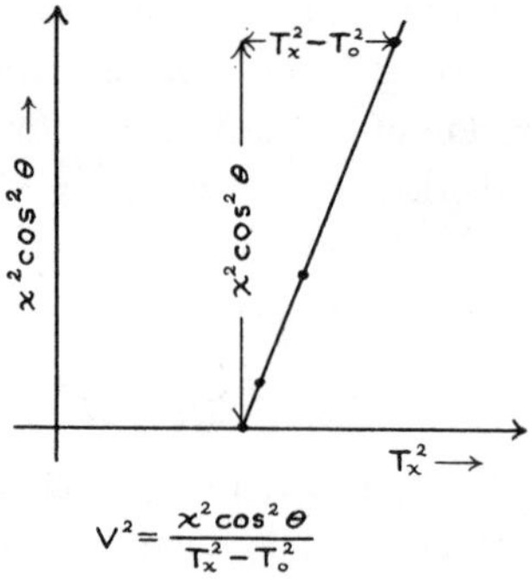

FIG. 7.12. Plot of ($x^2 \cos^2 \theta$) vs. T_x^2.

Regular reflection profiles have been used extensively for the determination of velocities and lateral variations of velocities.† The theory of this so-called "$T\Delta T$ Process" is illustrated in Figure 7.13. The first relation in the figure is the first relation in Figure 7.11 with T_x replaced by $T_o + \Delta T$. In the second relation we have squared and removed canceling terms. In

* Cecil H. Green, *Geophysics*, 3:295 (1938); L. W. Gardner, *ibid.*, 12:221 (1947).

† See W. E. Steele, Jr., *ibid.*, 6:370 (1941).

going from the second to the third, we note that $(\Delta T)^2$ is negligible compared with 2 $T_0 \Delta T$ and then we solve for $T_0 \Delta T$. For a range of reflections where V does not vary much, $T_0 \Delta T$ is nearly constant. Thus if we were to plot ΔT against T_0 we would get approximately a hyperbola as illustrated in Figure 7.14. By fitting a hyperbola to the datum points we can measure x/V as indicated in this figure. Knowing x we find V.

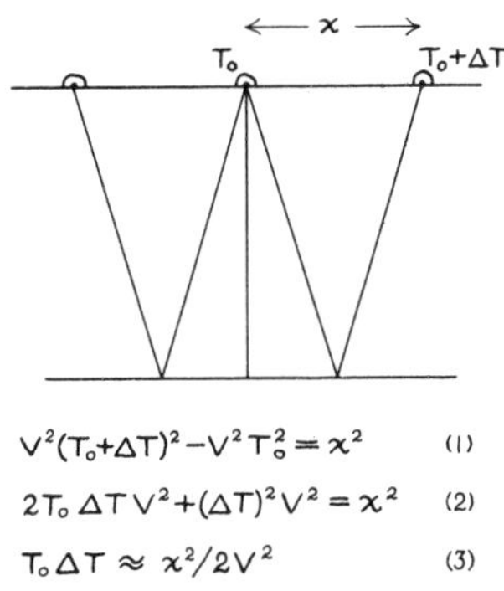

Fig. 7.13. Basis for $T\Delta T$ plot.

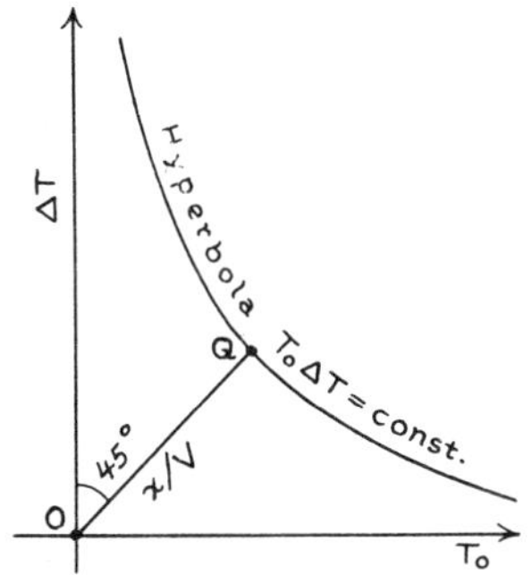

Fig. 7.14. $T\Delta T$ hyperbola.

The $T_0 \Delta T$ method is valuable for detecting lateral variations of velocity such as might seriously distort the seismic interpretation.

7.4. REFRACTION METHODS

Refraction methods have been used in the past to a considerable extent to determine velocities. However, they are seriously handicapped by the fact that they cannot detect a decrease in velocity with increase in depth. Since such decreases are quite common, this is a real handicap. But there are certain areas of the world where the decreases in velocity are negligible and in these areas refraction methods serve very well to determine the velocity distribution.

In this chapter we have outlined methods of determining the subsurface velocity. We are now ready to proceed to the use of this velocity information for the computation of depth, dip, and strike.

7.5. COMMENTS TO THE SECOND EDITION

Sonic logs in wells have been used since their development about thirty years ago and are still used today. They provide details not obtainable any other way.

We know that the velocity of propagation of a plane compressional wave (see Figure 14.1, equation (3); Figure 13.9, equation (8); and the comments for Chapter 13) is $V = \sqrt{(\lambda+2\mu)/\rho}$ where $\lambda+2\mu$ is an elastic constant and ρ is the density. For a sand, the elastic constant increases with overburden load much more than the density increases, so the velocity increases also. This was made beautifully quantitative by Fritz Gassmann in 1951 for a packing of elastic spheres. The greater the load, the nearer we come to the elastic property of the spheres themselves—they offer more and more resistance to squeezing from above and so do sand grains.

On the other hand, shales with very low sand content also increase their elastic constants under load, but they lose water, so the density increases. So the velocity increase under load may be expected to be less than for a sand.

Thus it was thought interesting to try to measure velocities using the reflection seismograph for the purpose of getting some lithological information before drilling. I had been an exploration geophysicist for only one week (1934) when L. W. Blau told me that velocities could be measured using reflection records. I immediately checked that this was correct, but I thought that the accuracy would be very poor. In 1941,

I saw that long spread (x^2,T^2) work was good enough to do better than measure "average velocities" and so I worked out a way to measure *interval velocities.*

I considered an approximating idealized model. The geological section was made of homogeneous layers of constant velocity, each layer bounded above and below by plane interfaces. I supposed the reflections observed came from these plane interfaces and that dips were zero. The velocity, V_1, and depth, z_1, of the first layer were found using Figure 7.11, equation (2), and then $z_1=V_1T_o/2$.

For the second layer, I used the second reflection to plot an x^2,T^2 graph. This graph is nearly straight, so that a tangent straight line can be drawn at any point on it with good accuracy. The equation of this tangent line can be differentiated with respect to x, to give at the point of tangency $dT_x/dx=x/V^2T_x=\sin\beta/V_1$ for the horizontal component of wave slowness. Since x, T_x, V_1, and V (from the slope of the tangent line at x^2,T^2_x) are known, we also know the angle β. β is the angle between the up-coming ray from the second reflection and the vertical at the tangency point. So we can draw this up-coming ray back down to the bottom of the first layer. The path in the top layer is of length $z_1/\cos\beta$, and the corresponding travel time is $z_1/(V_1\cos\beta)$. Also, the vertical time at $x^2=0$ is z_1/V_1. We can subtract these two times to remove the travel time through the first layer. The offset at the base of the first layer is $x-2z_1\tan\beta=x_2$. Thus, the x^2,T^2 graph for the second layer alone is

$$(T_x-2z_1/[V_1\cos\beta])^2-(T_0-2z_1/V_1)^2=x^2_2/V^2_2$$

where T_x and T_o are reflection times from the base of the second layer measured at the top of the first layer. Using this relation, we can compute V_2 and the thicknesses of the second layer z_2.

We continue this process to give single layer x^2,T^2 graphs, until all layers have been measured. We carried out this process also for dipping and/or wedging layers, as described in my paper in 1955. This work was done in the early 1940s when the available computers were very primitive. We sought as much accuracy as possible. We tried to make the interpretation internally consistent. For example, when we went from a higher to a lower velocity layer, we expected a negative reflection. We were careful to look for multiples of each good reflection. When we found them, that gave added precision to our assessment of both the single and the multiple. Sometimes we were able to make reflection velocity surveys just before well-shot surveys were scheduled to be made. The results agreed, on time versus depth graphs, to closer than 1 percent. The up and down trends of interval velocities were the same in general directions but which was the more accurate could not be decided.

None of the layers really had constant velocities and none of the interfaces were plane. We usually knew the dip and strike directions. We were more comfortable when we aligned our surface spread up or down dip. If we were not troubled by concave upward-interface possibilities, the strike direction would be the better choice, as this would require no correction for dip for small x^2's. To do the work properly, one must work so as to recognize interference effects due to concave upwards reflectors and then either make a successful analysis (separation of interfering parts) *or* recognize the impossibility directly if that is the case.

In many cases the work is most simply done by keeping x^2 small. This makes use of a well-known formula, which, as pointed out in my 1955 paper was first used and developed by John A. Legge, Jr. This makes the work much simpler than our more general approach.

Referring to Figure 7.15, we first write the trigonometrical formula (a). Since for small angles θ_1 and θ_2, the tangent and sine are approximately equal, we get the approximate equality (b), which by Snell's Law gives (c). But the pulse slowness component in the x direction is given by (d) (see the comments at the end of Chapter 5). Also (e) implies (f). Using (d) and (f) in (c), cancelling x and taking the limit as x approaches zero, yields (g). Thus solving for V^2_2 we get rela-

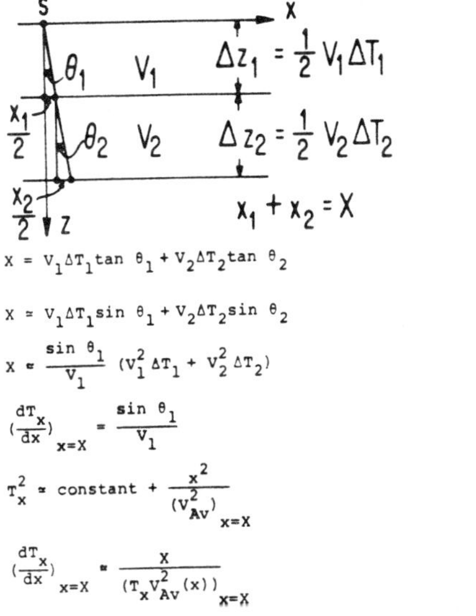

$$X = V_1 \Delta T_1 \tan \theta_1 + V_2 \Delta T_2 \tan \theta_2 \quad \text{(a)}$$

$$X \approx V_1 \Delta T_1 \sin \theta_1 + V_2 \Delta T_2 \sin \theta_2 \quad \text{(b)}$$

$$X \approx \frac{\sin \theta_1}{V_1} (V_1^2 \Delta T_1 + V_2^2 \Delta T_2) \quad \text{(c)}$$

$$\left(\frac{dT_x}{dx}\right)_{x=X} = \frac{\sin \theta_1}{V_1} \quad \text{(d)}$$

$$T_x^2 \approx \text{constant} + \frac{x^2}{(V_{Av}^2)_{x=X}} \quad \text{(e)}$$

$$\left(\frac{dT_x}{dx}\right)_{x=X} \approx \frac{X}{(T_x V_{Av}^2(x))_{x=X}} \quad \text{(f)}$$

$$(\Delta T_1 + \Delta T_2)\, V_{Av}^2(0) = V_1^2 \Delta T_1 + V_2^2 \Delta T_2 \quad \text{(g)}$$

$$V_2^2 = \frac{\Delta T_1 + \Delta T_2}{\Delta T_2} V_{Av}^2(0) - \frac{\Delta T_1}{\Delta T_2} V_1^2 \quad \text{(h)}$$

$$V_{Av}^2(n,0) \sum_1^n \Delta T_i = \sum_1^n V_i^2 \Delta T_i \quad \text{(i)}$$

$$V_n^2 = \left[V_{Av}^2(n,0) \sum_1^n \Delta T_i - V_{Av}^2(n-1,0) \sum_1^{n-1} \Delta T_i \right] / \Delta T_n \quad \text{(j)}$$

FIG. 7.15. Velocity Formula for Zero Offset.

tion (h). We generalize (g) to n layers to get (i). We can write (i) for $n{=}1$ layers and subtract this from (i) (for n layers) to get (j), which is what we were seeking.

In the actual measurement of the interval velocity, we must use a long spread of receivers. On the other hand, having made that measurement, we can look back at the whole problem more simply via the zero-offset formulation.

In my comments at the end of Chapter 8, I supply a simple program that can be used to make an ideal (x^2, T^2_x) calculation

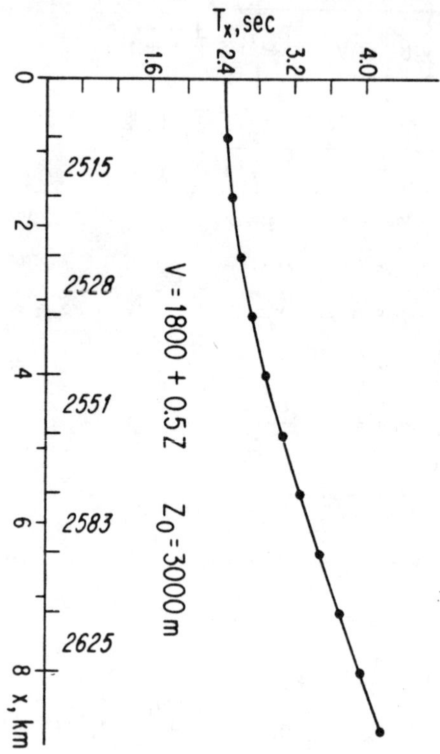

FIG. 7.16. Reflection Times, T_x to Depth 3000 m., and x^2, T^2_x Average Velocities for Increasing Offsets for Velocity 1800+0.5z

for the case where the velocity distribution is linear with depth, $V=V_D+az$. T_x versus x from that kind of computation is plotted in Figure 7.16 for a case that gives a very good fit to an actual distribution for a tertiary sedimentary basin, for example, for much of the San Joaquin Valley of California. Such an ideal calculation can, of course, be carried numerically to any accuracy desired. The (x^2, T^2_x) velocities for various x values are listed to show the way in which this varies with spread distance. The increase of the ideal (x^2, T^2_x) velocity with x is a simple consequence of Fermat's "least-time principle." The actual velocity measured with the zero spread is too low, but as you can see from Figure 7.16, this kind of error is sometimes acceptable. Also equation (j) of Figure 7.15 indicates that we have to consider the difference of two measures, each a little too low.

Reference is made to the just published SEG Monograph by Peter Hubral and Theodor Krey that covers interval-velocity determination very thoroughly.

CHAPTER 8

Depth and Dip Computations

This chapter is written primarily for seismologists, party chiefs, and computers engaged in direct interpretive work.

8.1. INTRODUCTORY REMARKS

It may be well at the outset to make a distinction between what might be called *routine computing* and *tailoring*. Routine computing is the kind that is done practically every day in a seismograph party's field office. Its main purpose is twofold; one is to find indications of interesting structures and the other is to eliminate unpromising areas from further consideration. The methods used have to be mass production methods that are somewhat oversimplified. This oversimplification is absolutely necessary if the work is to be turned out. *Tailoring* is used to obtain the closest possible approach to knowledge about potentially productive areas. The word "tailoring" is employed in this connection to imply a detailed piece of handwork especially suited to the individual problem at hand. The object of the tailoring job is to secure the closest possible approximation of the correct picture.

The present chapter is concerned with routine work rather than tailoring. This is true because here we consider only the components of dip in the line of profile. The cross component is completely neglected in this chapter. On the other hand, in the routine work we seek as high a degree of perfection as possible, keeping in mind that we also want to hold costs to a minimum. Thus, if there are two possible procedures and each costs the same amount of money, we shall select the one which most closely approaches the true picture.

8.2. *CONSTANT-VELOCITY CASE*

The simplest case with which we have to deal is that in which the velocity is assumed constant and is assumed known. The ray paths are straight lines and the pulse fronts are spheres or, in the cross section which we consider, they are circles. We now take up details of the calculations for the simplest case.

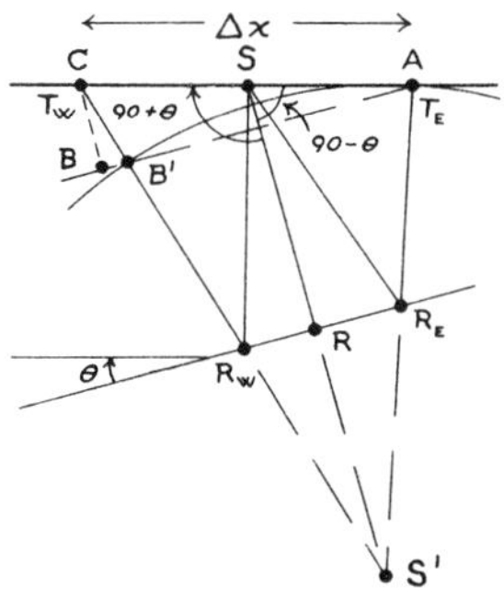

$$V^2T_W^2 = V^2T_0^2 + (\Delta x)^2/4 + VT_0\,\Delta x \sin\theta \quad (1)$$

$$V^2T_E^2 = V^2T_0^2 + (\Delta x)^2/4 - VT_0\,\Delta x \sin\theta \quad (2)$$

$$V(T_W - T_E)(T_W + T_E) = 2T_0\,\Delta x \sin\theta \quad (3)$$

$$\sin\theta = (V\Delta T/\Delta x)[(T_W + T_E)/2T_0] \quad (4)$$

$$\approx V\Delta T/\Delta x \quad (5)$$

$$\text{(For } \theta \text{ small)} \quad \theta \approx (57.3V/\Delta x)\Delta T \quad (6)$$

FIG. 8.1. Vertical section showing split-dip computation.

8.2.1. PLANE REFLECTOR

Refer to Figure 8.1, in which S is the shot point, and the point of reflection of the dipping reflector is the point R for the geophone at the shot point. S', the image point of the shot, is on the extension of the ray from S to R as indicated in the figure. This must be drawn so that $S'RS$ is perpendicular to the reflector. Also $SR = S'R$.

The calculation of the distance from the shot to the reflector SR is very simple. One has only to multiply the velocity of travel by the travel time from S to R. This travel time is evidently half of the reflection time.

A slightly more difficult problem is involved in obtaining the dip angle, θ. This is done as indicated below the diagram in Figure 8.1. In the first relationship the law of cosines is applied to the triangle $SS'C$, together with the identity $\cos(90+\theta) = -\sin\theta$. In the second the law of cosines is applied to the triangle $SS'A$ together with the identity $\cos(90-\theta) = \sin\theta$. The third relationship is obtained by subtracting equation 2 from equation 1, factoring out V^2, dividing through by V, and factoring the difference of the two squares of the times shot west and shot east into the product of the difference and sum, as indicated. Equation 3 may be directly solved for the sine of the dip angle, as indicated in equation 4. An approximation of equation 4 is given in equation 5. For small values of the dip, formula 6 is valid, since the sine of the angle is approximately equal to the angle in radians. The argument is stated more completely in the form:

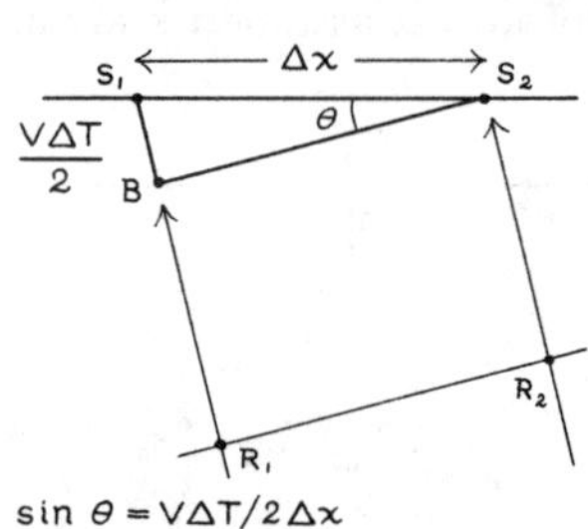

FIG. 8.2. Dip computation for two shot points.

(a) $\sin\theta \approx \theta$ (radians) $\approx (V/\Delta x)\Delta T$

(b) θ (degrees) $= 57.3\ \theta$ (radians) $\approx 57.3\ (V/\Delta x)\Delta T$

We have thus provided the means of computing the dip angle, θ. Knowing the dip, θ, and the distance SR, we can plot a little segment of the reflector, $R_W R R_E$.

We have just considered the calculation of depth and dip for a *symmetrical split spread.* Let us now proceed to consider what we shall call an *end-to-end spread,* as illustrated in Figure 8.2. Here S_1 is one shot point and S_2 is a second shot point; it is supposed that geophones are laid out on the line joining the two shot points. In the triangle S_1S_2B, the length of S_1B is equal to the velocity times half the difference in reflection time at S_2 and at S_1. The sine of the dip angle θ is given by the formula below the diagram. We may interpret this in terms of formula 5 in Figure 8.1. The difference between the two figures is easily observed by the presence of the factor 2 and by the fact that the "approximately equal" sign is not used in the equation in Figure 8.2. We may consider that the factor 2 occurs as we have indicated because the path S_1B is covered twice in the reflection so that ΔT has to be divided by 2, or we may suppose that the spread Δx is covered twice, as indeed it is, so that it has to be counted twice. Either viewpoint leads to the same result.

Plotting the data for the end-to-end shot requires a protractor in order to lay off the down-going ray with the proper dip angle, θ. A measuring scale must also be used to measure the depth.

8.2.2. Precision of Dip and Depth Computation

We consider here a number of influences on the precision of the dip and depth computation outlined above. We have already seen that formula 5 in Figure 8.1 is not precise. Thus a small error has been introduced in going from equation 4 to equation 5. It should be noted, however, that in normal circumstances this error is very small. For example, if Δx happens to be 1000 feet, the velocity 6000 feet per second, and the time 1 second, the error committed because of neglect of the last factor in equation 4 is a 0.3 percent error for small angles. Such an error is negligible compared with other types of error. But the error that occurs in going from equations 4 to 5 in Figure 8.1 increases as the reflection becomes shallower compared to the spread, as can easily be seen directly by inspection of the figure. Usually, in practice, the very shallow reflections are of no great interest and consequently the formula is more nearly valid in the cases where our interest is greater. We may in fact regard the approximation as entirely adequate and the effect of neglecting the error as of negligible importance.

Another effect is due to the fact that we do not determine the very first of the reflection. Usually we miss its beginning by approximately 0.030 second. This will lead to an error of 120 feet in the depth when the velocity is 8000 feet per second.

With modern seismic equipment designed, as it is, to eliminate the higher-frequency components of noise, the beginnings of reflections are almost always exceedingly difficult to pick up. Therefore, the determination of absolute depth is almost out of the question and errors of the magnitude just indicated are sure to be present. On the other hand, this is usually not serious because the *relative depth* from shot point to shot point is not so seriously affected. For example, in going along a line of profile we may be off uniformly 0.030 second on a given reflection. We cannot determine the constancy of this shift absolutely, but approximately the shift remains constant, so the error we introduce merely involves shifting the reflector a little deeper in the section than it actually is.

In connection with the error in depth mentioned above, we should say that a constant velocity very rarely corresponds to the actual facts. Consequently the *assumption* of a constant velocity in itself introduces errors which are likely to be much more serious than any that have been mentioned thus far.

We discuss the error in dip by referring to equation 6 in Figure 8.1. We can obtain some idea of the error in dip that results from making an error in ΔT. If we have an error in ΔT of 0.001 second with a spread of 1000 feet and a velocity of 6000 feet per second, we can make an error in dip of approximately $\frac{1}{3}°$. An accuracy of 0.001 second is better than we can usually obtain for ΔT. The error in ΔT is more likely to be of the order of ± 0.002 second. This larger error is likely because ΔT is the difference between two quantities, each of which is subject to a certain amount of error. It will be seen that if the error is ± 0.002 second in the above case, the error in dip will be $\pm \frac{2}{3}°$. Unfortunately such an error can be extremely serious in very low dip country like much of western Texas, the Texas gulf coast, and the Canadian plains region. The fact that we are working on the ragged edge of accuracy should, however, not blind us to the fact that we do not have perfect data and that there is a range of variation which can hardly be reduced by any field process or any juggling of the original data.

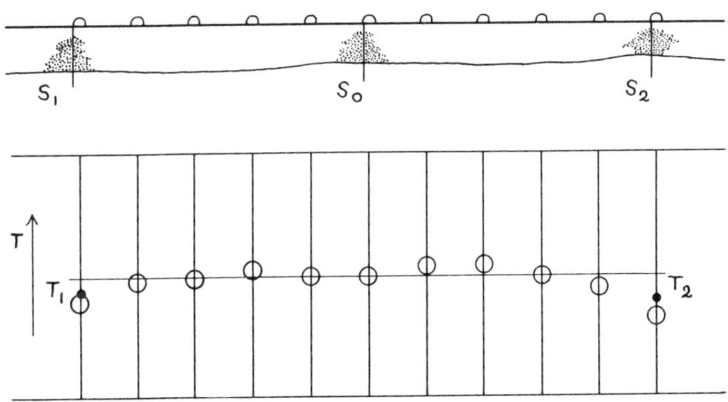

FIG. 8.3. Cause of irregularity due to shot hole drilling fluids.

The error of approximately $\pm 1°$ just referred to is not entirely due to error in reading time. The error in reading ΔT on the record is usually quite small and can be made extremely small by reading very carefully. But there is one error that cannot be reduced by the reading procedure; it is illustrated in Figure 8.3. This figure shows three shot points S_1, S_0, and S_2, just above which is an irregular line indicating approximately the base of the weathering. A series of geophones are laid out on the ground surface between the various shot points. The three little treelike shadings in the figure are in the weathered zone and are intended to indicate disturbances of the weathered section due to the penetration of drilling fluid. During drilling this fluid penetrates the walls of the shot hole and affects the velocity distribution in the neighborhood of the shot hole. This may have very serious consequences. The reason for the seriousness is that during the penetration of the fluid large quantities of air bubbles may be trapped with it. The trapped air has the incompressibility properties of air. The medium, on the average, has the density properties of the mixture of air, water, and soil. The velocity of such a mixture may become very low and thus there is a disturbance of the velocity in the shallow section near the shot point which may seriously distort the correction to datum. The danger of such distortion is much more serious in some areas than in others. For instance, it is very serious when the depth of the weathering is very great and the weathered region consists of sandy porous material.

In the lower part of Figure 8.3 an attempt has been made to indicate approximately how a series of reflection times may vary in going across the record. The upright lines represent the equilibrium positions of traces; the circular points indicate the times positions graphically. If the times were not disturbed they would all fall on the horizontal line in the figure. But because of disturbance, the times have a certain amount of scatter. It is this scatter that introduces the serious error referred to above. This error, which is very difficult to avoid, is often the major contributor to the error in the dip computation.

8.2.3. Normal-Move-Out

Normal-move-out is the term applied to the increase in reflection time due to an increase in the distance from shot to geophone when there is no dip. The normal-move-out can be calculated, for the case of constant velocity, very simply by reference to Figure 8.4. We wish to compare T_0, the travel time from S to C to S, and T_x, the travel time from S to B to A. We define the normal-move-out, N, as indicated in the first relationship. The next relationship (2) is obtained by applying the Pythagorean theorem to the triangle $SS'A$ (S' is the image point of shot point S). By applying to the definition the formula for T_x obtained from equation 2 (by extracting the square root of both sides) we obtain formula 3. An approximation of this formula is given by formula 4, which is approximately valid

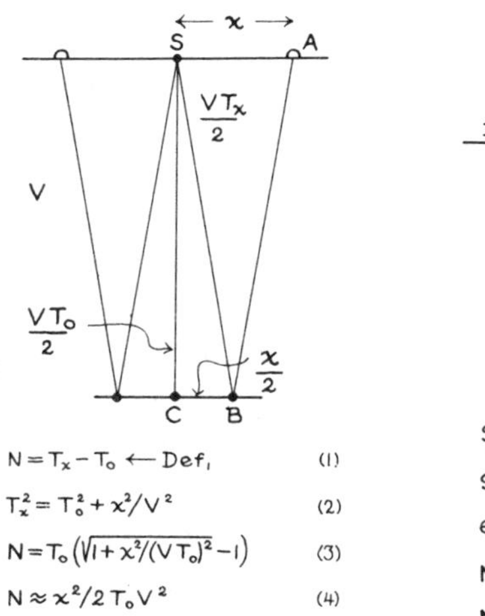

FIG. 8.4. Vertical section showing normal-move-out determination.

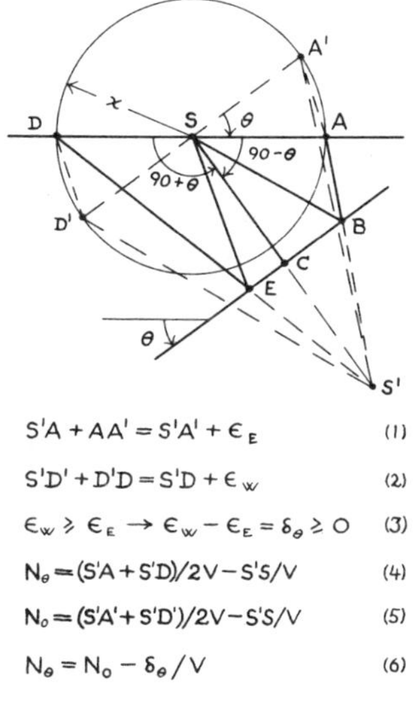

FIG. 8.5. Effect of dip on move-out.

when the spread, x, is small compared with twice the depth to the reflector, VT_0. To obtain formula 4 we expand the square root in equation 3 by means of the binomial theorem, keeping only the first two terms. Formula 4 gives the normal-move-out in terms of the spread distance, the center reflection time, and the velocity.

The normal-move-out is defined in the case where there is no dip. There is, however, a certain amount of move-out, or curvature of the reflection as it passes across the record, when there is some dip. It is now our intention to show the effect of dip on move-out. Refer for this purpose to Figure 8.5, in which S again represents the shot. But in the present case there is a dip of the reflector corresponding to the angle θ. We may suppose that as before the spread is laid out from A to D symmetrically on both sides of shot point S.

We shall now compare the situation in which geophones are located at A and D with the situation in which there is no dip, that is, if geophones are located at A' and D', as indicated in Figure 8.5.

The first relationship states that if we measure the distance from the image point, S', to point A and add to it the distance from point A to point A', we get the quantity $S'A'$ plus a small positive quantity ϵ_E. In the second relationship we have the distance from S' to D' plus the distance from D' to D equal to the distance from S' to D plus a small positive quantity ϵ_W. It is evident from the figure that ϵ_W is always greater than or equal to ϵ_E. This being the case, we can find the difference between ϵ_W and ϵ_E. This difference will be a positive quantity δ_θ, depending on θ. The move-out, N_θ, is calculated as indicated in the fourth relationship; that is to say, we take the travel time from S to E to D as one of the reflection times and the travel time from S to B to A as another reflection time, average, and subtract the travel time from S' to S. The normal-move-out for O dip is expressed in relation 5. Therefore we can express the relationship between the move-out for the dip of angle θ and the normal-move-out by equation 6. This equation expresses quite clearly that the move-out decreases as the dip increases. An explicit expression of the positive quantity δ_θ can be obtained in terms of the spread distance x, the time T_O, and the angle of dip, θ, but this expression is too complicated to be useful so we have omitted it. Furthermore, the planeness that we have assumed is an assumption of major importance not likely to be encountered in practice. In the next section we shall see qualitatively how this assumption affects the calculation of normal-move-out.

8.2.4. Curved Reflector

While the curvature of the reflector is neglected in routine work, most reflectors are actually somewhat curved. Hence the simplest qualitative aspects of the curvature problem are mentioned in this section.

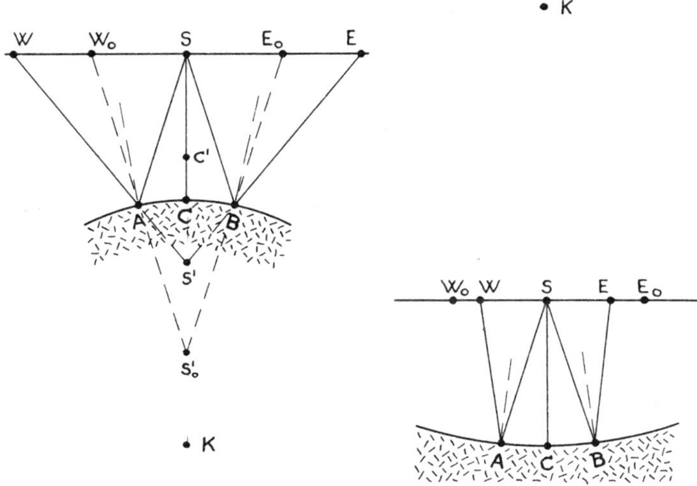

Fig. 8.6. · Effect of convex upward reflector.

Fig. 8.7. Effect of concave upward reflector.

Refer to Figures 8.6 and 8.7. In these figures K is the *center of curvature* of the reflector. In Figure 8.6 the seismic pulse energy traveling downward in a cone, $SACB$, is spread over the surface from W to E on reflection. Had the reflector been uncurved, the same energy would have been spread from W_0 to E_0 on reflection. Thus in the case of a convex upward reflector, the reflection is spread more thinly, i.e., the reflection is weakened. Similar considerations applied to Figure 8.7 show that the reflection is strengthened when the reflector is gently concave upward.

If the reader will imagine the reflector in Figure 8.7 to be moved deeper and deeper, he will see that points *W* and *E* approach each other until they coincide at the source, S. To achieve this, the reflector will have to be deepened a distance equal to *SK*. We say that the rays *focus* at S. If the reflector is lowered still further, the reflected rays will cross below S.

When the reflector is concave upward and deep enough so that the reflected rays cross or focus at or below the geophones, the problem of interpretation becomes far too complicated to be handled adequately by presently known routine procedures. We shall discuss this more fully in Parts III and V.

The main problem at the present stage of this discussion is to understand in a general way the effect of curvature on a reflection. Convex upward curvature increases move-out.

8.3. GENERAL CASE

In the general case the velocity depends not only on the position in the section but also on the direction that we move away from this position. That is to say, in general, the vertical velocity differs somewhat from the horizontal velocity at a given point. This is true of an *anisotropic* medium. In the case of an *isotropic* medium the velocity is independent of the direction from each given point.

Although we must recognize that the velocity of travel of the seismic pulses is often likely to have an anisotropic character, it is very difficult to make use of this in regular routine computing. Hence we shall *assume* from this point on that the velocity has an isotropic character.

Even when we have made this simplification, that is, the assumption of isotropy, a great deal of complication remains because we have allowed lateral variations of velocity. In the average geological section lateral variations of velocity do exist. In some cases they are so important that failure to take account of them distorts the interpretation to such an extent that the interpretation becomes meaningless. We are thus faced with a problem, which, although very difficult, is not entirely hopeless; however, it is beyond the scope of the present book. A paper by Dix and Lawler* gives a method of approximately taking account of the effect of lateral variations of velocity. This method was used in one case where, without it, the picture was one of hopeless confusion. With it, the picture made what one might call geological common sense.

8.3.1. Velocity Dependent on Depth Only

We assume in this section that there is no lateral variation in velocity and that the velocity is determined as soon as we know what the depth is. This corresponds to the most general case that is considered in routine computations.

Let us consider a ray traveling downward from the shot point as illustrated in Figure 8.8. This ray generally will curve as it moves downward and the curving will depend upon the velocities encountered as the ray passes through the section. The dependence upon velocity is given by Snell's law and is expressed by equation 1. The equation states that the sine of the angle between the ray and the vertical divided by the velocity is the same constant at each level. This constant, p, is characteristic of the ray. We can then write the sine of the angle i equal to p times the velocity at the corresponding depth, z. We can differentiate both sides of this equation to obtain equation 2, which gives the change in i corresponding to the change in z. Inspection of the drawing shows that the arc length AB is equal to the change in depth, Δz, divided by the cosine of the angle i_1. This is an approximate expression that would be exact only if the velocity were constant between A and B, so that the ray would be a straight

* C. H. Dix and R. Lawlor, *Geophysics*, 8:105 (1943).

line instead of a curve. But AB is also the radius of the curvature times the angle Δi in radians. By combining relations 3 and 4 we obtain an expression for the radius of the curvature, r, which is given by equation 5. In this equation we may substitute the expression for p from equation 1 and obtain the radius of curvature in terms of the velocity, the sine of the angle i, and the rate of change in velocity with respect to depth. Since the only variable along the ray is the

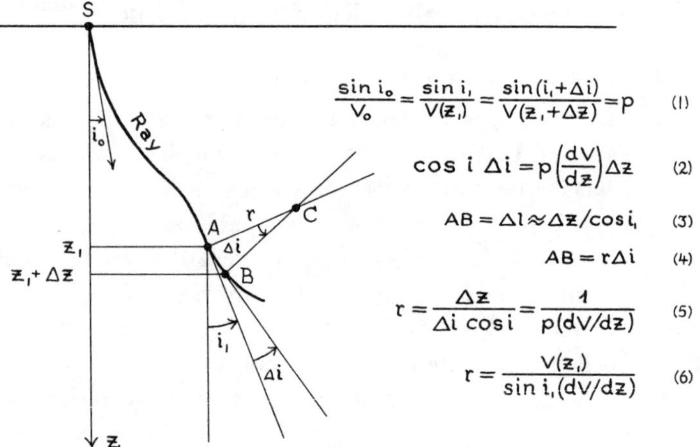

FIG. 8.8. Vertical section showing general down-going ray.

rate of change in velocity with respect to depth, it is clear that the radius of curvature of the down-going ray depends wholly upon this rate of change, once the ray is selected. However, when we go from one ray to another, we go from one value of p to another; therefore the radius of curvature varies according to equation 5 as we go from one ray to another.

We see at once that if the rate of change in velocity with depth is constant, the radius of curvature is constant. If the radius of curvature is constant, the ray is an arc of a circle.

In our study of the relationship of velocity to depth in the preceding chapter, we indicated the time-depth curves as well as the depth vs. average-velocity curves. Such curves obtained from well shooting or from velocity profiling may be taken as representing the relationship between velocity and depth. In order to use them, however, some simplification is usually necessary.

When all the dips are exceedingly small, we may consider that the simplification is automatically made for us. With an exceedingly small dip we may plot the depth points as if there were no dip; that is, these points are plotted either directly beneath the shot point or at a point halfway between the shot point location and the geophone location. This plotting procedure is much used in low-dip country. The principal objection to it is that even in low-dip country irregularities such as faults and lateral stratigraphic changes often make events on the record similar to those made by steeply dipping reflections. The interpretation of these latter events is practically excluded by adhering to such an oversimplified routine. Consequently this objection is serious.

It may be stated that if the velocity really were dependent upon the depth only, and if there were no important changes of density in the section, all the reflectors would indeed be of zero dip, but in such a case the section would completely lack interest.

The other simplification can be made if we assume that the velocity vs. depth graph can be fitted sufficiently closely, for practical purposes, by a straight line. In such a case the velocity either increases or decreases uniformly with increasing depth or remains constant. It is this case that we consider next.

8.4. VELOCITY A LINEAR FUNCTION OF DEPTH

A word or two of warning is appropriate in connection with this discussion. If we imagine that the true velocities of the section are available, it is of course highly unlikely that they will follow a strictly linear depth relationship. What we attempt to do is to fit a straight line to the velocity vs. depth relationship. This fit may not be perfect; in fact, it rarely is.

There is one point at which the imperfection of fit of the linear distribution of velocity with depth to the actual distribution may be troublesome. This is the datum point. In making this fit we use a certain datum velocity, but this may not be the velocity in the section at the datum. The reason is that if we require that these two velocities be coincident, the fit is not likely to be as good as if we allow ourselves greater flexibility at this point. Thus it must be understood that the datum velocity is not necessarily the same as the velocity used in making the datum correction described in Chapter 6.

Another warning should be issued to the effect that the linear distribution of velocity with depth should not be claimed to be much better, as far as accuracy is concerned, than certain other approximating distributions. For example, some companies prefer to employ what they call a "straight-line" computation method, using a variable average velocity. As long as the dips remain small, there appears to be no serious objection to this procedure. When the dips are large, however, the procedure is subject to serious question. In fact, any procedure is questionable for large dips when the cross component is neglected.

The main advantage claimed for the linear distribution of velocity with depth is that it is very simple to use. It can be arranged for routine computations so that it is almost as simple as the simplest process of interpretation. Furthermore, the amount of work required to produce a cross section is very small. In fact, it is about the same as that required to produce a cross section using straight-line methods.

8.4.1. Rays.

We see immediately that the rays are circular arcs; Figure 8.9 shows this directly. The first equation shows the linear velocity vs. depth relationship, where V_D is the datum velocity and z is the depth. a is constant and has the meaning indicated in the second relationship; that is, a is the rate of change of velocity with depth, a uniform constant rate in this particular case. If we refer to the result summarized in equation 6 of Figure 8.8, we obtain for the radius of curvature equation 3 in Figure 8.9. Thus the radius of curvature is a constant, since the datum velocity is constant, the value a is constant and the sine of the angle i_o is a constant. i_o is the angle that determines which ray we are considering.

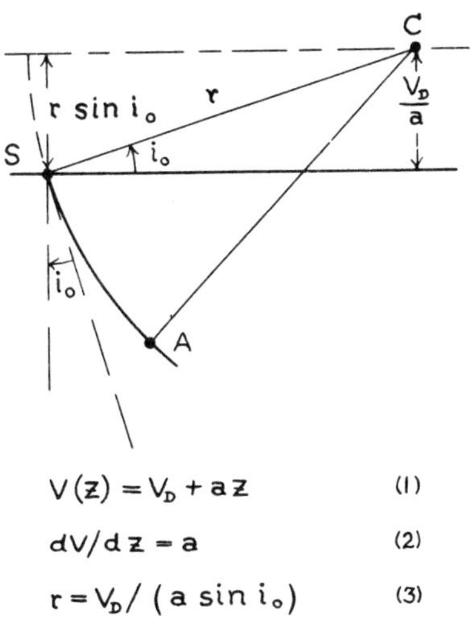

Fig. 8.9. Circular arc ray and line of centers.

We see immediately from Figure 8.9 that, since r is given by equation 3, the height of the center C above the datum line is $r \sin i_o = V_D/a$. Thus it is a property of ray circles that their *centers lie on a fixed line* a distance V_D/a above the datum. Note that this line of centers is the line at which the velocity becomes zero if the linear relationship is extended upward.

8.4.2. Depth and Dip

In order to use the results obtained in the last section in a practical way, we must be able to associate travel time with a given reflection point. For this purpose refer to Figure 8.10, in which S is the source and C is the ray circle center. A small elementary circular arc AB of length dl is shown. This corresponds to a change of angle, di, as indicated in the figure. The corresponding change of time is given by the first relationship, which may be rewritten in successive steps as indicated in the second equation. The details of this deduction are left to the reader. Equation 2 may be integrated as in the third relationship by means of a table of integrals. Thus the dif-

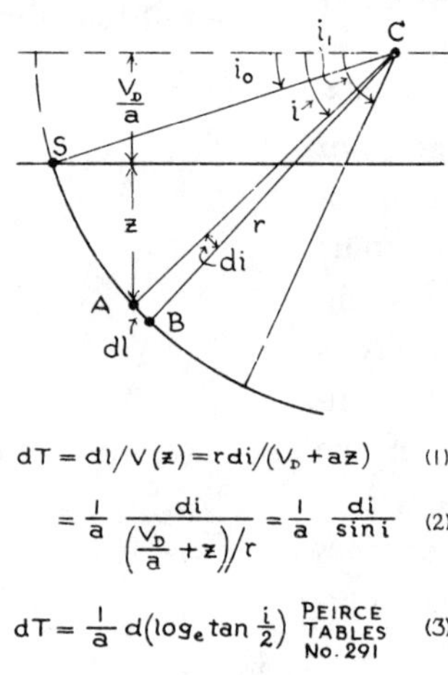

$$dT = dl/V(z) = r\,di/(V_D + az) \qquad (1)$$

$$= \frac{1}{a}\,\frac{di}{\left(\frac{V_D}{a} + z\right)/r} = \frac{1}{a}\,\frac{di}{\sin i} \qquad (2)$$

$$dT = \frac{1}{a}\,d\left(\log_e \tan \frac{i}{2}\right) \quad \text{PEIRCE TABLES No. 291} \qquad (3)$$

FIG. 8.10. Calculation of time along arc element.

ferential of the time is equal to $1/a$ times the differential of the natural logarithm of the tangent of half the angle, i. This important relationship we now proceed to use.

Figure 8.11 shows a source, S, and a reflector at P. The center of the ray circle is at C. The dip of the reflector is the angle i. We integrate the third equation of Figure 8.10, as indicated in equation 1 in Figure 8.11. The integration is carried out in the second equation. This equation may be rewritten as the logarithm of the ratio, as indicated in equation 3. The definition of the logarithm is used in order to go from the third to the fourth equation. We multiply the fourth equation through by the tangent of half of i_o to get the fifth equation. We thus have the tangent of half the angle i in terms of T and i_o. Thus we are in a position, at least theoretically, to compute the dip of the reflector using the time, the apparent dip at the surface, i_o, and the values of h and R, which we get from the time.

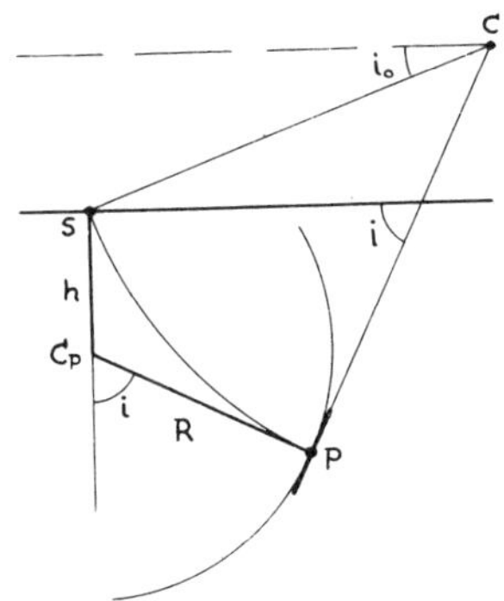

$$a\int_0^T dT = \int_{i_o}^{i} d \log_e \tan\frac{i}{2} \qquad (1)$$

$$aT = \log_e \tan\frac{i}{2} - \log_e \tan\frac{i_o}{2} \qquad (2)$$

$$aT = \log_e\left(\tan\frac{i}{2} \Big/ \tan\frac{i_o}{2}\right) \qquad (3)$$

$$e^{aT} = \tan\frac{i}{2} \Big/ \tan\frac{i_o}{2} \qquad (4)$$

$$\tan\frac{i}{2} = e^{aT} \tan\frac{i_o}{2} \qquad (5)$$

FIG. 8.11. Calculation of dip formula.

Formula 5 appears to involve only a and not V_D. However the V_D is still present in the fact that the center of the ray circle, C, is V_D/a above the datum.

8.4.3. Pulse Fronts

We have carefully avoided the quantitative use of the concept of a *pulse front* until this stage of our exposition. Referring briefly to Figure 8.11 again, focusing attention on the point P, the lower end of the ray SP, we note that (see Figure 5.1) the tangent to the ray at P and the tangent to the pulse front at P are mutually perpendicular, that is, they are orthogonal at P.

We now calculate the radius of curvature at P of the pulse front. For this refer to Figure 8.12 where $PC=SC=r$ is the

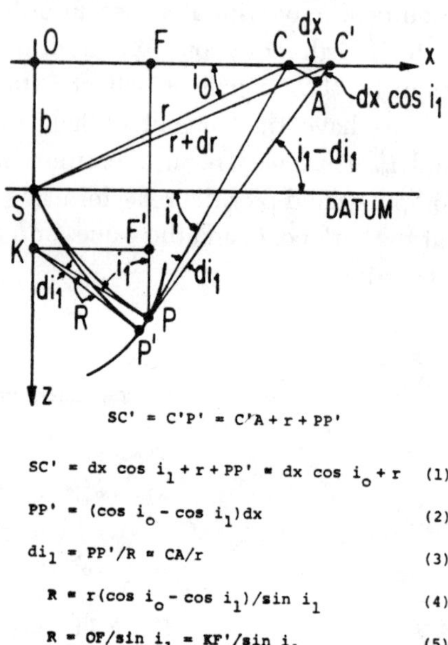

FIG. 8.12. Construction to Show that Pulse Fronts are Spheres.

radius of curvature of the ray arc, SP. We take the origin of rectangular coordinates, O, directly above S a distance $b=V_D/a$ so that the x axis is the line of centers for the ray circles. Move from C to the right an arbitrarily short distance, dx, to C'. Using C' as center construct a neighboring ray SP' where $SP'=C'P'=r+dr$ and the straight line $C'P'$ passes very close to P. Drawing the tangent to the ray at P' (which is perpendicular to $C'P'$), we see that the tangents to rays at P and at P' intersect very close to K in the figure. Because of the perpendicularities at P and P' the angles CPC' and PKP' are equal, say to di_1, as indicated. Also we see that equation 1 is valid. Notice that r may be eliminated from equation 1 to give 2. Taking R to be the radius of curvature of the little element of pulse front PP', we have equation 3. From 2 and 3, since $CA \approx dx \sin i_1$ we get equation 4. But in the limit as dx approaches zero we get relation 5, which places the intersection of the two ray tangents to P and to P' on the vertical z axis through S at K. Thus K is the center of curvature of the pulse front at P and $PK=R$ is the radius of curvature at P with K located on the z axis. It will be noted that there is nothing special about P; P can be any point in the x, z plane below datum.

In Figure 8.13 we have taken the same S, C, and P as in Figure 8.12. We have also used the same K as center and R as radius to draw the circle, *pf* through P. We want to establish that the circle, *pf*, is the pulse front through P. KP_1' is perpendicular to the z axis at K and $C_1'P_1'$ is perpendicular to the x axis at C'.

The arguments made above using Figure 8.12 to find the pulse front curvature at the point P can now be repeated for P_1' with the same results as for P. Thus the radius of curvature at P_1' will again be R for the special point P_1' where the reflector dip would be 90° whereas the reflector dip was i_l for P. It is fairly evident that, as long as we pick *any* point on *pf* we'll see that point has the same properties that P and P_1' have for a pulse front *at a point*. Since *all* points on the circle *pf* meet the requirement of orthoganality of ray and pulse front, this aspect of the problem is finished. There is however another requirement that is as basic as ray-pulse front orthogonality. This is that the travel time from S to P on the circle *pf* is the same for *all* rays between S and *pf*.

We begin by calculating the time T between S and P'. Formula 4 in Figure 8.11 allows us to write formula 1 in Figure 8.13. In triangle $OC_1'S$, we see that equation 2 holds.

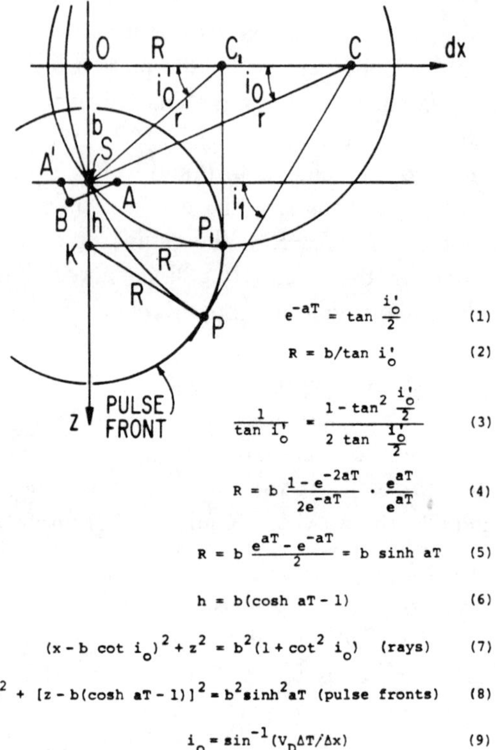

FIG. 8.13. Mutually Orthoganal Circles, Rays and Pulse Fronts, in a Plane.

Also the trigonometry formula 3 holds. Using 1, 2, and 3 we find 4, which leads immediately to formula 5.

Notice the equation 5 for R contains no angles, only T, a, and $b(=V_D/a)$. Furthermore the depth below datum of the center point K, which we call h, is seen to be $h=r'-b=\sqrt{b^2+R^2}-b=b\sqrt{1+\sinh^2 aT}-b=b(\cosh aT-1)$ or formula 6 in Figure 8.13, using a well-known relation between the hyperbolic sine and the hyperbolic cosine. Thus both h and R are independent of *dip*.

We leave as an exercise for the reader to show that the general equations for rays and pulse fronts are respectively given by formulas 7 and 8. This mutually orthogonal system of circles with parameters i_o and T form a net that can be used for a coordinate system, a bipolar coordinate system since the ray circles intersect the z axis not only at S but also at S' above the origin O.

We note that the ray plane, in which we have been working, can be oriented in any way about the z axis keeping S fixed. Thus properly we should speak of the pulse fronts as spheres. The angle giving the orientation of the ray plane may be used as the third coordinate.

Finally, in Figure 8.13 note the little right triangle $AA'B$, where the angle $A'AB=i_o$, $A'A=\Delta x$, $A'B=V_D\Delta T$. Thus we get $\sin i_o=V_D\Delta T/\Delta x$, which gives equation 9 in the Figure.

The relations thus far obtained are summarized in Figure 8.14. It is important to realize that the third relation in this figure gives angle i_o in terms of ΔT and Δx when V_D is given. Thus the relationships summarized in this figure are sufficient for our purpose. However, a warning should be issued in connection with the figure, to the effect that T is the *one-way*

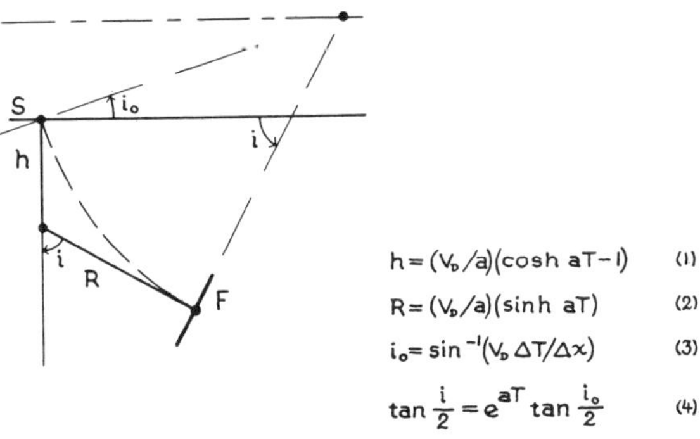

FIG. 8.14. Summary of useful formulas.

time, that is, half the reflection time. Δx is the spread distance over which ΔT is measured.

We have assumed thus far that there is only one shot point, that is, that we are dealing with a split spread. But we may be shooting an end-to-end shot from one shot point to another, as indicated in Figure 8.15. In this case, Δx may be taken as the distance between the two shot points. The natural procedure would then be to take the difference between the reflection time from shot point S_2 to reflection point A_2 and back to geophone g_2, and the reflection time from shot point S_1 to reflection point A_1 and back to geophone g_1. It must be observed, however, that if this difference is used, the path contributing the ΔT is covered twice; hence the ΔT thus obtained must be divided by 2 in order to be used in formula 3 of Figure 8.14. When this is done formula 3 is valid for the case illustrated in Figure 8.15.

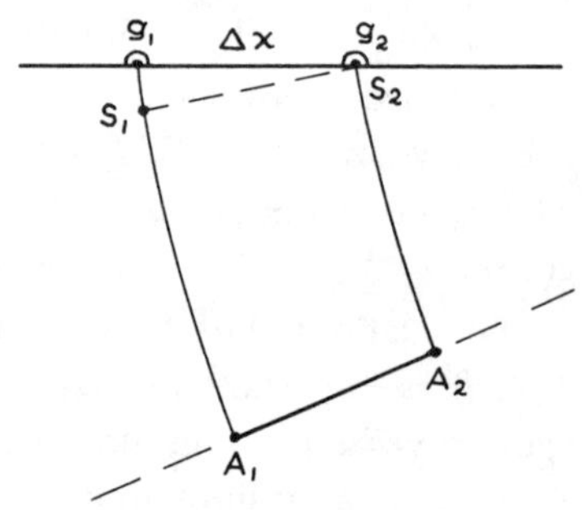

FIG. 8.15. Calculation of dip using two shot points.

At the present stage in our discussion we are theoretically prepared to use the linear increase of velocity with depth to compute depths and dips.

8.4.3. Tools for Routine Work

One of the most useful tools has been devised by R. H. Mansfield.* This is a circular slide rule containing many scales. It can be used to compute almost every quantity in-

* See *Geophysics*, 12:557-575 (1945).

volved when velocity increases linearly with increasing depth.

Let us consider first the part of Mansfield's slide rule that makes possible the easy computation of *dip*.

For this refer to pages 565-566 of Mansfield's paper, cited in the footnote. His equation 11 is equivalent to equation 2 in Figure 8.13 if we remember that T there is one-way time whereas T_o in Mansfield's equation is reflection time. Also, his equation 10b is equivalent to equation 3 in Figure 8.14. Thus if V_D, a, T_o, and $\Delta T/\Delta x$ are given, we can compute the dip, i. Since Mansfield's Figures 6 and 7 are difficult to read, their mode of construction is clarified below. The present treatment follows his closely. However, an alternative ordering of scales is described which may have minor advantages. One advantage is slightly greater ease of description.

Figure 8.16 gives a résumé of the theory and the construction principle. At the top appears the primary formula, obtained from equation 4 of Figure 8.14 by taking the logarithm of both sides. Secondary formulas 2 and 3 relate i_o *to* ΔT, Δx and V_D. Using these four formulas, we deduce formula 5, which is the basis for the construction.

The diagram in Figure 8.16 shows the construction. The scales are set as they would be if T_c were zero. This simplifies formula 5 so that our problem is reduced to matching $\log_{10} \tan i/2$ to:

$$\log_{10}\left\{\frac{1-[1-(V_D/\Delta x)^2(\Delta T)^2]^{\frac{1}{2}}}{(V_D/\Delta x)\Delta T}\right\}.$$

We must specify $V_D/\Delta x$ numerically. We do this by setting $V_D/\Delta x=5$. This is equivalent to Mansfield's specification $V_o/2x=10$. With this specification the dip scale is *measured* according to $K \log_{10} \tan i/2$ but *labeled* according to i. The ΔT scale is measured according to

$$K \log_{10}\left[\frac{1-(1-25\Delta T^2)^{\frac{1}{2}}}{5\Delta T}\right]$$

PRIMARY FORMULA

$$\log_{10} \tan(i/2) = (aT_c/2) \log_{10} e + \log_{10} \tan(i_o/2) \qquad (1)$$

SECONDARY FORMULAS

$$\tan(i_o/2) = (1-\cos i_o)/\sin i_o = (1-\sqrt{1-\sin^2 i_o})/\sin i_o \qquad (2)$$

$$\sin i_o = (V_D/\Delta x)\Delta T \quad (3) \qquad \log_{10} e = 0.4343 \qquad (4)$$

DEDUCED FORMULA FOR SLIDE RULE CONSTRUCTION

$$\log_{10} \tan\frac{i}{2} = 0.2171\, aT_c + \log_{10} \frac{1-\sqrt{1-(V_D/\Delta x)^2(\Delta T)^2}}{(V_D/\Delta x)\Delta T} \qquad (5)$$

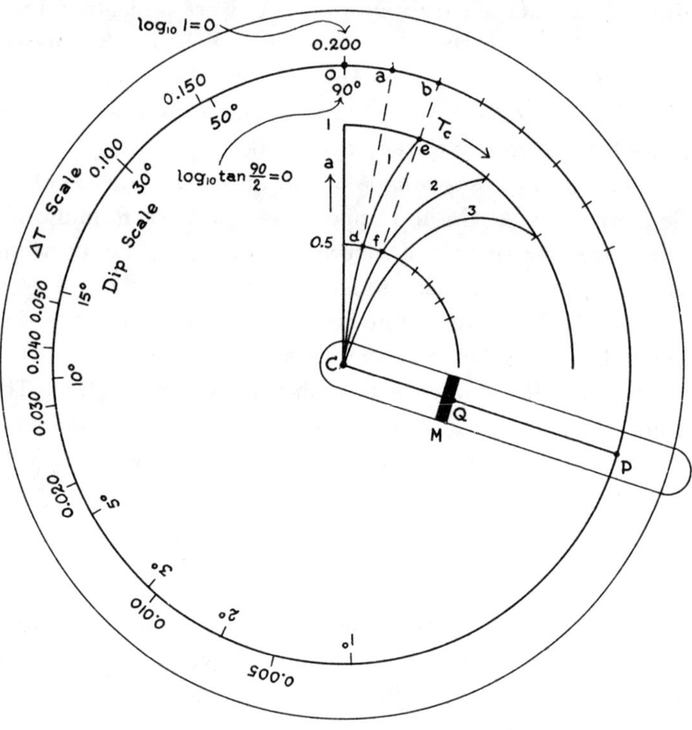

FIG. 8.16. Construction of circular slide rule for dip computations.

but *labeled* according to ΔT. Here K is a *scale factor* that is a fixed number as soon as the size of the scale is selected.

The origins of the ΔT and dip scales are at point O in Figure 8.16. Note that for $\Delta T = 0.2$,

$$\log_{10}\left[\frac{1-(1-25\Delta T^2)^{\frac{1}{2}}}{5\Delta T}\right] = \log_{10} 1 = 0.$$

Also

$$\log_{10} \tan 90°/2 = \log_{10} \tan 45° = \log_{10} 1 = 0.$$

Thus we start measuring both dip and ΔT scales counterclockwise from O.

As an example, consider a dip of 30°. It may be verified that 30° corresponds to $\Delta T = 0.1$ when $V_D/\Delta x = 5$. Thus $i = 30°$ corresponds to $\Delta T = 0.1$. A few other points on both the dip and the ΔT scales are included in Figure 8.16. Naturally, in a rule that is made for actual use more complete scales will be required.

An a scale and a T_c scale are also included. Measure a from the center, C, outward along a radius. Thus $a = 0$ at C. Circular arcs corresponding to $a = 0.5$ and $a = 1$ have been drawn. The T_c scale must be *measured* so that the clockwise rotation of the ΔT scale will be $K \times 0.2171\ aT_c$, measured on the arc where the ΔT and dip scales adjoin. If, for example, $a = 0.5$ and $T_c = 1$ sec., the clockwise rotation will be $K \times 0.1086$ measured on arc Oab in Figure 8.16. But the time has to be read on arc df, corresponding to $a = 0.5$. Oa corresponds to the rotation $K \times 0.1086$. Hence if we connect a and c by a straight line we determine point d where T_c reads 1.0 sec. Point b likewise determines point f on the $a = 0.5$ arc (where $T_c = 2.0$ sec.), as well as point e on the $a = 1$ arc (where $T_c = 1.0$ sec.). Thus it will be observed that a network can be drawn on the face of the rule composed of circular arcs where $a =$ const. and of curves where $T_c =$ const. If we use the reference rod CP with the mask M (perhaps of black masking tape) we thereby fix reference to a particular a value, as is usually done in any given area. Then if $a=0.5$ and $T_c=1$ we rotate rod CP until Q coincides with d. Then P will determine a and the ΔT scale may be rotated until the $\Delta T=0.2$ sec. point coincides with a. If $\Delta T=0.100$ sec., the dip will be about 37.5°. Thus it will have increased, as it should, from 30° to 37.5° for fixed $\Delta T=0.100$ sec., when T_c increases from $T_c=0$ to $T_c=1.0$ sec.

The a, T_c-scale network can be placed at any convenient location on the inner disk. For example, it can be rotated rigidly counterclockwise about 90°, thus keeping the dip part on one side of the rule, as Mansfield has done.

Note the following minor differences from Mansfield's rule. The a scale in Figure 8.16 is drawn outward, whereas Mansfield drew it inward. The virtue of Mansfield's arrangement is that the T_c scale can be compressed to make space for other scales. A universal rule seems to require Mansfield's plan. Also the T_c scales have been reversed. This is a consequence of reversing the order of the ΔT and the dip scales. This exchange was made with the idea that perhaps the operator could set the reference rod CP and hold it in place with respect to the inner disk with the thumb and forefinger of his left hand while rotating the outer disk to the reference point P. This would be facilitated even further if the inner disk were on a fixed base and the reference rod were fastened to the outer disk permanently so that when $T_c=0$, CP will coincide with CO. The outer scale (ΔT scale) would be rotated with the reference rod line (CP) until it reached the proper T_c on the proper a circle.

No claim is made here that this rule arrangement is significantly better for dip computations than Mansfield's rule is. Exactly the same principles are involved in both cases. The writer knows from experience that Mansfield's rule is exceedingly convenient to use.

The table on page 557 of Mansfield's paper gives an extensive list of quantities capable of being calculated by his rule. The rule is very convenient for all these calculations. Though not always the most convenient device, the rule is certainly far superior to numerical computation for many purposes.

Other tools that make the computation systematic are shown in Figures 8.17 and 8.18, which represent forms on which the data may be recorded. Figure 8.17 is a form for a

Profile No. ________ Dir. ______ Spread __(Δx)__ Date ______

Reduction Factor ________ Datum El. ______ V_D ____ a ______

No.	T_c	ΔT	Grade	$(\Delta T)_{corr.}$	Dip	No.
1						1
2						2
3						3
4						4
5						5
6						6
7						7

FIG. 8.17. Table for one-shot dip computations.

regular split shot. Figure 8.18 is a form for an end-to-end shot. The *profile number* must be given in each case for identification. Also stated is the *direction* along which the profile is shot, particular care being taken that it is the direction in which dips are positive. The reverse direction of course will correspond to the direction in which dips are negative. The *spread distance* must be recorded and it is wise to record the *date* as a further identification check. The next item that should be entered is the so-called *reduction factor*. The forms have been planned so that a circular slide rule can be used for computing the dip. This slide rule construction is based on the assumption that the ratio of the datum velocity to Δx is 5. In general, the ratio will be different from 5. The reduction factor is the factor by which the observed ΔT must be multiplied in order to bring ΔT to the value we would have, using a ratio of V_D to Δx equal to 5. For example, if the datum velocity is 6000 feet per second and Δx 1000 feet, the actual ratio will be 6. In this case we will have to increase the spread

Profile No.______ to Profile No.______ Date ______

Dir. ______ Spread (2Δx) Reduction Factor ______

Datum El. ______ V_D ______ a ______

No.	T_{C_1}	T_{C_2}	ΔT	Grade	$(\Delta T)_{corr.}$	Dip	No.
1							1
2							2
3							3
4							4
5							5
6							6
7							7

FIG. 8.18. Table for two-shot dip computations.

to 1200 feet in order to bring their ratio down to 5. We then have to increase ΔT by multiplying by 6/5. Therefore the *reduction factor* in this case will be 6/5. It is also wise to record the *datum elevation*, the *datum velocity*, and the *a value* for further identification purposes. It will be observed that both Figures 8.17 and 8.18 have numbers down the left and right sides. These numbers signify the number of the reflection. In Figure 8.17 the *reflection time,* ΔT, and the grades are recorded directly from the seismic record. ΔT is multiplied by the correction factor to get the *corrected* ΔT. Using this corrected ΔT and the reflection time in conjunction with the dip slide rule, we read the dip and record it in the column headed "Dip." In the case of end-to-end shots, the table is very similar except for the spread. The spread is *counted twice*; hence ΔT does not have to be divided by 2. In other words, we put the necessary correction in the reduction factor itself so that we can take as ΔT merely the difference of the two reflection times. From here on, the end-to-end computation is the same as that for a split spread.

For a given value of a and V_D we may use formula 1 of Figure 8.14 to compute the value of h, the depth of the center of the pulse-front circle below the source. It should be remembered that the time in this formula is the one-way time, which is half the reflection time. By using a table of hyperbolic cosines we can easily compute a table of h values. This table can be used directly, or an h scale may be constructed from it, the scale showing times corresponding to the distances recorded on the scale.

The R scale can be constructed in a manner very similar to that used for the h scale. This is shown roughly in Figure 8.19, in which S is the shot or source at datum, and C is the center of the pulse-front circle. Attached to the scale is a protractor which is used for measuring the dip angle. The point of reflection is the point B, and AA' represents a segment of the reflector. The scale may be put in a clear transparent holder with clamps at both ends. The zero point of the scale is at B. R values are measured along the scale and the *corresponding reflection times are written on the scale* so that we have to use only the reflection time in conjunction with point C. Point C is located below S by the h scale or table referred to above. The R scale is placed so that C corresponds with the point on the scale corresponding to the reflection time. The protractor is then adjusted so that the proper dip is registered and the segment of the reflector, e.g., AA', is drawn on the cross section.

The clamps are put at both ends of the R scale for stretching purposes. It is usually good practice to make the R scale itself a little bit shorter than the scale of the paper on which the plotting is to be done. It is then easy to stretch the R scale so that its scale matches the paper scale. In this way the meas-

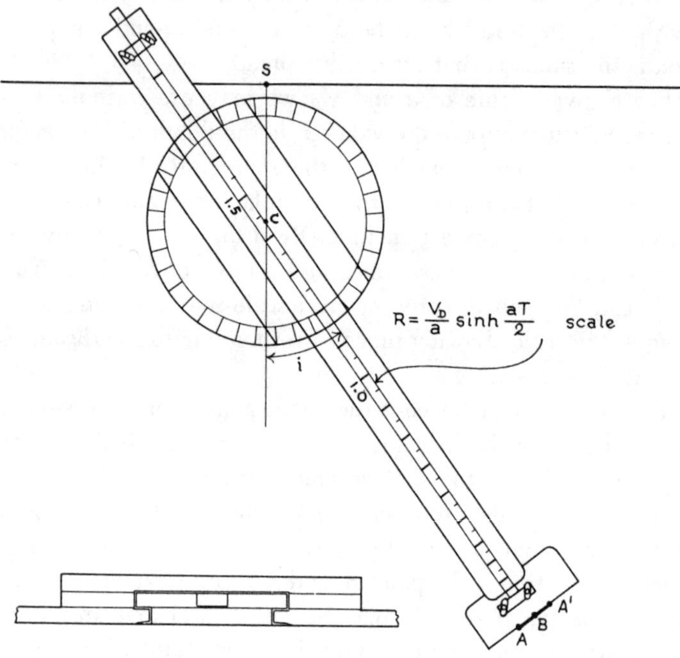

Fig. 8.19. Dip plotting machine.

uring scale or R scale is kept in proper adjustment with the cross-section scale for various changes of temperature and humidity.

The insert in Figure 8.19 shows how the scale can be constructed and the protractor attached. The form shown here has three layers of clear plastic. The kind of plastic that at present appears best suited for this purpose is vinylite, because it is clear and strong, and not subject to rapid deterioration or much temperature expansion.

The *R* scale illustrated in Figure 8.19 may not be the best type to use for regular routine computations. The angle is not adjusted as easily as it might be.

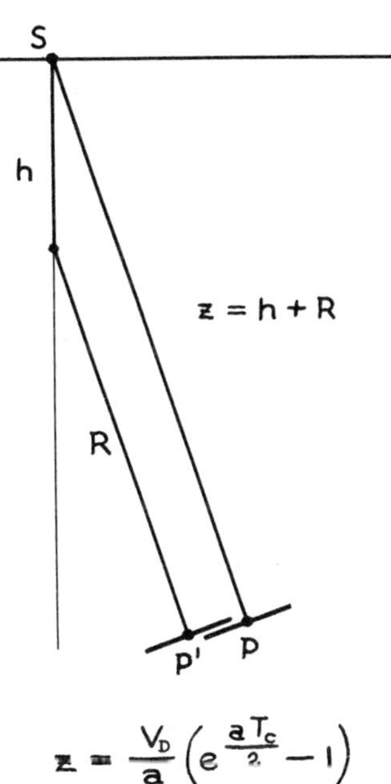

FIG. 8.20. Straight-line vs. curve ray paths.

A more convenient arrangement has been described by John W. Daly.* Daly's instrument contains both *h* scales and *R* scales as well as the necessary protractor. The author has not used it but has seen it in use and he believes it to be very convenient where much routine plotting has to done.

For small dips another procedure employing a so-called *z* scale may be used. A *z* scale is the scale obtained by adding the *R* and *h* scales. Its use eliminates the necessity for using an *h* scale and thus makes it possible to use what is essentially a straight-line plotting procedure. Using a *z* scale is thus equivalent to using an average velocity which varies in accordance with the linear velocity vs. depth relationship. For a better understanding of the use of the *z* scale, see Figure 8.20, on which two plots are made. One plot is made with a *z* scale, giving the reflection point *P*; the other is made by

* See *ibid.*, 13:153-157 (1948).

the h and R scales, giving the reflection point P'. The difference for small dips is negligible. The value of z in terms of the reflection time can be reduced to the formula shown under the drawing by adding equations 1 and 2 of Figure 8.14, making use of expressing the hyperbolic functions in terms of exponentials, and converting from one-way time to reflection time.

The cross-section paper usually used is millimeter paper 50, 75, or 100 cm. wide, depending on the depth of the cross section and the scale used by the party. The vertical and horizontal scales are always equal. It is often helpful to do the plotting on a drafting table that can be tilted quite steeply. It is of course desirable that the eyes be at normal working distance from the paper and approximately perpendicular to the board at the reflection to avoid parallax errors.

8.4.4. Organization of the Work

The picking of reflections and their marking have to be done directly after the datum corrections have been made. The next step is to transfer these data to the computation forms and compute the dips. The completed forms are then used in connection with the plotting machine to construct the cross section. The main extra step required is plotting the depth, h. This increases the amount of work somewhat, but not very much. When it is not important to plot this depth point a z scale may be used on the plotting machine, which makes the work very simple.

Different companies use different grading techniques. Those who use none at all object to grading on the basis that it is overdone. But the companies who use an extreme grading technique may contend that a certain amount of grading is useful. The author has used a system in which no letters are placed on the cross section, the grades being indicated automatically by the method of plotting. The average grade reflection which is regarded as *reliable* and which represents a reflection that carries all the way across the record is drawn as a solid segment that is of average heaviness and whose length is half the surface coverage. The *very excellent* reflections, which usually are few in number, are emphasized on the section not by means of a longer segment but by one twice as heavy. These may be regarded as A-grade reflections. The C-grade or *lower-grade* reflections are drawn in as light solid segments much shorter than half the spread length. These reflections represent reflections that apparently do not continue all the way across the record but are fragmentary, consisting of one or more parts. They are generally regarded as unreliable in themselves but very important to show on the cross section because in conjunction with other reflections they may give important information.

8.5. COMPARISON OF ROUTINE METHODS

One of the simplest routine methods for making seismic cross sections is merely to *plot* the reflection times directly beneath the shot points. This method has some merits when the dips are so small as to be almost negligible. One advantage is that it is the simplest of all possible methods, requiring almost no computation. However, this may not be an advantage if this oversimplification leads to wrong conclusions. The method has the disadvantage of introducing a varying vertical scale distortion. Thus the horizontal and vertical scales are not only not equal, but the vertical scale varies with depth. The most serious disadvantage arises because the method fails to account for apparent steep dips such as are often generated by faults, lateral changes in lithology, etc., by diffraction (see Chapters 11 and 14).

Another commonly used method is the *varying average-velocity method*, in which, for each time, a certain average velocity is taken, and it is assumed that the path is a straight line. Plotting the straight ray path is easy and the method has much to recommend it. A great many seismic cross sections have been interpreted on this basis. Objections to the method are that steep dips plot too shallow and overturned dips cannot be plotted.

In a special case of this method the z scale is used, the average velocity being selected so as to correspond to a linear increase of velocity with depth.

The next method is the circular arc or curved ray method which was outlined in the immediately preceding sections. This method requires an extra step because of the h scale. This is about the only addition required in comparison with straight-line methods. The curved ray method is especially valuable when the dips are higher. It should be remembered, however, that this advantage is somewhat diminished because, when the dips are higher, it is a serious error to neglect the cross component, as we have done entirely throughout this chapter.

Perhaps we may say that the circular arc method of computing has the advantage of being a single routine procedure which is easy to carry out in practice, involves little work, and covers the widest possible range of situations without leading to serious distortions. Whenever straight-line methods are used, it is necessary to limit the dip to 90°. Indeed as the dip approaches 90°, the depth is very poorly determined. This is not necessarily the case with the circular arc method. On many occasions overturned dips have been encountered which gave cross sections that made good geological sense. Thus, the circular arc method reduces the chance of getting into trouble because of introducing a distorted picture that is misleading. On the other hand the method should not claim too much accuracy. Its major claim is that it is easy to use and can be employed very efficiently in a field office.

We have considered reduction to datum, determination of velocities for converting from observed time to depths and dips, and the processes of making this conversion. But we have neglected the fact that the seismic profile line may not always be perpendicular to the strikes of the dipping interfaces. We have computed depths and dips *as if* there were no cross component of dip. This neglect is not serious when all the dips are small. However, it is likely to be serious when the dips are above 30°, as they often are.

So let us now see what can be done to remedy this defect. We need to calculate the *strike* and the *total dip* by simple routine methods. These methods we now consider in detail.

8.6. COMMENTS TO THE SECOND EDITION

This is the chapter where the differences between 1950 and 1980 are great and important, due to the explosive developments in electronics and computer technology. Aims are not very different, but emphasis has shifted. The growing technology, which is a truly remarkable *means*, has become so important in this competitive world that it sometimes tends to overshadow the *ends*. This requires much balance and understanding on the part of the management people. Their problem is difficult.

The technology is made up of a large number of simple parts, simply interrelated. By keeping in mind the aims of the various parts, starting with the major assemblies and working slowly toward but not to the nitty details, one can achieve a sufficient understanding at some adequate level without having a specialized technical background.

If a nonspecialist wanders into a geophysical prospecting seismograph office and asks to see what they have to offer of importance, that person may be shown a large print, called a seismic record time cross section (Figure 8.21). The careful host will state that each of those little points along the top of the section represents positions along a line on the surface of the ground where a sudden impulsive energy source is

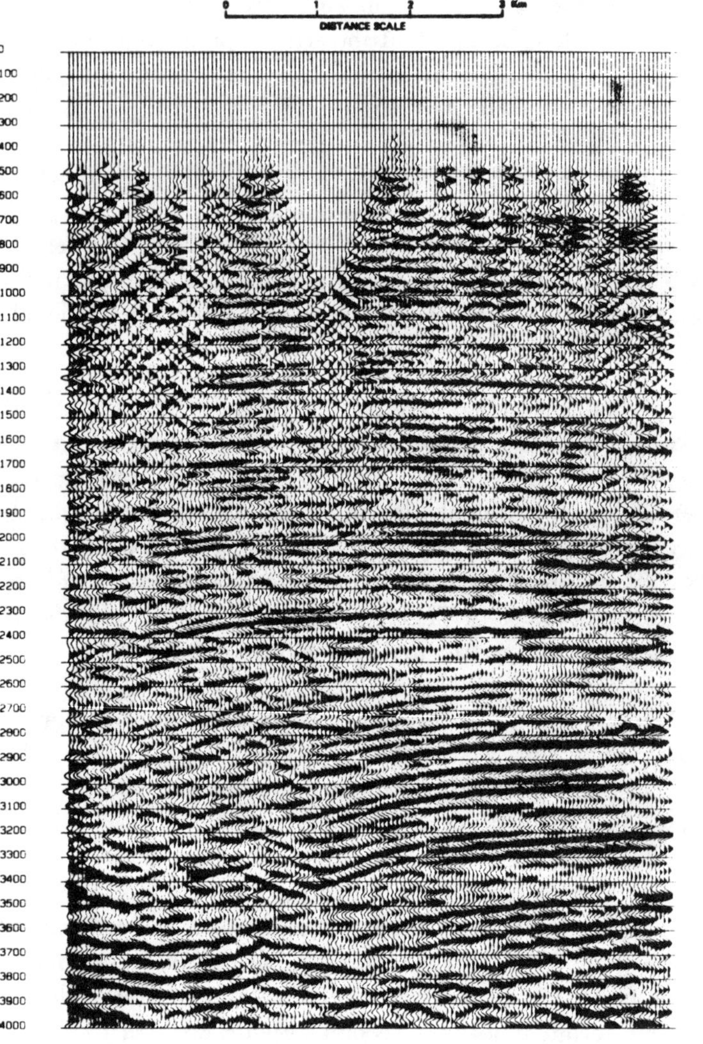

FIG. 8.21. Seismic Record Section. Unmigrated Example. Courtesy of Chevron Oil Field Research Company.

released, generating seismic waves that travel into the rock below, to places that reflect energy back to a receiver which is very close to the energy source. Then, directly below this source-receiver (SR) point are various shades of gray (white to black) along a narrow vertical strip, which indicates, by the variation of shade, *when* the reflected pulse arrives back from the reflector; the longer the time, the farther down the reflection indication appears. Our nonspecialist-geophysicist conversation might go something like this. "So all you need is a source and one receiver?" "Well, no—we have to use a hundred receivers and gather the reflected energy together, so that it won't be too small to detect. These hundred receivers are spaced out, one at each of the little points you see along the top line on the surface." Then several questions and answers until . . . "Yes, I can see that if the simpler, cheaper, single source and single receiver could have been used, the competition would have forced its use." "Correct sir." "So what you see on this one line may come from places all around this vertical line under the source-receiver point?" "That is exactly correct for this one source-receiver pair. But look at the other source-receiver pairs along the line on the ground surface. Notice that right here, at reflection time 2.835 seconds, the dark shading, nearly black, is repeated for many pairs, suggesting that this darkening may come from a reflection sent back to the ground surface from a nearly plane reflector, that appears level, relative to this line of source-receiver pairs. If you had another line of source-receivers running across this line, you could probably determine whether the reflector is dipping across this line or not." As our manager edges toward the door, he remarks, "I'll drop by another day." "Yes, please do."

Our manager sees a positive value and a problem—how much data are piled on this narrow strip below this *SR* pair, how confused does it get, how well can these fellows that use these seismograph sections distinguish them from real geological exposed sections, like road cuts, through sedimentary rocks, do they sometimes fool themselves in weak moments, and so on.

In Chapter 8 we took care of most computational problems using the velocity distribution of section 8.4, namely the linear increase of velocity with depth $V = V_D + az$. We now emphasize the usefulness of this same linear approximation, applied to present problems. Look at Figure 8.22. S is a source at an

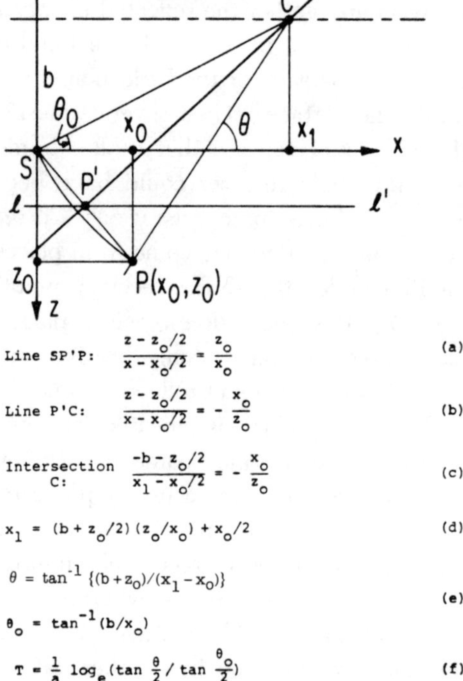

FIG. 8.22. Calculation of Travel Time between Two Given Points, for a Linear Velocity-Depth Dependence.

origin of (x,z) coordinates. A receiver, P, is located at (x_o,z_o). We know V_D and a. We wish to calculate the travel time, T, from S to P.

The line $11'$ is parallel with the x-axis half-way between S and P. The equation of the line $SP'P$ is given by (a). The line drawn perpendicular to $SP'P$ through P' has the equation (b). That line intersects the line of centers of ray circles at C. Call the coordinates of C, x_1, and $-V_D/a=-b$. Using these coordinate values in equation (b), we get equation (c), which, solved for x_1, gives relation (d). Having (d), we can get the angles θ and θ_o from relation (e). Finally, the one-way time, T, is given by (f), which comes from Figure 8.11.

We make a simple arithmetical program to compute T, given a, b, x_o, and z_o.

For the purpose of illustrating a property of (x^2,T^2_x) velocity relations when $V=V_D+az$, I let x_o be replaced by nx_o, where $n=1,2,\ldots,13$ and $b=3600$ meters with $a=0.5$. These results appear in the comments to Chapter 7.

In 1950, given T_o, $\Delta T/\Delta x$, a, and V_D, we used R and h scales (Figure 8.13, (5) and (6)) and the dip slide rule to calculate and plot the small segment representing an element of the reflector on the seismic reflection cross section.

Today this could all be mechanized, but we would have to have previously measured the record to get T_o and $\Delta T/\Delta x$. We use the "roll along" and "stacking" procedure to provide quantities that can lead to the same result, but by a different process.

Refer to Figure 8.23 to get some background for the technique so that we can use the same numerical example, that is, $V=(1800+0.5z)$ meters/sec with dip possibility included. Two reflection points P_1 and P_2 fall on a single reflecting plane with dip, θ. The circular arcs P_1D_1 and P_2D_2 are used to plot D_1 and D_2 using the R and h functions, just as we did thirty years ago. But now we use the stacked record time section to

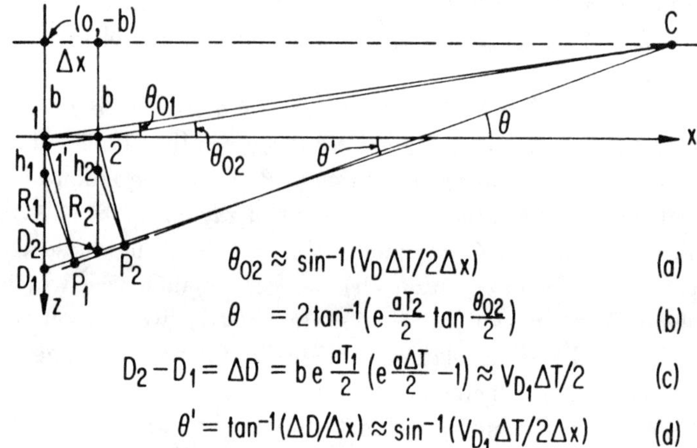

FIG. 8.23. Reflector Dip and Record Section Dip.

get $\Delta T/\Delta x$, and relation (a) to get θ_{02}. Then from (b) we get θ. In equation (c), I have given a relation between ΔD, ΔT, and the velocity at depth D_1. Dividing (c) by Δx, the small distance between stations 1 and 2, we get relation (d). The small dips, θ and θ' do not show how poor θ' is as an approximation to θ, on Figure 8.23. If you go to higher dips, say $\theta = 60°$, it becomes obvious that plotting "depth points" directly below the source-receiver point can be very misleading. Henceforth, I shall regard the times sections of the present day as very compact "bookkeeping" devices for exploration seismologists, potentially very valuable but, in some cases, possibly misleading.

Imagine an ideal situation, where there is *no noise* disturbance, and we proceed down a line of *SR* points, shooting identical shots and receiving a single trace seismic record at each station. We assemble all of these traces, side by side, on a single wide record. This would constitute a modern reflection seismograph time section (see Figure 8.21).

In the actual case, we aim at this ideal by gathering and stacking (see Figure 2.12). Each gather has to be made *coherent*, before it can be stacked. We outline processes for making these corrections, usually called normal-move-out (NMO) corrections. Just to have a definite example, we suppose the velocity distribution is given by $V=1800+0.5z$) m/s. We focus attention on a reflector at 3000 meters depth. I calculated $T_x=2T$ according to Figure 8.22, and listed it for offsets $x=0$, 800, 1600, . . . , 8800, and 9600 meters. These T_x's are used to calculate the corresponding NMO's, which are listed in order as follows: 0, 0.021, 0.082, 0.181, 0.312, 0.471, 0.652, 0.851, 1.063, 1.286, 1.517, 1.753, and 1.993 seconds.

It is customary to select a series of constant average velocities, for example, 2000, 2150, 2300, . . . , 3950, and 4100 m/s, and calculate NMO's as if those velocities were constant for

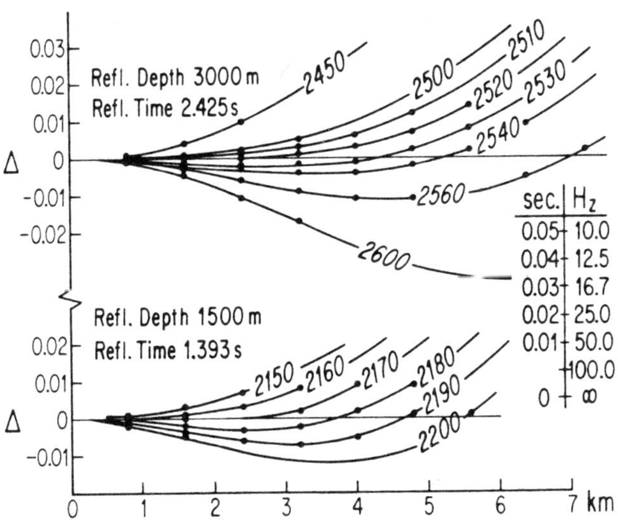

FIG. 8.24. Coherence in Stacking Corrections for Constant Velocities Minus "True" Velocity Assumed (1800+0.5z) m./s.

all reflection times. For constant velocity, the reflection times fall on an exact hyperbola, that is, the (x^2, T^2_x) graph is really straight. So NMO $= T_x - T_o = x^2/[(T_x + T_o)V^2]$ for this case.

On Figure 8.24, I have plotted Δ (NMO), which is NMO for constant velocity minus NMO for $V=1800+0.5z$, for stations 800 meters apart along the abscissa. These are plotted for two depths, $z=1500$ and 3000 meters, and for a set of constant velocities. For example, for $z=3000$ meters and $V_{constant}=2500$ m/s, as the offset increases, the incoherence increases until at about $x=4000$ m, a 50 Hz spectral component of our signal is just exactly out of phase and so is cancelled. Lower frequencies are still reinforced to some extent in the stack from zero to 4000 meters. If we want higher resolving power, we must either shorten our offset range (and reduce our signal-to-noise ratio) or base our NMO on a better velocity choice. I would select $V=2530$ m/s for the constant NMO velocity out to $x=4800$ m. This would handle the signal spectrum up to about 100 Hz. We have indicated a scale showing the relation of Δ(NMO) to the first cancellation frequency on the right edge of Figure 8.24.

Quantitative aspects of making the NMO are apparent from Figure 8.24. A very good stacking velocity can be selected for offsets up to the reflector depth, but if you rely on constant velocity panels 150 m/s apart, as for example 2450 and 2600 m/s, the part of the spectrum above 30 Hz has to come from the part of the spread offset less than half the reflector depth, or 1500 m. So if we were to mute the panels for offsets greater than half the depth, one should be able to make a fairly good selection of a constant velocity for stacking. But try to note in this case, that the panels for 2450 and 2600 show better high-frequency response than do those for 2300 and 2750 near 2.425 seconds. Perhaps if one can see, in this case, that 2450 and 2600 are equal choices, we might try the average 2525 m/s, which should be much better.

When very large numbers of channels and long lines are used, if one demands very high-fold numbers, it seems necessary to use a good approximation to the correct velocity distribution. I would try to fit a linear distribution, V_D+az to the problem. NMO tables for various values of V_D for a fixed a value, can easily be made and stored for computer use. The selection of a is not critical. The important thing is to get an increase in V with increasing z. For this purpose, $a=0.5$ can be used almost universally. 0.5 may be too high or too low, but it is better than zero. For very long offsets and shallow reflectors, muting is the only practical solution for the problem. For finer tuning, the linear velocity approximation will usually allow one to gain large-fold numbers.

If you have a large number of channels, you can use a small channel interval—say 10 meters. We have here a problem of balancing various costs, for example, field operations, computer, storage, and so on and a value estimation. I believe that that judgment should be made by a person or persons who have good technical understanding of the problem—see the comments at the end of Chapter 4.

In case dip needs to be taken into account for stacking, the question is, how can we achieve coherence? This can be done over small collections of channels using NMO's with panel sampling. As there are violations of *normalness*, inconsistencies may arise. We should try to be reasonably precise about the meaning of the particular stacking velocity. This may involve a measurement of velocities, in a complicated situation, that may not be worth the expense.

Now we look at the *migration* problem. Because we plot our time sections straight down under *SR* pairs and because we are thus storing data that may not actually come from straight down and because these time sections are widely thought to give at least a qualitative picture of what is below any surface point at the top of the time section, we are in grave danger of misleading a good part of the industry people. So we must restore the picture presented, to give a picture with any falsifications reduced as much as possible within reason. We migrate.

The recent book by Robinson and Treitel (1980, pp. 374–398) discussed the problem of migration, concisely and clearly, in what they call the two-dimensional case. In my effort to always try to keep in mind the fact that, even when we really place our seismograph line strictly perpendicular to the strike of all reflectors, even then the process is really (and obviously) three-dimensional so far as the wave propagation is concerned. The strike direction coordinate is really there, but + and − symmetry in that direction permits us to ignore that direction in formal mathematical manipulations, but a truly two-dimensional problem would require a whole series of *line sources* laid out along the strike direction, a much more difficult problem than the actual one with a simple line of point sources. I am sure the authors agree but think the extra explicitness unnecessary.

The record time section is made of shades of gray plotted on narrow strips directly below *SR* stations, using *CDP* stacking procedures. These narrow strips and the time sections they constitute are regarded here as compressed "bookkeeping" devices. No one really considers that such events, thus plotted, really refer to reflections just below the *SR* points concerned. I shall insist that this linear distribution has by itself an unknown direction or set of directions. If we use the constant velocity, *c*, then we can consider that the distribution of greys is not only on the vertical line below the *SR* point, but throughout the solid sphere centered at the *SR* point. Since it is difficult to illustrate such a sphere, with degrees of greyness dependent only on radius and *not* on direction, we restrict the discussion to a vertical plane section along the line of *SR* points, so that the spheres appear as circles. We are thinking of these circles as instantaneous wave fronts, so that ct has the distance dimension that may be measured with the same scale as used for measuring horizontal distance along the top of the section.

So our time section is expanded to a set of circles of radii given by the "bookkeeping" time section. This is illustrated in Figure 8.25, where these circles have the equations (a),

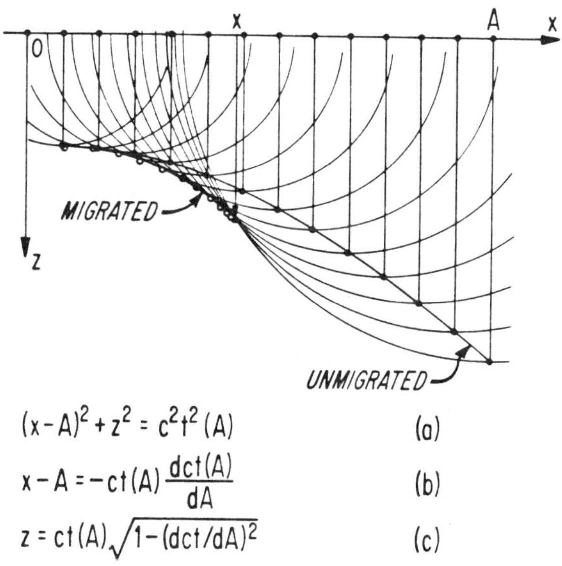

FIG. 8.25. Migration Illustrated.

where the parameter A is the x coordinate of the "bookkeeping" data strip, that is, of its *SR* point. The solid points at the bottom of the vertical lines represent unmigrated reflection events, which are read off the "bookkeeping" time section that is labeled by t, not ct.

What we want is the *envelope* of this family of circles. To find this, we differentiate (a) with respect to the parameter, A, to get relation (b). (Refer to any calculus book under envelopes in the index.) Then we eliminate $(x-A)$ between

(a) and (b) to get equation (c), where cdt/dA is the tangent of the "dip" angle appearing on the "bookkeeping" record section. Notice that when $x=A$, equation (a) says that $z=ct(A)$. Thus formula (c) tells us to reduce this z by a factor $\sqrt{1-(cdt/dA)^2}$ in order to "time migrate." For correct results it is necessary to keep the same units for x, z, and ct on the time section.

When the unmigrated time section shows a dip of 45°, the reduction factor is zero, (90°). For 0°, 10°, 20°, 30°, and 40°, the respective reduction factors are 1.00, 0.98, 0.93, 0.82, and 0.54. Also, the horizontal shift of the reflection point is given by (b).

If you want to do a better job and can find the proper velocity distribution, that can be done. However, in most cases, I would find a linear approximation to the actual velocity and use that. This can be done easily, using Miller's method as given in Figure 7.6. We want the x and z scales to be equal. We relate the reflection time scale, T, to the depth, Z (and x scale by $Z=R+h=b(e^{aT/2}-1)$. The mapping of a time section on a depth section is easily carried out on the vertical strips. Then wave fronts are spheres just as they were for the constant velocity case, but the centers of these spheres now go below the datum in accordance with $h=b$ $(\cosh aT/2-1)$. We have the same envelope problem, except that now equation (a) becomes $(x-A)^2+(z-h)^2=R^2$, where h and R are functions of A. If this linear velocity versus depth approximation is to made practical, some younger man will have to work through the details. I have seen remarkable fits to quite complicated structural problems, using the circular ray plots, even in cases where many overturned reflectors could be seen from both top and bottom. It was interesting to map a productive anticline that we were not allowed to shoot on, both from top and from bottom, the latter coming all the way across under the anticline.

CHAPTER 9

Strike-Dip Determinations

This chapter is primarily for seismologists, party chiefs, and computers. It is a technical chapter of necessity. As we have pointed out in earlier sections, neglect of the cross component of dip is legitimate only in the case of a small dip or when we know the strike. When we know the strike and can shoot perpendicular to it the methods outlined in the preceding chapter may be used with considerable confidence. In other cases the methods discussed in the present chapter must be used.

The present chapter gives methods for computing the strike, dip, and depth from field observations. These methods will be presented in such a way that they can be used with a high degree of efficiency in a field party office.

9.1. FIELD TECHNIQUES

The field techniques mainly used are illustrated in Figure 9.1, the upper one being the *cross spread* and the lower one the *L spread.* The cross spread is perhaps used more than the L spread because it is easily added to the line shot when it appears to be particularly needed. The L spread has some advantages over the cross spread, the principal one being that it is easier to lay out an L spread with a regular set of cables designed for use with split spreads. In some areas, however, three-dimensional information is of such importance that the special cables used with cross spreads amply justify their cost. In complicated regions the cross spread offers some advantages over the L spread, because the information is obtained on the

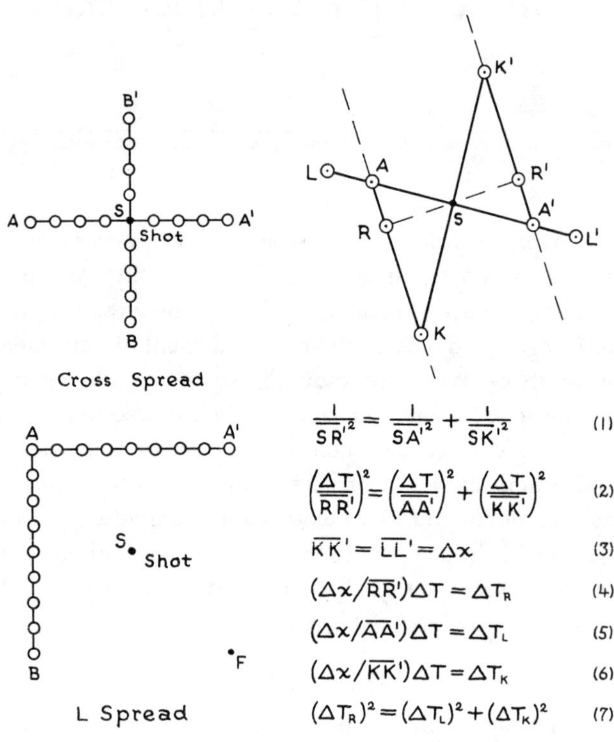

FIG. 9.1. Cross and L spreads.

FIG. 9.2. Calculation of resolved dip and strike.

cross spread from a more compact area of the earth's surface than is true with an L spread. This compactness may be of considerable importance especially when the reflections are fragmentary. A reflection may not carry far enough over the surface of the earth to be recorded on a full L spread, whereas it may be recorded on the whole cross spread. The cross spread should, if possible, be laid out so that AA' is at right angles to BB', though this is not absolutely essential.

9.2. MOST DIRECT COMPUTATION

Figure 9.2 shows a cross spread in which LL' may be regarded as the line direction and KK' as the cross direction. S is the shot point. The two parallel lines $A'K'$ and AK represent the strike direction, as explained earlier (see Section 2.4). Thus the reflection comes to A at the same time as it comes to K and the same reflection comes to A' at the same time as it comes to K'. The method of utilizing the field setup illustrated in Figure 9.2 was explained in Section 2.4.

It is our purpose to transform this fundamental idea into a formula that is easy to use. For this purpose we start with relationship 1 in Figure 9.2, which may be deduced by applying the Pythagorean theorem to triangles $SR'A'$, $SR'K'$, and $SA'K'$. This relationship is multiplied by $(\Delta T)^2$ and divided by 4 as indicated in the second formula. Formula 3 defines Δx. We are now ready for the operations shown in formulas 4, 5, and 6. These operations consist of finding a relationship between ΔT for the imagined spread RR' and ΔT_R for the entire spread corresponding to the distance Δx. We assume in each of these three formulas that ΔT is proportional to the spread distance. Substituting formulas 4, 5, 6, and 3 in the second formula and removing a constant factor, we obtain formula 7.

Formula 7 is one of our key formulas. It states that if, when ΔT_L is along the line and ΔT_K is in the cross direction, we square them and add their squares, we obtain ΔT_R in the resolved direction for the same spread, Δx. This formula is easier to use than formula 1 because we are given directly ΔT_L along the line and ΔT_K across the line.

We can carry our investigation one step further by reference to Figure 9.3. Here our object is to determine the strike. When we measure the strike by angle α between the line and the strike direction, it will be evident from the figure that the tangent of this angle is given by relation 1. We also have relations 2 and 3 just as we saw them in the preceding figure. We can then write angle α as the angle whose tangent is the ratio of the two ΔT's as in equation 4. Thus we have a formula for the angle which determines the strike direction in terms of ΔT alone.

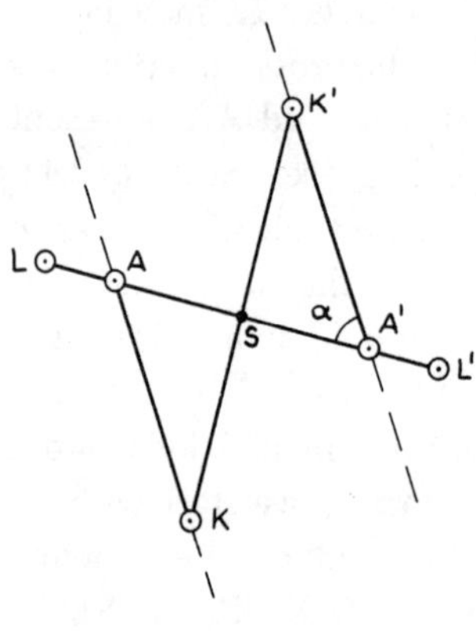

$$\tan \alpha = \overline{KK}'/\overline{AA}' \quad (1)$$

$$\overline{KK}' = \Delta x\, \Delta T/\Delta T_K \quad (2)$$

$$\overline{AA}' = \Delta x\, \Delta T/\Delta T_L \quad (3)$$

$$\alpha = \tan^{-1}(\Delta T_L/\Delta T_K) \quad (4)$$

Fig. 9.3. Calculation of strike.

The results obtained in Figures 9.2 and 9.3 are summarized in Figure 9.4. This figure shows a coördinate system in which ΔT_K is plotted upward and slightly to the right, and ΔT_L is plotted to the right and slightly downward. The resolved ΔT_R is constructed by completing the rectangle, as indicated. Angle α is the angle between the strike and the line direction. The formulas we have obtained are given in equations 1 and 2. Another way of writing these formulas is shown in equations 3 and 4, which are sometimes useful.

Before proceeding to the construction of an instrument for determining strike and dip, let us consider the slightly more general problem in which the cross spread is *not* perpendicular to the line spread; this is shown in Figure 9.5. We could use formulas 1 and 2 in this figure and the fact that α and β are related angles and that the relationship is known.

On this basis we could handle the entire problem, but it is

much simpler to handle it as indicated in Figure 9.6. In the upper left part of the figure we have drawn a ΔT_K axis and an origin. We have plotted ΔT_K and have indicated by the line OR the ΔT_R value at which we shall eventually arrive. This portion of the figure may be interpreted in terms of a cross and a line spread at right angles to each other. The same is true of the upper right-hand portion of the figure. Here we have plotted ΔT_L and have indicated how the situation would appear if

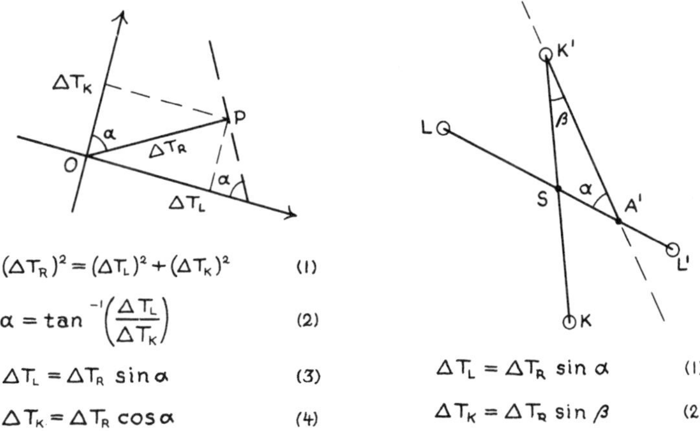

FIG. 9.4. Summary of strike-dip calculations for perpendicular cross spreads.

FIG. 9.5. ΔT relations for nonperpendicular cross spreads.

the cross had been perpendicular to the given line. These two parts are superimposed on the lower left portion of the figure. We have only to remove the extra perpendicular axes that were in the upper figures to obtain the lower right-hand portion of the figure. Once we have obtained this construction we see how to get the resolved dip by measuring the ΔT components along the directions of the corresponding profiles. When we have made these measurements we draw perpendiculars to these lines which intersect at point *R*. This point in conjunction with the origin *O*, or the center point of the spread, determines

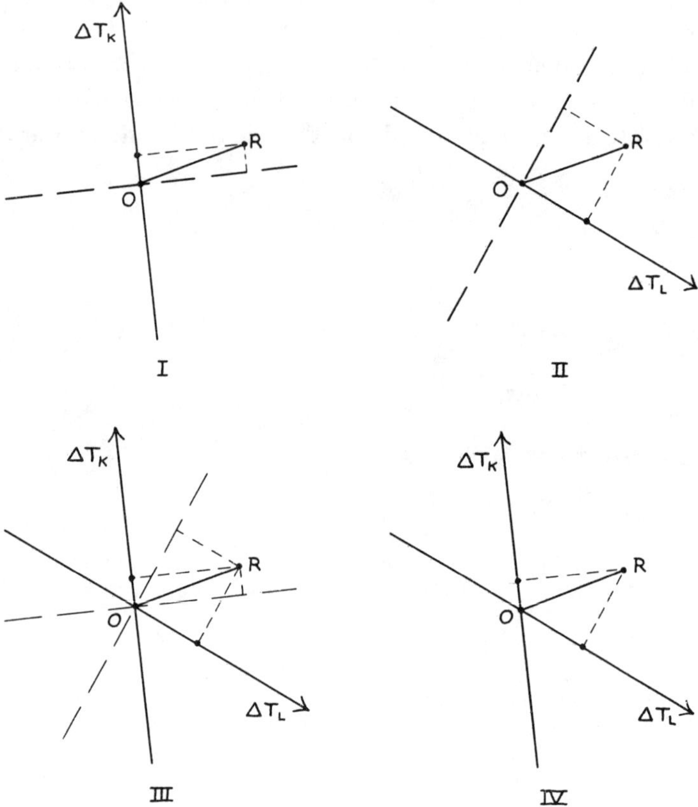

FIG. 9.6. Synthesis of method of computing dip and strike in non-perpendicular cross spreads.

ΔT_R for the resolved dip. It also determines the direction of strike.

With this background of understanding we are now ready to consider actually doing the computation.

9.3. SIMPLE MECHANICAL COMPUTATION

Figure 9.7 shows the genesis of an instrument that can be used to compute the strike and the dip from ΔT information. In Part I of the figure appears a base with the cardinal directions, a center, and the necessary protractor to measure the angles. In Part II is a ruled transparent disk and in Part III is a similar disk oriented at right angles to the disk in Part II. These two transparent disks are superimposed over the base, as indicated in Part IV. In addition there is a measuring rod in order to determine the resolved ΔT_R. Referring to Part IV of the figure, we see that we measure ΔT_N in the north direction and ΔT_E in the east direction. The point whose coördinates have these ΔT values is marked and the resolved ΔT_R is thus measured on the rotating bar.

Part V of Figure 9.7 illustrates the case previously considered in which the cross spread is *not* at right angles to the line spread. In this case the *points* show how the ΔT's are measured. A good characteristic of this instrument is that the strike and dip direction can be measured with it just as easily when the cross spread is not perpendicular to the line spread as when it is. However, a perpendicular cross spread gives the most precise determination.

In Part VI of Figure 9.7 is shown the mapped result of the computation illustrated. S is the shot point and P is directly over the point of reflection. The *strike-dip symbol* indicates the direction of the strike. The dip is of course toward the shot point.

The foregoing discussion has not gone into the problem of relating the direction of dip to the sign of ΔT. It will be evident, however, that positive ΔT's have been related with the south and west components of the dip in the example which has been used. This relationship is a purely arbitrary matter and has to be selected by the party chief or computer.

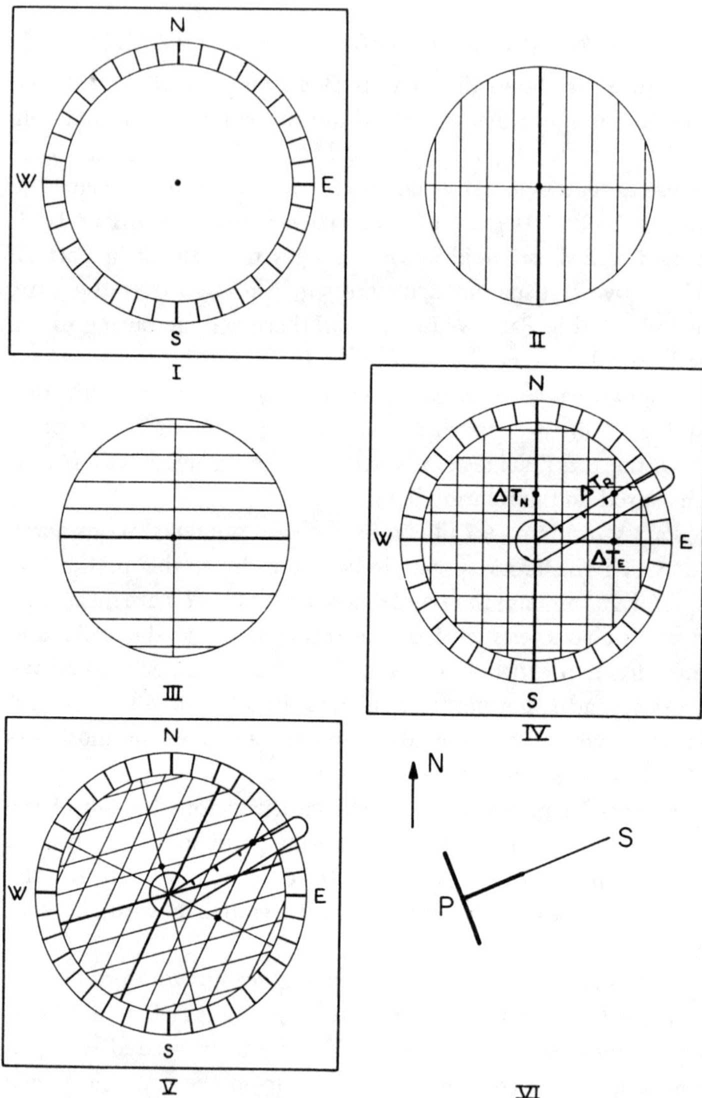

FIG. 9.7. Synthesis of strike-dip resolver. (Courtesy United Geophysical Company.)

In order to facilitate the use of the dip resolver shown in Figure 9.7 we should record the data in such a way that the results are ready for computation, and we should record the results of the computation. Figure 9.8 shows a form designed for this purpose. At the top are entered the general data. The part at the left gives the line data which include Δx, the direction of positive ΔT, and the reduction factor for use with the

Company ______ Shot Point ______ Date ______

LINE					CROSS			V_D ______, a ______				
Δx ______ Dir. ______					Δx ______ Dir. ______			Red. Factor ______				
Red. Factor ______					Red. Factor ______			______				
No.	T_1	T_2	ΔT	$(\Delta T)_c$	T	ΔT	$(\Delta T)_c$	$(\Delta T_R)_c$	Dip	Dir.	Depth	Offset

Fig. 9.8. Form for strike-dip computation.

dip slide rule if this computation is to be made. The middle part gives the cross information and the right-hand part gives the resolved dip information. The line columns have a place for both T_1 and T_2. Thus the form may be used for an end-to-end shot as well as for a split spread on the line. Occasionally both types of computation must be made. The observed ΔT occurs with the corrected ΔT. On the cross, provision is made only for the time, assuming that the cross will be a split spread. The line ΔT and the cross ΔT, after correction to a standard spread suitable for use with the dip slide rule, are resolved on the dip resolver shown in Figure 9.7. This gives the first column in the right-hand portion of the form. From this resolved ΔT_R, by means of the circular dip slide rule (see Fig. 8.16), the resolved dip can be computed. The direction of the dip is read directly off the dip resolver protractor. Usually it is desirable to record also the depth and offset. These may be measured on a blank piece of regular cross-section paper like that used for plotting cross sections. In case the velocity increases linearly with the depth, the h and R scales described in Section 8.4.3 should be used.

From the data in the right-hand group of columns in Figure 9.8 we can construct a strike-dip map. This is constructed as illustrated in Part VI of Figure 9.7. The distance *SP* is the scaled offset. The direction of the dip is obtained from the direction column. The depth must be written in, along with the strike-dip symbol, which should also, of course, include the corresponding dip angle.

A great many strike-dip maps have been made in the exploration industry, but they have generally been prepared in cases where the structural problem was difficult. This difficulty usually makes correlation of reflections impossible. This in turn makes it difficult to tell whether a given strike-dip symbol applies to a given horizon or not. Without this latter knowledge a great deal of confusion can arise, especially if unconformities exist. In Section 10.1.3 we shall discuss a means of avoiding this confusion.

9.4. APPROXIMATIONS

We have shown how strike and dip can be easily determined by means of a strike-dip resolver, the theory of which has been presented. It is our purpose in the present section to note some approximations.

In Figure 9.9 we see that angle θ, in radians, is given by the definition in relation 1. This may be compared with the definition of the sine of this angle given in relation 2. It is evident from the drawing that, for small values of θ, the sine of θ is very nearly equal to the value of θ, as indicated in relation 3. This relation is, of course, only approximate. We recall that the ratio of ΔT_L to Δx is approximately equal to the sine of the angle divided by the velocity. In this case angle θ_L is the component of dip along the line direction computed, the cross component being ignored. A similar remark applies to formula 5, which relates the cross ΔT_K to the cross component of dip, θ_K. Relation 6 is of course the same one for the resolved dip. We have shown relations 4, 5, and 6 as approximate to emphasize the fact that they are based on the assumption that the reflection goes across the record *straight*. Actually, as we have seen, every reflection has a slight *curvature* as it moves across the record. Therefore these three relations have to be regarded as approximate. We substitute these relationships in formula 7 for the resolved ΔT_R. This gives formula 8 in terms of the sines of the angles. Relationship 9, also approximate, may be obtained from equations 8 and 3. Relation 9 then expresses the resolved dip θ_R in terms of the line component and the cross

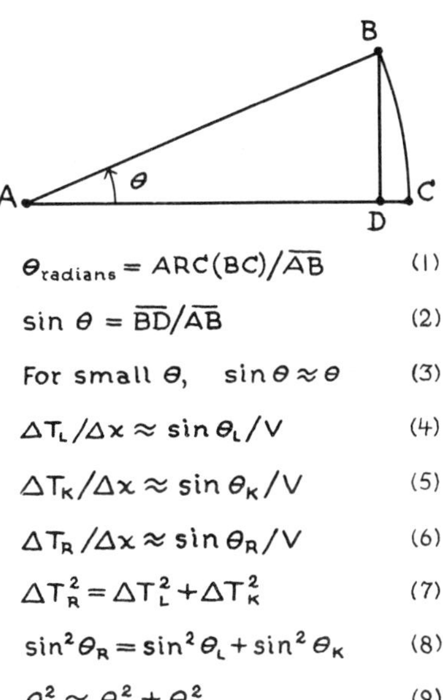

FIG. 9.9. Approximate resolved dip formula.

component and is valid as an approximation when the component dips are small enough.

Relation 9 is mentioned because it has been used a great deal. However, there is not much need to use it, for relation 7 can be used just as well and it remains accurate even for overturned dips.

In closing this chapter it may be remarked that in any complicated structural area where dips are high the methods discussed in this section are required in order to obtain interpretable results of real value. In such cases we must not ignore cross dip components. Ignoring information may be seriously misleading. Thus if we happened to shoot parallel with the strike, to take an extreme case, a dip of 90° might show only as a deep undipping reflection on the corresponding cross section. Such information would be decidedly misleading and misleading information is often extremely expensive for it may lead to further costly operations.

We have considered in detail about all the routine computing necessary for practical operation of a seismic party in the field. An effort has been made to make clear the assumptions used. These assumptions limit the applicability of routine methods and for this reason should be understood. We have carried the practical aspect of the work far enough so that the party chief, seismologist, or computer who understands the processes described can do very good routine interpretive work.

But he must impart his information, thus gained, to others working on the exploration problem, to geologists, managers of exploration, and also other geophysicists.

Moreover, the direct worker's information is still very fragmentary. He has analyzed details. Now he is ready to put them all together. This synthesizing process is just as important for him as for his associates. Whereas others can see the forest, only the investigator of details can see both the forest *and* the trees.

9.5. COMMENTS TO THE SECOND EDITION

This chapter gave the basic considerations for three-dimensional (3-D) work of 1950 vintage. Today there is an almost desperate struggle to automate this kind of work. Such work can be done now only at high cost.

The obvious way is to use a network of lines crossing each other on a "roll along," "stack," "migrate" basis; then resolve dips, lean depth sections away from vertical appropriately to get a second approximation map of the major reflectors with a set of 3-D models, using techniques of this and of the next chapter.

CHAPTER 10

Presentation of Results

This chapter is written primarily for party chiefs, seismologists, computers, and geologists. If geologists understand something of the methods used in preparing seismic results for presentation, they can make better use of these results.

Fortunately the writer's task has been much simplified by the recent appearance of the *Geophysical Case Histories,* Volume 1, published by the Society of Exploration Geophysicists under the editorship of L. L. Nettleton. This volume includes many examples of presentation of results and should be studied in connection with the present chapter. It also contains many interesting cross sections, contour maps, and dip maps, which may serve as examples. Also important are the reports that accompany the maps and cross sections. The principal difference between the reports in the *Case Histories* volume and the reports of primary interest here is that hindsight rather than foresight is involved and so, of necessity, some of the Case Histories have a clarity which the original reports probably did not have. Perhaps some day a case history will be written that includes all the original reports, together with the poor guesses and mistakes. Many of the Case Histories are remarkably sincere and unbiased. Space limitations, loss of detail because of imperfect memory, and the generally inarticulate nature of early stages of exploration probably account for part of the incompleteness. Even so, the volume is a truly remarkable achievement just as it stands.

The present chapter has two main purposes. One is the *synthesis* of much scattered data into a significant whole. This purpose is partly aimed at giving the original geophysicist who works up the information a chance to see it all at once and to make sure that the various parts have a certain mutual consistency. The second purpose is to make a *permanent record* in the form of cross sections, maps, and reports, which may be referred to in later years in attempts to return to the position occupied by the interpreting geophysicist and associated geologists when his work was completed. This purpose is by no means a trivial one. It is of course very difficult to accomplish—so difficult, in fact, that it should be fully realized that this purpose is essentially impossible. During the passage of time, information is bound to be lost, forgotten, misplaced, confused, etc. The individual charged with presenting results should have this fact in mind so that he will realize what a great task he faces when he starts his work. Such a presentation constitutes a real challenge to any individual. Unless treated with great care, information keeps no better than other fruits.

The chapter is divided into three main sections, the first dealing with *cross sections*, the second with *maps*, and the third with *reports*.

It is not the object here to give instructions for constructing cross sections, maps, and reports, for these of necessity will differ from company to company in accordance with variations in basic policy and philosophy. Little attention is given to the simpler types of presentation now commonly used. Instead, most of the space is devoted to the presentation of what may be regarded as improvements on the routine types commonly used. These improvements are not always essential, but in certain complicated cases they are very valuable.

10.1. CROSS SECTIONS

Cross sections are presented in order to give depth and dip relationships found from the seismic records. They vary from the very simple to the very difficult, such as vertical cross sections.

10.1.1. Time Sections. Small Dip, Zero Offset Plotting

A great many cross sections have been made as *time cross sections,* that is, the reflection time is plotted along the section, giving a time profile. The time is shorter for the high places and longer for the low places in the section; hence there is a certain structural resemblance between a time section and a dip cross section. A fair percentage of all the cross sections now used consists of time cross sections. They are quite adequate when the dip is very small and the depth is medium. Even with relatively small dips, at great depths the *neglected offset effect* may be important enough to make the time cross sections inadequate.

A cross section closely related to the time cross section is that in which depths are plotted directly below shot points. These depths are proportional to the times according to some time-depth relationship. Thus in a certain sense such plots, which neglect offset effects, are structurally equivalent to time cross sections. They are valid in the same circumstances in which time cross sections are valid. Their advantage over the straight time cross section is that the amount of closure across a high can be estimated more quickly. An example of such a cross section was shown in Figure 2.4.

It is sometimes remarked that the time cross section has an advantage over any other type, in that no assumptions are made regarding the velocity distribution. Observed times are plotted directly. It is usually wise to keep assumptions to a reasonable minimum, but there is such a thing as keeping their number too small. Certain risks have to be taken in the exploration business and certain risks prove valuable. The risk involved in making reasonable assumptions regarding velocity distributions is, in most cases, one of the valuable risks that ought to be taken.

10.1.2. Cross Sections with Offset and Neglect of Cross Component

In this section we consider the type of cross section now most used in the industry, at least in all areas where there is a chance that dips will be high. As was explained in Chapter 8, such cross sections are almost as easy to make as those in which the depth is plotted directly below the shot point. Consequently these cross sections are also used to advantage in low-dip country. One of their principal advantages is that they give an approximately proper offset to the deeper reflections and thus often produce a more consistent-looking picture. The method of constructing these cross sections was outlined in Chapter 8. An example of one may be seen in Figure 2.2. That one happens to be a cross section plotted with split dips, but a similar cross section could be plotted using end-to-end dips.

The title of this section indicates that we are neglecting the cross component. This is true, and very important, especially where the cross component has a chance of being large. It must be emphasized that neglecting a cross component is not quite the same thing as finding the line component. The line component of dip may be various quantities. It may be the component of dip in a vertical section parallel to the direction of the line. If this is so, it is not generally computed in making the cross section and is not the quantity which we computed in Chapter 8. The quantity computed there is the dip when the cross-component is neglected.

We may perhaps get a better picture of the situation if we follow a regular cross section around a sharp bend. In order to do this, refer to Figure 10.1. At the top are two cross sections. Below at the left is the corresponding map, showing contours and shot point locations. It will be seen that line BA makes a right-angled turn into AA'. Let us now consider the problem of going around this turn. In order to show the problem more clearly, the ends of the cross section corresponding to point A have been enlarged in the lower right-hand por-

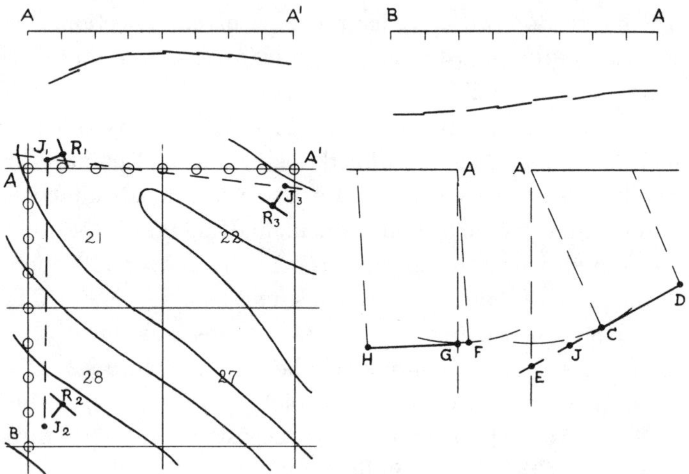

Fig. 10.1. Carrying a seismic cross section around a corner.

tion of the figure. It will be observed that no point on HGF, the plot of the reflector on line BA, has the same depth as CD, the plot of the reflector on AA'. Thus there is a discrepancy in depth in turning the corner. What we should remember is that, when we turn such a corner, we have a *common time* for the reflection, but *not a common depth*, calculated with neglect of the cross component of dip. Later, in Section 10.2.1, we shall consider this problem again; but for the present it is used only to bring out the fact that cross sections, thus made, may not tie with respect to depth when we turn a corner. Such a tie is not to be expected and should not be forced. Only when the dip is very small, in all directions, can a profile thus constructed be continuously tied with respect to depth in turning a corner. The tie is continuous with respect to time but not depth. Inadequate treatment of this problem could be made the basis of an argument in favor of simple time cross sections. However, in the cases when such sections are not valid either, this type of argument is not good.

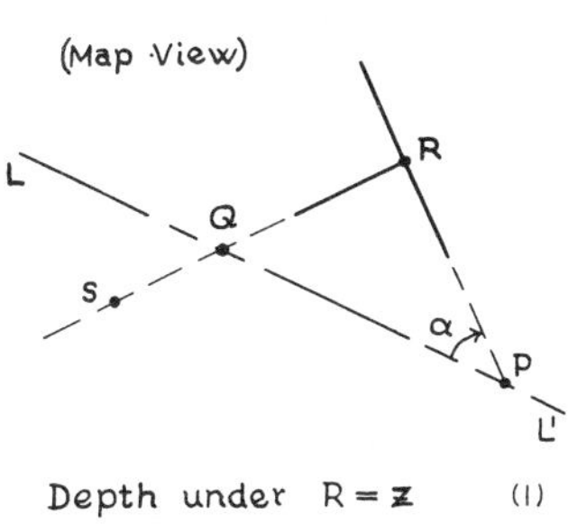

Depth under $R = z$ (1)

Total Dip $= \theta$ (2)

Depth under $P = z$ (3)

Depth under $Q = z_Q$ (4)

$QN = z_Q - z = QR \tan\theta$ (5)

$z_Q = z + QR \tan\theta$ (6)

$\tan\phi = \frac{QN}{QP} = \frac{QR \tan\theta}{QP}$ (7)

$= \sin\alpha \tan\theta$ (8)

Fig. 10.2. Basis for constructing vertical cross sections.

There are several ways of solving this problem, one of which we consider in the next section. The other is considered in Section 10.2.1.1.

10.1.3. Vertical Cross Sections

Vertical cross sections are made from strike-dip data. Let us refer to Figure 10.2, which shows a shot point S and a strike-dip symbol at point R. Suppose we wish to make a vertical cross section along the line LL', as shown in the map view in the upper part of the figure. This line, LL', makes an angle α with the strike. We are going to assume that, at the buried point of reflection, underneath R, the reflector is approximately plane. Thus, if we make a projection of the reflector, as if it were plane, onto the vertical section through LL', we shall obtain approximately the intersection of this vertical section, LL', with the actual reflecting interface.

Clearly the assumption that we have just made is only partly valid. In most of the cases where vertical cross sections are valuable, the reflectors depart seriously from planeness. It is therefore important that we keep the fact of this assumption well in mind so that we will not blindly produce a seriously distorted picture. Our object is to reduce distortion to a minimum; and to do this, we have to keep the distance we project from our point of reflection to the vertical cross section as low as possible. We shall return to this point a little later.

The first four equations in Figure 10.2 give definitions of symbols. The third one merely states that if we go along the strike from *R* to *P* we do not change the depth. The depth under *Q* is denoted by z_Q, as indicated in the fourth relationship. The relation between the depth under *Q* and the depth under *R* is shown in the first vertical view. From it we see that the distance *QN*, which represents the additional depth below the depth of the reflector at point *R*, is given by *QR* times the tangent of the dip angle θ. Equation 6 is obtained by solving equation 5 for the depth under *Q* by adding z to both sides of that equation.

At the present stage we have two possibilities of finding the depth of the projection of the reflector in the vertical plane through *LL′*. One is to use point *P* and relationship 3; the other is to use point *Q* and relationship 6. Either way will give the depth. Evidently the simpler procedure to use is relation 3 and this relation will be used in almost all cases. Sometimes, however, relation 6 must be used. For example, if the strike of the reflector is parallel with the line *LL′*, no such point as *P* exists, so *Q* has to be used. When the strike is almost parallel with line *LL′* on the map, it is usually advantageous to use point *Q*, because *P* will be remote from the place where the segment of projected reflector is plotted on the vertical cross section.

We may carry the theory of the vertical cross section one

step further by looking at the second vertical view in Figure 10.2. This is a view in a vertical plane directly under QP, along the line of the vertical cross section. Angle ϕ is the component of dip of the reflector in the vertical cross section. It is evident that ϕ has the tangent shown. Referring to relation 5 we can complete relation 7. Relation 7 may be written in the form of relation 8 by referring to the map view in the figure, where it is seen that the sine of α is the ratio of QR to QP. Formula 8 gives the component of dip in the vertical cross section making an angle α with the strike, when the total dip angle is θ. This formula was derived in Figure 5.2; it will be recalled that the tangent of an angle is the reciprocal of the cotangent of the same angle. So far as the simple projection of the dipping reflector into the vertical plane is concerned, all the needed information is given in formulas 3, 6, and 8 of Figure 10.2. If these relationships are used, all the remaining work can be done graphically. One item is still lacking, and that is the means of making the *distance of the projection a minimum.*

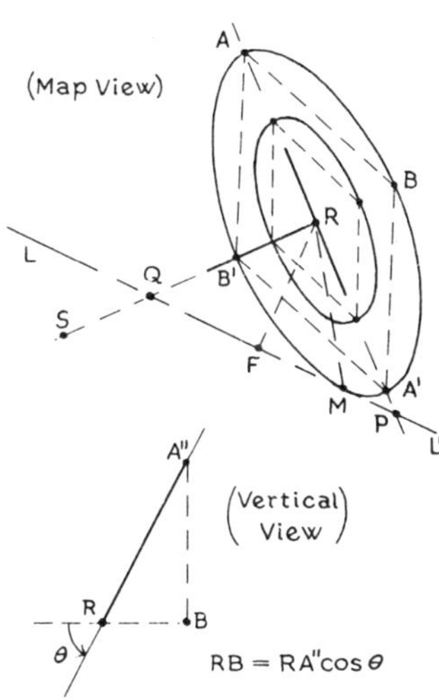

FIG. 10.3. Vertical view of circles drawn in dipping reflecting plane.

We can consider one aspect of this problem of minimizing the distance of the projection by referring to Figure 10.3. The map view shows the strike-dip symbol at R and the trace of the vertical cross section LL'. The shortest distance of projection happens to be the projection from R to M, as shown. Let us see how this point M is constructed.

The ellipses in the figure, centered at point R, represent vertical views of circles drawn in the reflecting plane. Since the reflecting plane is dipping at a considerable angle, the vertical view flattens the circles, as indicated. Thus the distance RA is the radius of the circle. The distance RB is a radius which is foreshortened because of the vertical view. The relationship between RB and the dip angle θ is shown in the vertical view; RB is the radius times the cosine of the dip angle. No lengthy explanation of the fact that RM is the shortest distance, in the dipping reflector, to the vertical plane section, will be given here. The reader is advised to study the situation carefully, so that he understands that this is indeed the case.

We now proceed to make the situation a little more quantitative. Refer to Figure 10.4. The equation of the ellipse is relation 1. The equation of the tangent to the ellipse is equation 2. This may be computed by the usual methods of analytic geometry and differential calculus and will be found in most books on elementary differential calculus. Relations 3, 4, and 5 show the coordinates of points P, Q, and M. The sixth relationship shows the tangent of the dip angle; this is dy/dx, and is given by the relationship at the right of this equation. The angle between the strike and the direction RM is angle β and its tangent is given by formula 7. This, indeed, is the relationship that will probably be used in most cases. This is the formula by which β can be computed, since the angle θ and the angle α are both known. If desired, a slide rule can be made for this computation by using logarithmic formula 8. However, so few computations of this sort are needed in practice that a slide rule hardly repays the work that goes into its construction. After a person computes a dozen or so of these shortest projection distance points, such as M, he gets a very good intuitive picture of the situation. The computation is then no longer necessary, for the position of M can be guessed each time with sufficient accuracy.

In this connection, a warning should perhaps be issued that, if the resolved dip gets very close to 90°, the position of M has a great effect upon the depth of the projection. About the only conclusion that can be drawn here is that when the dip is very close to 90°, the depth of the projection is very uncertain.

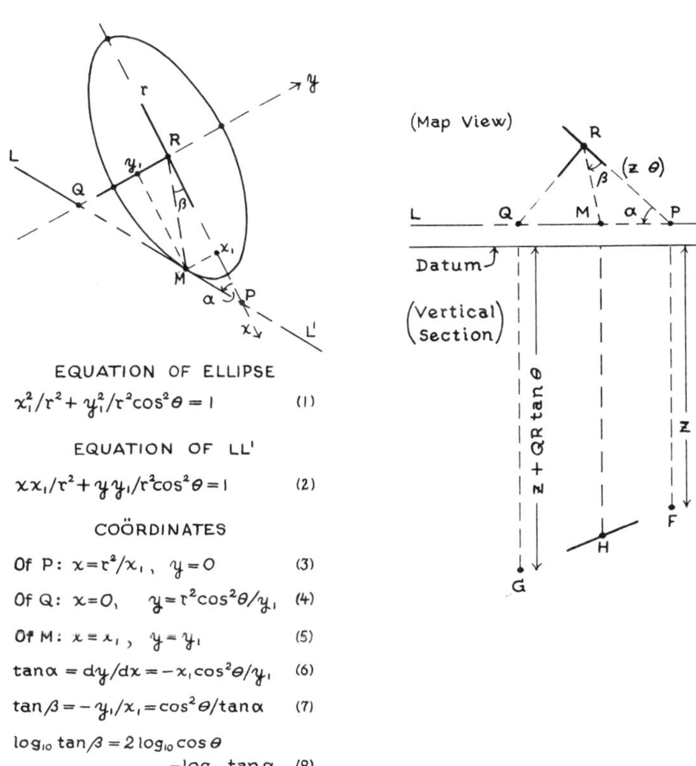

Fig. 10.4. Calculation of shortest offset point M.

Fig. 10.5. Summary of construction procedure for vertical cross sections.

We are now ready to summarize the more practical aspect of making a vertical cross section. For this purpose refer to Figure 10.5. Suppose that a strike-dip map has been constructed and that a small piece of it is shown in the upper view in the figure. The trace of the vertical cross section is the line LL' on the map. Since the strike is not parallel with LL' but intersects it at point P, we have a simple means of determining the depth. We drop downward from the datum at point P to a depth z, to determine point F. Then we turn the angle ϕ, which may be determined from equation 8 of Figure 10.2. We then locate point M and project down to the point H, which represents the *center of the dipping segment in the vertical cross section.* The dipping segment is drawn about this center with the proper angle, ϕ. If on the other hand the strike had been parallel with LL' we could have used point Q and determined the depth of G by the formula in the figure. In this case we should have had to draw the line using point G as a depth basis. The dip would have been zero or almost zero, unless the resolved dip were nearly 90°, in which case it is doubtful whether we would have projected it at all, since it would have to be projected such a long distance.

In the actual construction, the strike-dip map should be complete, showing strike-dip symbols and the associated depth and angle of the resolved dip. When this information is given on a strike-dip map one can complete the operation with a set of triangles, a protractor, and a scale. In place of the triangles and protractor, a detail drafting machine properly oriented with respect to north and south directions will prove very valuable. If much of this type of work has to be done, a small drafting machine is an investment that is well worth while.

The construction in Figure 10.5 is drawn on paper to the same vertical scale as that on the map. After a set of cross sections are thus constructed on paper, the transparent vertical cross sections may be constructed as illustrated in Figure 10.6. At the left are two flat parts with half-cut slots, cut so that the parts will fit together egg-crate fashion, as shown at the right. This cellular network of transparent plastic, when placed over the strike-dip map, provides a means of showing three-dimensional structure that is very valuable. On the transparent sheets are traced in India ink the segments of the reflectors that have been projected on the paper work sheet to the same scale. When all these tracings have been made, the cellular structure is assembled and placed over the map in the proper position. The details of structure can then be inspected directly.

The writer has made and supervised the making of several vertical cross-section interpretations of complicated structures.

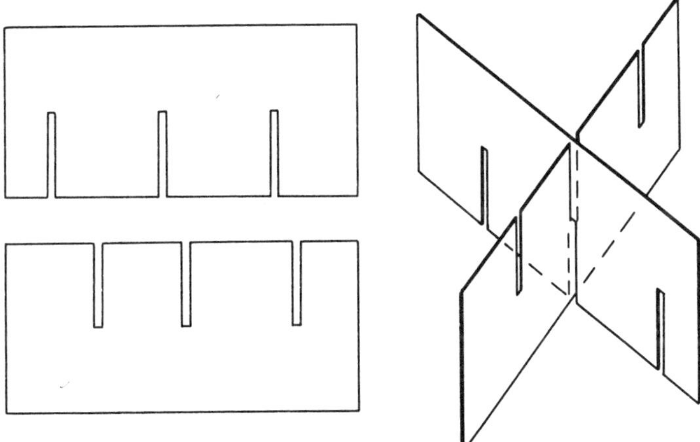

FIG. 10.6. Transparent vertical section construction.

The procedure is simple in principle, but difficult to execute. However, when such vertical cross sections are needed at all they are extremely useful because they show structural details very clearly. The actual construction is difficult enough so that the computer is usually justified in using a regular cellular structure in which the spacing between lines is almost uniform and the directions are always north-south or east-west. This procedure keeps his orientation problem simple. When such directions are not used, the orientation problem becomes complex and requires very close attention to the meaning of each operation.

In spite of the large amount of work required to construct the vertical cross-section interpretation, it contributes so much information regarding the structure that it is well worth while in complicated structural problems.

Since one has to look directly at the vertical cross sections on the map, it is helpful to use the simple grading procedure used by the writer (see Section 8.4.4).

The method of making vertical cross sections is, of course, by no means foolproof. It is possible to make ridiculously long projections which introduce serious distortion into the picture. It is also possible to produce a distorted picture without introducing long projections. An example of such distortion is shown in Figure 10.7, with the projection of strike-dip symbols *A*, *B*, *C*, *D*, and *E* shown in the vertical section below. If we blindly project *F*, *G*, *H* into the vertical section we get the type of distortion shown. Such distortion does not spoil the relative structural picture here, unless an unconformity is present. This might have been the case, and we might have had at the top of the structure a section which showed practically no dip from this top up to the datum surface. In this case projections *H* and *F* would contradict the low-dip evidence. It should be understood that this contradiction was introduced in error by the interpreter and should be removed. Thus we see that the method can be

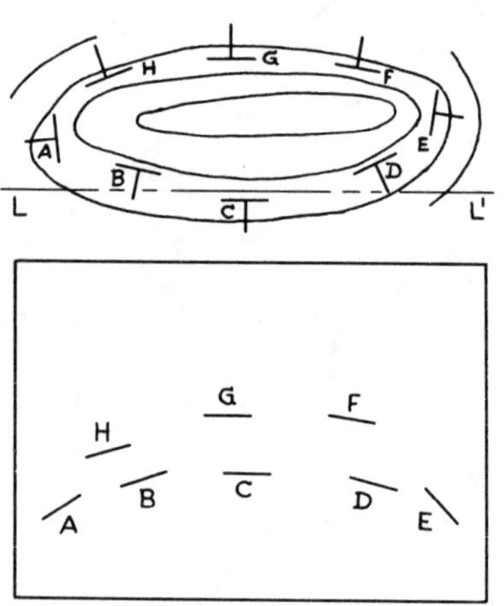

FIG. 10.7. An error to be avoided in making vertical cross sections.

misused just as any other instrument of interpretation or production can be misused.

10.1.4. Size and Accuracy. Grading

The size of a cross section varies greatly from one company to another. Some companies use very large cross sections because the plotting is easier on them. However, large cross sections are harder to handle than small ones. Very large tables are required for their study. Also such cross sections may be plotted to such a scale that the random errors in reading reflections, reduction to datum, and computing depths and dips appear very large.

There is something to be said in favor of using a cross section which can be plotted and read with about the same degree of accuracy as can be obtained in the interpretation. One advantage of such a scale is that it does not lead anyone to exaggerate the accuracy of the work. Unfortunately, accuracy is not always expressed in terms of accuracy of timing the records, reading ΔT, reduction to datum, and calculation of depth and dip. Sometimes it depends on the assumptions involved in making the calculations. Sometimes it depends on what is neglected in making them. For example, when the cross dip is large, neglecting it introduces a serious error. This error can never be expressed in terms of the scale of the cross section. All that can be expressed in these terms is the minimum random error under optimum circumstances.

Another advantage of smaller cross sections is the fact that they can be handled and used more easily. They are much more likely to be unrolled and looked at than very large ones.

The question of grading is exceedingly important. We have already touched upon it in several places. Opinions differ in this regard, for some people object to grading altogether, whereas others are in favor of a rather extreme grading system. In favor of the extreme graders is the fact that the grader probably has closer contact with the raw data than anyone else, and for this reason his judgment can probably be trusted better than anyone else's. His grades will probably depend to some extent upon his emotional condition at the time, to some extent upon his environment. Thus, grades made in October should not be compared with those that were made in July. On the other hand, on a given cross section the grades can claim to indicate only relative values. Thus a person who is studying the cross section can decide that certain reflections are better than other reflections. This type of information has undoubted value even though one may legitimately claim that the grading is sometimes overdone.

The author's system of using only three grades was explained in Section 8.4.4. A more elaborate system that has been used with considerable success is that devised by Phil Gaby.* Gaby's system, and others like it, are highly recommended.

A few words in favor of the people who do not like any grades to appear on cross sections. The principal justification for this is that if grades are omitted, the records have to be studied in connection with the cross sections if any adequate evaluation is to be made. Whether grades are included or not, the process of studying records in conjunction with the cross sections is highly recommended. In fact, this is usually the only adequate means of properly evaluating a survey. Therefore, if the system requires the use of records along with the cross sections, it may be regarded as a good system for that reason.

No one should reach the conclusion that the quality of reflections is proportional to their value or importance. Sometimes the poorest evidence is the most important.

The widespread practice of excluding all but the reliable reflections that make a consistent picture cannot be justified on any objective basis.

* See *ibid.*, 12:590-617 (1947).

10.2. MAPS

We shall now consider the various types of maps that are used in presenting results.

10.2.1. Contour Maps

It is not our purpose to describe in detail the making of contour maps. This has been ably done by several authors.* We may say, however, that while contour maps serve the very valuable purpose of synthesizing a great deal of information for the interpreter and for the reader of the report they are occasionally somewhat misleading. Especially are they misleading when they are based upon wholly inadequate data. The practice of making so-called phantom cross sections and basing maps on them has been valuable in some cases and seriously misleading in others. These phantom cross sections, which are not based upon actual ties of reflections or upon adequate correlations, may pass right across faults and thus give an erroneous structural picture in exactly the places where the structural picture should be adequate in order to secure the most important part of the information sought. On the other hand it should be remembered that all exploration work is to some extent a gamble and that making phantom cross sections involves taking a chance. This chance is usually taken because of the feeling that it is necessary if any results are to be obtained. Cross sections made on this basis yield contour maps which are very much open to question.

Where depth contour maps are based upon reflections which are correlated on the basis of either character or time ties, the corresponding contour maps may be relied upon to give good relative structural pictures. Distortions still appear, usually because of the oversimplifications that have had to be made

* For example, see F. H. Lahee, *Field Geology*, 4th ed., pp. 383-394 (1941).

regarding the distribution of velocity. These distortions can usually not be avoided in the early stages of the exploration work.

10.2.1.1. Transferring Data

The problem of transferring data from the cross section to the map in making contours is often simple. A reflection horizon is selected to be contoured and depths are written on the map at selected places along the profile corresponding to the measurements on the cross section. This can usually be done simply and adequately in cases where the dip is small.

In cases where the dips are larger, the transfer of data from the cross section to the map presents a greater problem. It is of course possible to make vertical cross sections and transfer the data from them to the map; but so much information is gathered in making these cross sections that a contour map seems almost superfluous.

R. A. Peterson* has presented a method of transferring data from cross sections to contour maps which we now describe.

Refer back to Figure 10.1. On the map at the left will be observed three strike-dip symbols, R_1, R_2, and R_3, corresponding to shot points *A*, *B*, and *A′*. Peterson's method of transferring data makes use of the fact that is illustrated in the lower right-hand part of this figure. Observe that if *DC* is projected to *E*, we arrive at the point *J*, which is at the depth corresponding to the reflection time when no dip is involved. Point *J* is approximately, but not exactly, midway between *E* and *C*. Thus we can find the map position, or approximate map position, of the point above *J* by taking the mid-point between the strike-dip symbol and the shot point. This has been done at each point and we have thus located J_1, J_2, and J_3. When these three points are located, we draw the dashed lines as indicated. We then slide the cross section from the map position say *AA′*

* Unpublished.

perpendicular to this line AA' so that it coincides with the line J_1J_3. We take off the depths on the cross section and put them at the proper places on the dashed line.

This process is not entirely exact, but it is an improvement over the procedure of regarding the cross section, as regularly made, as a vertical cross section. It will be seen, by studying the example in the figure, that since this process is approximately correct it certainly has the great virtue of not placing the depth point very far from its correct position. It will be evident that Peterson's method of transferring the data is well suited to most cases. The cases in which his method breaks down are those in which the dips are very high, so that we are practically forced to make vertical cross sections.

10.2.1.2. Contouring

As we said above, contouring will not be described in detail, since other authors have already done so. However, the contouring of seismic data introduces some problems that do not arise in contouring either topography or geologic data.

An example is found in the case of contours that attempt to present a surface of very low relief. If the dip is extremely small and the seismograph lines are laid out on a grid network of say five miles between lines, we encounter a curious situation. The regular contouring procedure requires that we give greater emphasis to data near the neighborhood where our contours are being drawn than to data far away. In the case we are discussing, this emphasis has the effect of making the contours wiggle back and forth along the line of profile. Thus it will be seen that contours drawn on this principle tend not to cross the open squares at all, but that each open square appears as a little high or a little low. Such a case is evidently very seriously influenced by the position of the data. It is also apparent that the presentation cannot be regarded as correct.

Evidently, such a map is so highly influenced by the location of the lines as to be quite valueless. We can only conclude from looking at it that the map failed to uncover anything of interest. Every little high can be regarded as on one side or the other of the line. Thus the little highs only indicate possible highs in the uncovered areas. The situation is encountered fairly often in conducting semireconnaissance seismograph surveys. One way of avoiding it is to introduce a tilt into the whole picture. This tilt makes the contour lines run across the open spaces. After the contouring has been drawn with the tilt, the next step is to remove the tilt. This has the effect of producing another type of contour map in which the contour lines do not follow along the profile lines. But we are still in the dark, however, regarding the way the contours should go in between the seismograph profile lines. Our conclusion is that we need more observational data or that the region is not suited to seismic prospecting.

The picture need not be entirely hopeless, however. If we have adequate data so that we can draw a reasonably good contour map, we can incorporate in the map a certain amount of *smoothing*. The philosophy behind this is that we make our picture as simple as possible. Lack of simplicity may correspond to the facts of the case, but unless we have definite evidence of this we have no basis for making our picture complex.

In addition to smoothing we also introduce a certain amount of bias from our knowledge of *regional trends*. It is sometimes objected that regional trends should not be introduced in contouring unless the trends automatically come from the seismograph data. On the other hand, if we have valid information regarding regional trends, we can use it to advantage in many cases. The question is whether we wish to use valid evidence or not. There is no very compelling reason for restricting our information to that made available by the seismograph.

10.2.2. Dip Maps

In the earlier days of reflection seismograph work a good deal of plain dip shooting was done. This shooting was presented on a dip map. We do not present any such maps; instead, we refer the reader to pages 70, 80, 90, and 329 of the *Geophysical Case Histories* volume for examples of them. The first three of these are ordinary dip maps such as were used to indicate structure without contours. In the fourth one, contours have been drawn as well as the dip arrows. The rough utility of such maps is self-evident and requires no further comment.

10.2.3. Strike-Dip Maps

We have already mentioned strike-dip maps in the discussion in Chapter 9 and Section 10.1.3.

The reader is referred to the strike-dip map on page 534 of the *Geophysical Case Histories* volume (see also our Fig. 1.3). This map includes the principal information that led to the discovery of the Wilmington oil field in the Los Angeles basin area. This is perhaps the most important and valuable strike-dip map ever made. It had to be constructed under difficult conditions because the area is highly industrialized and the noise level is high. Fortunately there was just room enough on the land side for the operators to obtain most of the information necessary to indicate a closure. Doubtless, had they been able to move a short distance out to sea, they could have completed the picture and made it somewhat more definite. Even so, however, this particular operation represents an outstanding achievement.

As was mentioned above, the principal objection to a strike-dip map is that it tends to collapse a great deal of three-dimensional information on two dimensions. This may make the picture very confusing unless the structure is relatively simple.

In making contour maps it is desirable to include strike-dip data corresponding to the level contoured, wherever such data are available. The value of doing so is that it gives a direct check on the strike that is shown by the contouring. Whenever the two methods of presentation are inconsistent a careful rechecking of the data should be made in an effort to find out which kind of data are more reliable so that greater emphasis may be placed on them.

10.3. REPORTS

The purpose of writing a report is usually twofold. There is the immediate purpose of synthesizing, gathering together all the outlying information in a unified whole, so that the entire picture may be seen at once. Another purpose that is usually emphasized is the storing of information for later reference. This purpose is generally considered more important than the first one, but the relative importance of the two is very difficult to assess.

Available information, like available energy, tends to be lost with the passage of time. Whatever the effort to make a *complete* report for the storage of information, it should be realized that completeness is an unattainable goal. The only thing that can happen to information is for it to diminish, unless work is put into the process of obtaining new information. All we can do therefore is to make a report as complete as reasonably possible.

10.3.1. Check Lists

In making reports, relatively full check lists are of great value. The following check list might be used in writing a report covering seismic information:

1. Time of work.
2. Personnel in various capacities.
3. Surveyors' data, notebooks, plane table sheets, maps, roads, accessibility, permits.

4. Drillers' logs, drilling times, equipment reports.
5. Shooters' reports, dynamite and caps used, hole notes.
6. Observers' reports.
7. Records.
8. Cross sections, maps, computation sheets.
9. Descriptive part. Method of shooting, spread, filters, mixing, method of datum correction, method of depth and dip computation, general quality of results, main structures of interest, validity measure and means of determining it, simplifying assumptions.
10. Résumé of present status, conclusions regarding findings.
11. Problems still to be solved and suggested means of solving them.

It cannot be claimed that the above check list is complete. The writer of reports has continually to alter a check list on the basis of his experience. He adds some items and occasionally omits one that he finds either covered in another part of the check list or somewhat irrelevant.

10.3.2. The Descriptive Part of the Report

The descriptive part of a report is mainly concerned with what was done and what was found. It has a sort of skeletal background which we may regard as the *structure* of the description. This structure or form of the description ought to follow, to a certain extent, the form or structure of what was actually done and found. Figure 10.8 will illustrate our meaning. At the upper left is a circle labeled "Start"; this indicates the beginning of the work. At the beginning we can embark on a large number of *possible alternative procedures*. We select one of these alternatives and follow it for a certain length of time. The one chosen is labeled "Path taken"; the others issuing from the circle are labeled "Alternatives." After we have decided upon a path, we reach a stage where we have to reconsider. This "Stage of Reconsideration of Path" is indicated by another circle. Continuation of the path taken up to this point is labeled "Continuation." Suppose we decide that this path is not the one on which we wish to continue. We wish to

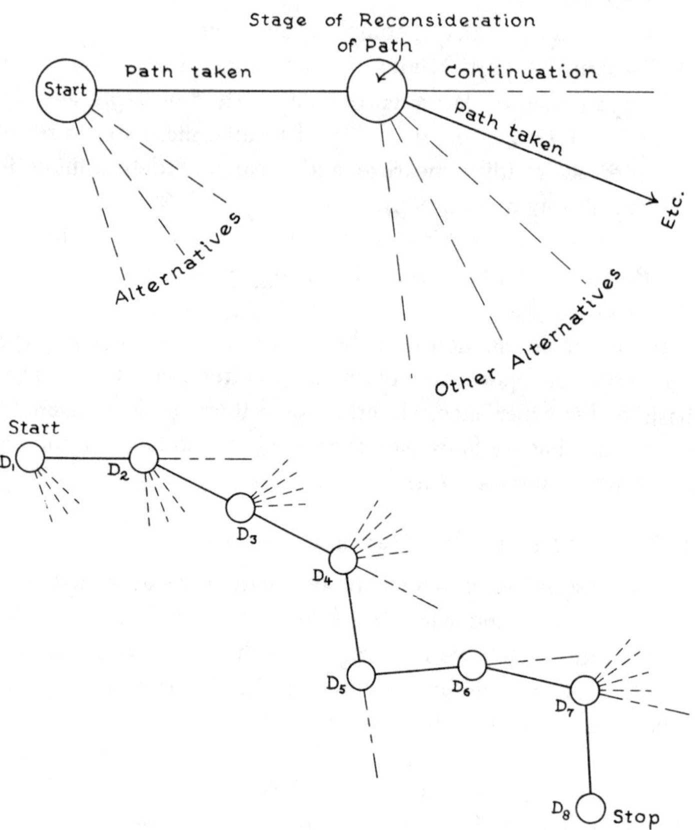

FIG. 10.8. Structure of a seismic exploration program.

take a slightly different path, so we change our course, as indicated in the diagram.

Let us consider this idea in greater detail; refer to the lower part of the figure. Our start location is labeled D_1. We travel on our chosen path over to D_2, then change direction and go down to D_3 and D_4. It will be noted that we stop at D_3 for reconsideration. Here we decide that the other alternatives are not the ones for us to take, so we continue along the same course that we decided upon at D_2. However, at D_4 we take an alternative path down to D_5. At D_5 we reconsidered our position and took another path over to D_6. Thus we follow our course of action from the start at D_1 to the stop at D_8. This may illustrate diagrammatically the type of course our exploration has followed. The *D*'s may be considered as labeling *points of decision.* For example, at D_1 we decided to proceed, perhaps with a reconnaissance seismic survey that consisted of placing cross shots on a squared grid with 5-mile intervals. After making this reconnaissance survey we decided, at D_2, to join these reconnaissance points with continuous profile lines. About a third of the way through the joining procedure we reconsidered the situation and decided to continue the joining procedure down to decision point D_4. At D_4 we decided to fill in intermediate profile lines crossing the 5-mile squares in order to get greater detail within the squares themselves. We did this until we reached decision point D_5, at which time we decided that we needed to cover a slightly greater area; hence the original reconnaissance program was extended outward until we reached point D_6. Here we decided that we had covered a large enough extended area, so our extended section was connected with continuous profiling. At decision point D_7 we decided on further work filling in the 5-mile squares with information. Finally we decided that enough information had been gathered and the crew operation was stopped. Stopping involves as much decision as starting.

At any point in this course of action a report may be written covering the procedures taken. Progress reports are usually written periodically covering procedures over a certain period of time. Generally, in a progress report, emphasis is placed on the information obtained; but when the work has been completed, as at D_8, a full report should be written in order to describe the entire procedure from start to stop.

It may be mentioned that at each of the decision points a choice of alternative paths is made in much the same way as was described in Section 4.4. Such a decision involves very little logic. It usually calls for imagination, attempts to picture the likely outcome with various alternative courses, and a sort of sampling process. We go down each alternative path a little way, so to speak, just to get a small sample to help us make our decision. If we base the actual selection on the information at hand and on our judgment, the decision has little to do with deductive logic. Deductive logic may be very important in taking us along the path from one decision to the next, but it is of relatively minor importance in the decision itself. Its use in decision making is that it sometimes helps us to take a slightly larger sample of various alternatives in our imagination.

Although in our example the work was stopped at D_8, this means only that the seismograph crew operation was stopped; in other words, the gathering of this kind of information has come to an end. The *use of the information* is another story. It involves a continuation of the exploration process; in no way does it involve a stopping.

A few remarks should perhaps be made regarding one item in No. 9 of our check list; this is the *validity measure* and the means of making it. This is a very important part of the report and should always be included. The measure of validity is usually, of necessity, a rather unprecise part of the work. The seismologist merely states whether he regards the picture as reliable or not reliable. He may be willing to give odds regarding its reliability. If such an estimate is possible, an attempt should be made to secure it; because however poor this estimate may be, it is nevertheless the carefully considered opinion of the man who has had the closest contact with the records, the operations, and all details of the interpretive work. Consequently, up to this stage of the exploration, he is the individual best qualified to give such an estimate. The fact that his estimate may be wrong should not prevent him from making it. The best that can be done under the circumstances is to get an estimate from the individual best qualified to make it.

10.3.3. Dating

One very important part of every report is the *date* on which it is submitted. This date gives notice that the report was written on the basis of information on hand at this time. Naturally as time passes and more information is accumulated, a better picture of the situation may be expected. It is for this reason that the date has so much importance.

In closing this section the author should like to mention Payne's discussion of the writing of reports.* Although his discussion covers reports primarily in the field of oil property valuation, it may very well be extended to the field of seismic prospecting reports. His description and check lists are the best that the author knows about and they can be recommended without reservation.

We have now covered the main items of routine interpretive work required for the applied seismologist and seismic computer. Many of these items are also vital to the geologist charged with, or happily engaged in, directing a broader exploration program including seismic prospecting. From the technical fragments can be formulated a synthesized interrelated unified presentation of results—reports, sections, maps.

But haven't we depended on an experienced party chief

* Paul Payne, *Oil Property Valuation*, Chap. 7, especially pp. 180-189 (1942).

or seismologist to pick our reflections? Yes, we have. And probably we shall have to continue to do so. Days, months, and years of experience in picking reflections in a great variety of circumstances cannot be transmitted via a book. So the author finds himself in a difficult position. He feels that to stop with the description of routine techniques would merely introduce the more interesting aspects of seismic exploration. The need for pushing beyond simple routine becomes more and more evident as discoveries become more and more difficult.

So let us embark on a difficult journey. Let us consider seismic records—not the simple clear-cut records that require no comment—but the disturbing factors that make the seismologist either a routine skimmer or an exploration research man in the best sense. Let us look at these records with the assumption that every event has a significance that we can find and that is worth finding. Thus every success will make the next investigation easier and will make us keener and more resourceful explorers.

10.4. COMMENTS TO THE SECOND EDITION

The 3-D models discussed in this chapter were useful and beautiful, when, as always, they were constructed with care and pride of workmanship.

Part III

INTERPRETATIONS

CHAPTER 11

Picking Reflections

This chapter is written for seismologists, party chiefs, and geologists. It is especially for seismologists and party chiefs, but geologists who are on their toes and are interested in using the results of seismic exploration will find the material of considerable interest to them.

The general problem of picking reflections is, of course, extremely important. It is also one that is somewhat difficult to describe. The reason for this difficulty is that, throughout the short history of reflection seismic work, the actual process of picking reflections has been a kind of private art, each picker having his own little tricks and ways of looking at the problem. The purpose of the present chapter is to discuss these matters and to attempt to bring out some of the important features of the problem.

11.1. NOISE vs. SIGNALS

Noise is the first thing we see on the average seismic record. Noise can be defined a little more precisely. We may regard noise as (1) uninterpreted pulses and waves, plus (2) mistakenly interpreted pulses and waves, plus (3) interpreted parts which cannot be correlated with the geological structure.

Much of the noise comes from the shot itself. This apparently is due to a complicated refracted system of pulses in the stratified section. It is probably also due, in part, to a great many little reflections, so small individually that they cannot be used as reflections.

We may illustrate the second kind of noise by Figure 11.1, which represents mistakenly interpreted pulses and waves. At

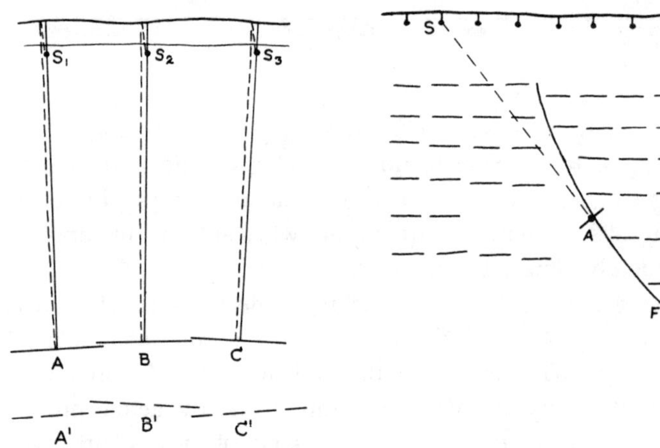

FIG. 11.1. A method of making reflection noise.

FIG. 11.2. Noise arising from potential misinterpretation.

the top of the figure is the surface of the ground. In cross section are shown three shot points, S_1, S_2, and S_3. When the shots are fired three reflections are obtained which give rise to segments *A*, *B*, and *C*. We suppose now that these three segments represent approximately the geological interface. They do not fit together perfectly and thus form only an approximation to this interface. Below *A*, *B*, and *C* are dashed segments *A′*, *B′*, and *C′*. These could also be plotted on the cross section, but they represent noise since they correspond to a mistaken interpretation. The true interpretation is assumed to be that *A′* should coincide with *A*, *B′* with *B*, and *C′* with *C*. It is supposed that *A′* is due to a pulse which leaves S_1, travels upward to the surface of the ground, is reflected downward to reflection point *A*, and then returns to the surface of the ground. The delay between *A* and *A′* is just twice the up-hole time. This type of process occurs fairly often and a great many reflections have been plotted that really represent noise, as indicated in Figure 11.1.

Another type of misinterpretation is illustrated in Figure 11.2, which shows a cross section with a line of shot points and a set of reflections. A fault is indicated in the section and there appears to be a reflector at point *A* which is plotted as indicated from shot point *S*. Interpreting this segment at point *A* as a reflector is likely to be erroneous in this case, because there exists no surface parallel to or coincident with this segment that would give reflections from this point. Instead of an actual interface, point *A* is more probably just an irregularity in lithology that acts as a scatterer, the energy from *S* being scattered back from *A*, or the region around *A*, as if the point were an approximate new source. Thus the dip indicated by the segment is without significance in this case. The only significant feature is point *A* itself, which gives the approximate location of an irregularity. This sort of thing happens quite often in connection with faulting, as we shall see in greater detail later in this chapter.

One type of noise that is particularly obnoxious in some cases consists of multiple reflections. These are reflections corresponding to, first, a single reflection at a geological interface, followed by a reflection at either the surface of the ground or the base of the weathering, or both, this in turn being followed by a reflection at a geological interface which returns the energy to the surface where it is picked up as a reflection. This is particularly troublesome because there is usually no way of telling from the records themselves whether a reflection is multiple or not, without a very detailed study of the situation involving shooting velocity profiles. By shooting velocity profiles, the multiple reflections can be separated. Occasionally multiples can be detected in terms of their depth, for the shallow section may show a very small dip, whereas the deeper section shows a larger dip. The multiples appear in the deeper cross sections with very small dip and interfere with the true single reflections coming from the deeper geological interfaces. Another means that can sometimes be used to pick out multiples depends on the fact that the multiples tend to show increased dip in a very regular way; that is, doubles have approximately double dip, and triples approximately triple dip, compared with the dip of the corresponding single.

Other sources of noise which are probably familiar are the noises introduced by the ground unrest due to people walking, wind blowing, and traffic on roads; sometimes there is a more basic ground unrest that is very widespread and is probably related to what the earthquake seismologists call *microseisms*. Another source of noise is sometimes the instruments themselves. This, of course, is particularly troublesome and every effort should be made to avoid it. With proper engineering care this type of noise can be largely eliminated.

The term *signal-to-noise ratio* is commonly used to describe the situation. A signal-to-noise ratio, however, does not have much meaning except in a very rough sense. We can measure the maximum amplitude of a reflection, make an estimate of the background noise amplitude, and then form the ratio and call this the signal-to-noise ratio. This is what is meant by the term, but its significance is not very sharp. Sometimes the signal is at the same level as the background noise and yet is detected without great trouble. This is due to the fact that the detection of the signal is in terms of the regular or approximately regular line-up of troughs and peaks across the record, whereas the noise lines up irregularly across the record, at least the part of the noise that does not correspond to reflected and refracted pulses originating from the shot point.

As is well known, *filtering* is used to reduce the noise level. For example, high-pass filters may be used to reduce the low-frequency *ground-roll* noise. Low-pass filters may be used to reduce the *wind* noise. Another means of reducing wind noise that is often used is burying the geophones below the surface of the ground. Another important way is to design the geophones so that they have no sharp corners and are as insensitive as possible to little rotational motions of the phone.

Without exaggeration, noise may be said to be the most serious problem in exploration seismology.

11.2. SIMPLEST REFLECTED PULSE

We now wish to look at a signal, the *reflected pulse.* We may classify reflected pulses according to whether they are *positive* or *negative.* This we explain by means of Figure 11.3. At the left of the figure is shown, in profile cross section, a shot below which is located a geophone *g* which feeds into an amplifier *A*, which drives a recorder *R*. The result of firing this shot may be determined in this way, and we will imagine that the result appears below at the left. This gives the *original pulse* with its dependence upon time. Thus it measures the *shape* or *character* of the original pulse.

Now let us consider the same situation except that we insert a perfectly rigid incompressible basement. This is, of course, an extreme case, but it is the simplest one of its type. The geophone is at the surface and is connected with the amplifier *A* and the recorder *R*. The shot is fired, the pulse travels downward, is reflected by the unyielding basement interface, and is returned to the surface. It is fairly evident, from the figure, that the initial pulse, with a compression on the front of it, will run into the immovable interface and be reflected with a compression on the front of the reflection. In a later chapter we shall show that the reflected pulse has the same shape as the original pulse. We refer to such a reflection process as a positive reflection process and to the corresponding reflection as a *positive* reflection. Thus the positive reflection has the

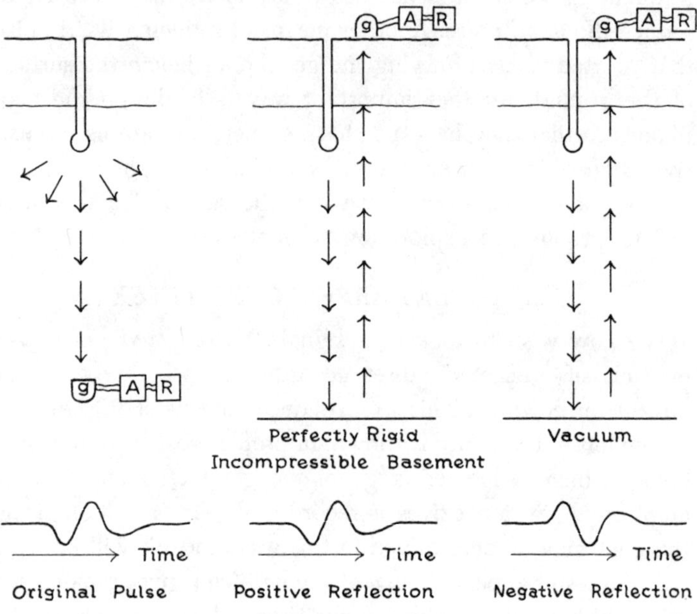

FIG. 11.3. Initial pulse with positive and negative reflections.

same shape as the original pulse, as indicated in the lower middle part of the figure.

We now consider the opposite extreme case, illustrated at the right side of Figure 11.3. Instead of a perfectly immovable interface, there is a free surface with a vacuum on the other side. This free surface is perfectly movable. If it swings down a bit, it will be brought back by the elastic properties of the medium above it. Imagine a compression pulse coming down from a shot and striking the free surface. The material is under compression at this instant and the free surface itself offers no resistance, so the compression is relieved by an expansion in which the particles at the free surface swing down into the vacuum. They are of course pulled back by the elastic properties of the upper medium, but in this downward swing they generate a rarefaction instead of a compression; hence the first part of the reflected pulse appears as a rarefaction instead of a compression. In a later chapter we shall show that such a process generates a reflection which is just the reverse of the original pulse; that is, where we had a compression we now have a rarefaction, and where we had a rarefaction we now have a compression. This reversed reflection we call a *negative* reflection.

Whenever the density and velocity increase suddenly as we pass downward from one geological layer to another across an interface, we get a positive reflection. Whenever the velocity and density decrease suddenly downward, we get a negative reflection. Thus our seismic record contains both kinds of reflections. However, because a geological section usually increases its velocity with increasing depth, we have many more positive than negative reflections.

It will be worth while to look into this reflection process in a little more detail. This may be done by means of Figures 11.4 and 11.5. Figure 11.4 concerns the process of forming a positive reflection. It shows seven motion-picture frames, *A*, *B*, *C*, *D*, *E*, *F*, and *G*, each of which has a heavy vertical line with a series of points on it. These points represent particles in the material, with their proper position at the particular instant corresponding to this frame of the picture. At the left of this line is a lighter parallel line that has another series of points with circles around them. These points are intended to present the effect of the incident pulse. The incident pulse is included within the bracket. This pulse is traveling down toward the interface, which is at the end of the lines. The first two particles at the lower end of the left line are not displaced. The third particle has been displaced downward, as have also the fourth, fifth, and sixth. The seventh particle, however, is at the very back of the pulse and at the present instant is not displaced. The quiet part behind the pulse, correspond-

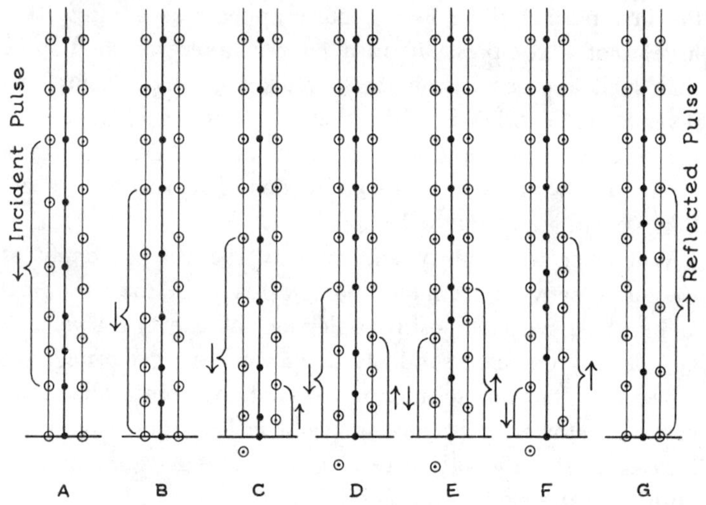

FIG. 11.4. Positive reflection process.

ing to the eighth and ninth particles, likewise shows no displacement. It is apparent that the displacements are such that there is a compression at the front of this pulse and a rarefaction at its rear. Since the pulse has not reached the reflecting interface at this instant, the total displacement of the particles shown on the heavy line is the same as that of the particles corresponding to the incident pulse. The third line, at the right of the heavy line, also shows a line of particles but without displacement. This line will be used for representing the reflected pulse. In the next frame, *B*, the incident pulse is just reaching the reflecting interface. There is as yet no reflected pulse.

The third frame, *C*, shows the first process involved in the reflection. The displacements of the incident pulse have to be carried down another step. This means that a particle which was originally at the interface is now displaced below it. But such a displacement is evidently impossible if the interface is immovable. Therefore this displacement, which had to be inserted, has to be completely canceled in the corresponding displacement in the reflected pulse. This cancellation is indicated by the fact that the resultant displacement shown on the center line is zero at the interface. This is made possible by introducing a displacement in the reflected pulse that is equal and opposite to the displacement in the incident pulse, as shown. This process is carried out with each detail until finally, in frame *G*, the incident pulse has completely disappeared and the reflected pulse is on its way back up. In each frame the points on the heavy line represent actual particle positions. These positions should be thought of as representing the sum of the displacements of that particular particle due to the incident pulse and the reflected pulse. The displaced particle's position has been determined in this way.

Let us now attempt to describe the negative reflection process in some detail, as indicated in Figure 11.5. Here, as before, at the left is shown an incident pulse which has not quite reached the interface. In *B* the incident pulse is reaching the interface. The third frame shows the incident pulse with its corresponding displacement into the vacuum. In this case the interface offers no resistance whatever, so that the displacement of the reflected pulse is the same as the displacement of the incident pulse. This introduces a separation of particles that creates rarefaction. It will be observed that, in the final stage, this process gives rise to a reflected pulse which has a rarefaction at its front and a compression at its rear. We may also observe that, in accordance with our procedure of adding displacements of the incident and reflected pulse particles to get the resultant displacement, the surface shows a *double displacement,* as indicated by the dashed lines in frames *C, D, E,* and *F*.

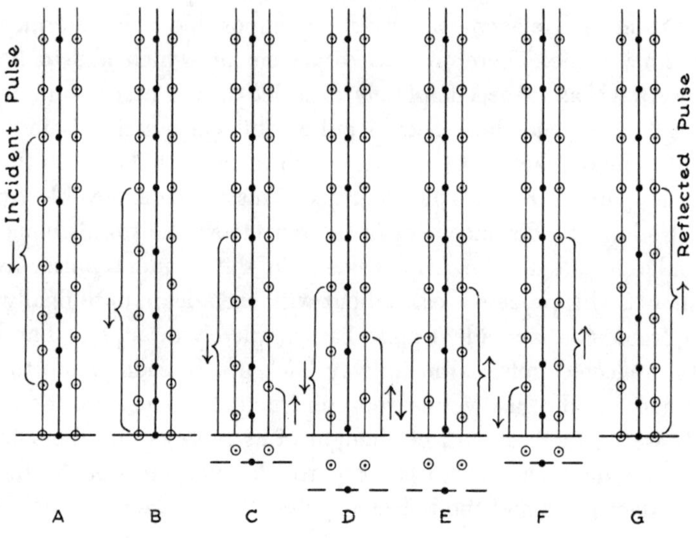

FIG. 11.5. Negative reflection process.

Of course the most outstanding example of negative reflections comes from the surface of the ground itself. When a reflection comes up to the surface of the ground, the situation is very similar to that illustrated in Figure 11.5. If a geophone is located at a receiving point, the displacement of the geophone will be not simply the displacement of the particles within the earth, but double that displacement, as indicated in Figure 11.5. This doubling is due to the superposition of the reflected and incident pulses. Thus from the point of view of good amplitude, it is good practice to place the geophone on the surface of the ground instead of burying it. This can be easily demonstrated in the field by digging with a hand auger holes perhaps 6 or 8 feet in depth for burying the geophones. The first foot makes very little difference because the pulses are fairly long. But the difference between a surface plant and one 6 feet deep is readily observable.

If the reader has a certain amount of background in physics and will study Figures 11.4 and 11.5 in some detail, picturing to himself the process we are attempting to illustrate by means of these figures, he will obtain a physical picture of the reflection process that is particularly valuable. We have not proved in any detail what we have shown in these two figures. This proof will come in a later chapter when we discuss the theory of the reflection process. However, what is shown in these figures is in accordance with the sound part of the theory and may be established by direct consideration of simple physical principles in conjunction with the study of these diagrams.

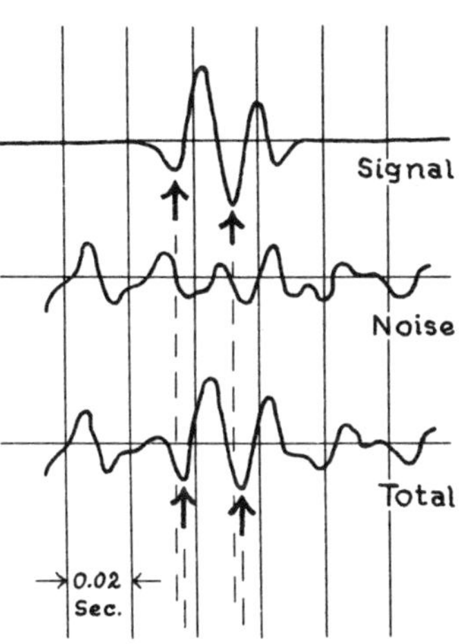

Fig. 11.6. Effect of superimposing noise on a signal.

It may be desirable at this point to illustrate pictorially how background noise can affect the signal. This is shown in Figure 11.6, the upper part of which shows a signal without noise. Underneath is shown a random background noise. At the bottom of the figure we have attempted to add this noise to the signal to show the signal with the noise background. Our purpose in doing this is to make clear that the presence of background noise in a signal in most cases changes the position of the troughs and peaks in the signal, as is indicated by the arrows. Since records are timed by reading the time at a trough, or peak, on the reflection, we should remember that this time has in it the *effect of variation due to background noise.* Thus although we can easily read reflection times to 0.0005 of a second, it is ridiculous to do so in most cases, primarily because true reflection times contain this random background noise error. The average value of such an effect appears to be of the order of ±0.0015 of a second. In difficult and bad situations this error may be much larger. In very favorable situations it may be somewhat smaller. It is quite evident from Figure 11.6 that the background noise distorts the original shape of the pulse, usually called its *character.* Thus this noise makes the character of the reflection difficult to see.

11.3. VARIATIONS ALONG THE LINE OF GEOPHONES

In some areas, when we inspect the records, we find the reflections lining up very regularly across them. These are usually areas where the reflections are considered very good. But there are other areas where the reflections are not so good but are disturbed by various causes. These variations we consider in the present section.

11.3.1. Variations Due to Noise

If the noise level is high, this will of course introduce variations along the line of geophones, as we saw in Figure 11.6. A trough will be shifted sometimes backward by noise and sometimes forward, giving rise to an irregular-appearing reflection.

11.3.2. Weathering Variations

When a weathered layer is irregular with respect to velocity or to depth, or to both, as is often the case, this causes a *shallow* irregularity in the reflections. This irregularity, however, is not very serious, since it affects all the reflections in the same way. In most cases, shallow irregularities in the weathering also introduce corresponding variations in the first breaks, and this may be used as a clue in detecting irregularities in the reflections. This clue, in conjunction with the templates described in Section 11.4, may be used in getting around the difficulty. However, it is the purpose of the datum correction to remove irregularities due to weathering.

11.3.3. Shallow-Velocity Variations

Occasionally we find important variations in the shallow velocity of material beneath the weathering. In this situation some correction should be made, for otherwise the deeper reflections will be very irregular on the plotted cross sections and no geological interpretation can be given. This effect can usually be corrected by selecting a shallow reflection—shallow, but still deeper than the region where the velocity varies seriously—and making an *isopach section* from this reflection to all deeper reflections. This procedure is usually very satisfactory, since it also automatically corrects for weathering irregularities that might otherwise be missed.

11.3.4. Variations Due to Curvature

In areas where the reflectors are almost planes with small dip and small variations of dip, the reflections carry across the records with little change in character or amplitude. On the other hand, where there are changes of dip, either gradual or abrupt, these changes introduce irregularities into the record. The effect due to curvature is shown in Figure 11.7. The shot point is point B. Extending downward from it are rays going to points A', B', C', D', E', F', and G'. These rays are reflected to the corresponding surface points A, B, C, D, E, F, and G. The distribution of energy in the pulse is fairly uniform as the pulse travels down to the reflector. But when it is reflected, because of the curvature of the reflector, it tends to be concentrated in certain parts and spread out in others. Thus it tends to be concentrated at D''' and C''' and spread out at G''' and F'''. Hence geophone G will receive a very weak reflection, whereas geophones D and C will receive a strong one. Thus the appearance of the reflection will vary greatly as it carries across the record.

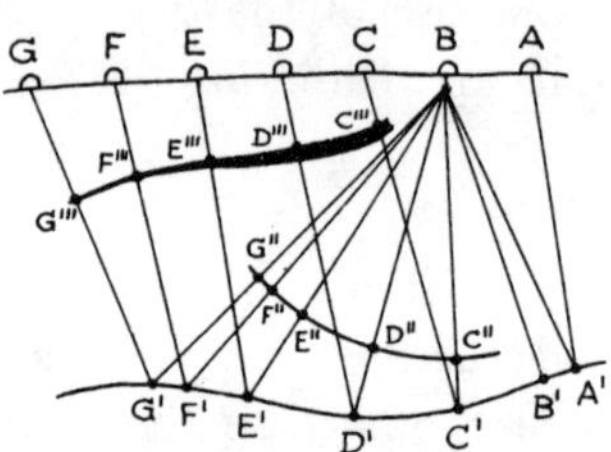

FIG. 11.7. *Top:* Effect of curvature on the amplitude of a reflection.

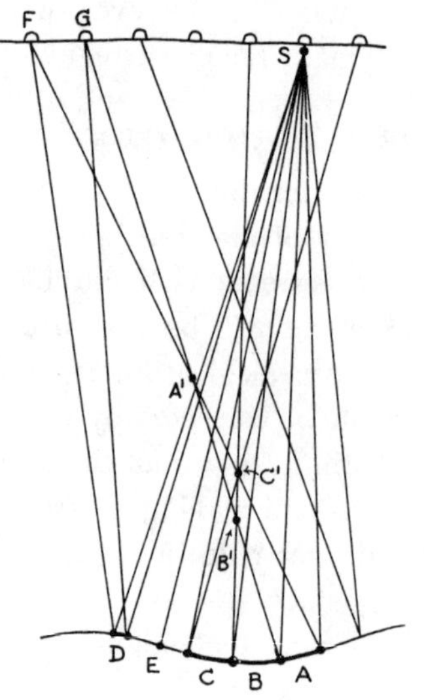

FIG. 11.8. *Right:* Disturbing effect of curvature and buried foci.

The situation illustrated in this figure is not extremely acute, but it becomes very serious if the reflector is moved deeper, as in Figure 11.8. When the reflector is deeper, the reflected rays tend to cross one another as illustrated at points A', B', and C'. These crossing points may be called *foci* of the reflected pulse. When the foci are buried, they introduce particularly difficult situations on the corresponding record. As we shall show in a later chapter, this is due to the fact that the pulse, in passing through a focus, tends to undergo a serious change in its shape or character. Thus, although we try to pick a given trough on a record, in certain parts of the record this trough has a different character due to the presence of buried foci; hence the timing is very different on this part of the reflection from what it is on another part. This effect has been observed many times. It is particularly disturbing when it occurs near one end of a split spread that is being used for dip computations, because it makes the dip computation invalid.

11.3.5. Interference Effects

In this section and its subsections we consider the effects of the superposition of pulses and waves of different kinds. Interference phenomena have perhaps been first encountered by the reader in his study of optics in his first course in physics. The interference and reinforcement patterns that occur in diffraction phenomena and in Newton's rings between plane thin plates are commonly described in elementary physics textbooks and are demonstrated in the elementary physics laboratory. A similar phenomenon occurs in sound, but the demonstration experiments are more difficult to perform and consequently are seldom presented in the elementary course.

The person who studies seismic records will encounter interference phenomena before he has studied many such records. The superpositions of different pulses can take place in many different ways. Since it would be futile to try to discuss all of them, we consider only particular cases in the subsections.

11.3.5.1. Interference Effects Due to Variation in Thickness of the Reflector

Refer to Figure 11.9, in which are shown four different situations. Case *A*, at the left, shows two reflections from a very thin layer. The figure has been drawn in such a way that the thickness of the layer is a very small part of the length of a whole pulse. It is also supposed, in each case, that the velocity of travel in the thin layer is twice the velocity of travel both above and below this layer. This being the case, we get a positive reflection from the top of the layer and a negative reflection from the bottom of it. This is indicated by the positive pulse on the left and the negative pulse on the right. In each case we have supposed that the negative pulse has the same amplitude as the positive pulse. This, of course, is in general not quite correct, but it is often roughly so. We have oversimplified the situation in another respect by neglecting the effects of internal

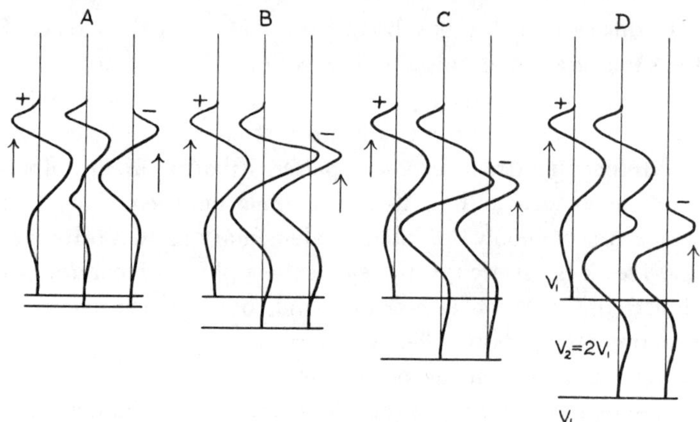

FIG. 11.9. Effect of reflectors of varying thickness.

multiple reflections. This, however, is not too serious because these multiples are quite weak in actual cases. It will be observed that the sum of the two pulses in *A* gives another pulse of somewhat irregular shape. This resultant pulse is of a type that may well be confused with background noise. This is especially true if the thickness is allowed to vary. Suppose, for example, that the difference in reflection time from top and bottom is 0.004 of a second, that the velocity, V_1, is 8000 feet per second, and the velocity, V_2, is 16,000 feet per second; the thickness of the thin reflector in *A* will be 32 feet. This thickness could easily vary by 100 percent either way over short distances, thus giving rise to wide variations in the character of the reflection formed from top and bottom pulses.

It is evident, without actually drawing the figure, that as the thickness of the reflecting layer approaches zero, we approach a condition where the sum of the reflections of the pulses from the top and bottom is zero. Thus the thinner layer is a poorer reflector.

Let us now consider a slightly thicker layer, shown in *B*. Here we see that the two pulses add up to give a stronger pulse than either one would give separately. In this case the thickness of the reflector is 104 feet and the time difference in reflections from top and bottom is 0.013 of a second. It is quite evident that the *sum character* of the pulse due to reflections from top and bottom is markedly different from the character of the individual pulses themselves; therefore, if a layer of this thickness were widespread over an area, we could identify it to a certain extent by the character of its reflection.

We now consider case *C*, in which the thickness is 200 feet. Here the character has changed so that, if we look closely, we can see the two parts of the sum reflection. This we could not do in *B*. It is clear that in case *C* there is still a strong reflection of the sum.

In *D* the thickness of the reflector is 320 feet. Here we see that the two reflections are beginning to separate, so that, if we were careful, we could make a distinction between the top and bottom reflectors. In order to do this, it is necessary to find out what the positive pulse character is and interpret the combination as a positive reflection followed by a negative one of the same character.

A reader who has studied seismic records for a certain length of time will recognize in Figure 11.9 situations which he has seen repeatedly. Sometimes an analysis can be made in terms of the simple reflection pulse shape, or character and an interference pattern can be separated into a couple of component parts. Usually, however, such an analysis is impossible, largely because the background noise is usually too strong to allow the character of any single reflection to be determined with sufficient accuracy. In order to separate two reflections it is necessary to know this shape and to have the top or earlier reflection reasonably well separated from the later component at the front of the composite pulse. Unless this is the case, the analysis cannot be made, because the unknown amplitude of the early component makes it impossible to subtract this component to find the later component of the composite pulse. In general, when there is a composite pulse, all we know is that two of them have been added together and give an interference pattern. They can be separated in many ways, each of which gives two components that can be added to produce the observed composite reflection or pattern of reflections. Thus in order to make an analysis, we have to have pretty good information regarding the shape of positive or negative reflections, uncomplicated by compositing with other reflections and also—and this is sometimes very serious—uncomplicated by interference effects due to the curvature of a single reflector.

It is worth while to analyze an interference pattern, but it is exceedingly important not to force an analysis when one is impossible.

11.3.5.2. Interference Effects Due to Faults and to Sudden Changes of Dip

We shall return to faults in Section 11.5.4, but since they give an interference effect, we mention them here. When there is a fault, reflections from different levels, possibly with slightly different dips, combine and give interference patterns. The same situation occurs if there is a sudden change of dip. For example, if on one side of the line of geophones we have a 0° dip, and we change to a 1° dip at the shot point and we continue with 1° on the other side of the shot point, we have a sudden change of dip which combines two reflected pulses that interfere with each other. Analysis in this situation is sometimes difficult. However, it is well worth while in spite of the difficulty, because often such information is of vital economic importance.

11.4. TEMPLATES

In the foregoing sections we have discussed in some detail the aspect of reflections that is referred to as *character* or *pulse shape*. We now proceed to another aspect that can be studied by means of what are called *templates.*

11.4.1. Directions for Making Templates

This section gives directions for making templates like that shown in Figure 11.10. Part *A* is a reflection, *B* is a series of points on one side of a clean transparent draftsman's triangle, and *C* shows the completed template. In addition to the triangle, a china marking pencil is required. This is a dark brown or black wax pencil that will mark readily on the slick surface of the triangle.

We select a reflection, if possible an *intermediate reflection* with respect to the cross section we are interpreting. For example, if we can use reflections between 0.5 second and 1.5 seconds on our records, we will try to pick a reflection at around 1 second. This is the case in the record that we have used as an illustration; our reflection is just below the 1-second line (see part *A* in Figure 11.10). This intermediate reflection ought to be a good, reliable reflection, and as strong as possible. Having picked this intermediate reflection, we now select a good trough or a good peak, as *far forward* as possible. On this particular example we have selected a trough. The peak in front of this trough was not selected because there was some evidence that the end of a shallower reflection was disturbing it to a small extent. We try to pick as far forward on

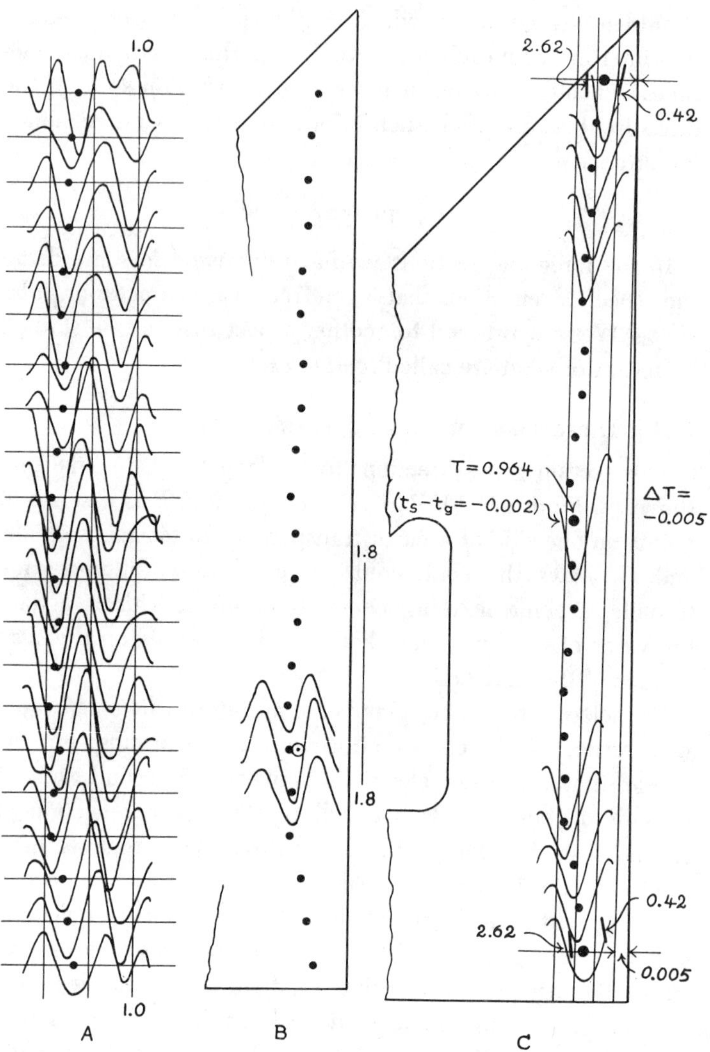

FIG. 11.10. Construction of a reflection template.

the reflection as possible, because usually the forward troughs and peaks have a slightly higher-frequency character with the instruments in common use today and generally are less disturbed by interferences than later peaks and troughs. After the trough has been picked, we mark a series of equilibrium points, as illustrated in the figure. It will be observed that each of these points is placed at a position over the bottom of the trough where the trough would be if the trace came to rest. The purpose of placing these points at the *equilibrium positions* is to reduce the effect of variations in amplitude. The more common practice of putting the points, or little vertical dashes, at the bottoms of the troughs themselves is satisfactory when the amplitude is constant. However, some means is needed for correcting for the variation in amplitude, when one occurs, and the means used here has proved very useful in practice. Having placed the points at the equilibrium positions on the record, we put the transparent triangle over the record and with the china marking pencil mark dots on the triangle directly over those on the record.

We then move this series of dots on the triangle to various other reflections in an effort to determine whether the *trend* we have used is *characteristic of the reflections in general on this record.* Of course, if we move to shorter times there will be more curvature of the reflection, or more *move-out.* But this is a continuously varying quantity and can be taken into account. Likewise as we move to deeper reflections, there is less move-out; that is, the reflection goes straighter across the record with less curvature.

The purpose in comparing the first series of dots with the other reflections is to find the places where the original reflection was disturbed by *noise that does not disturb other reflections.* Such a case is shown in *B*, where the 6th trace from the bottom is found to be offset because of a disturbance. We could have guessed this from a direct look at our first reflection which shows a disturbance very clearly. Although the presence of a disturbance may be detected, its exact nature is usually difficult to determine without reference to other reflections, and this is what is illustrated in *B*. Here only the 5th, 6th, and 7th traces from the bottom are shown, together with the correction that must be made to our original spot. The original spot is shown as an open circle, the corrected spot as a black dot.

Having made this correction, we transfer the template back to the original reflection and very carefully determine ΔT for it. It is assumed here that this is a symmetrical split shot, which it is, and that ΔT is the time difference between the top trace or top group of traces and the bottom trace or bottom group of traces, depending upon the procedure used for obtaining ΔT. When ΔT has been determined the dots are wiped off the triangle with a paper towel and the triangle is placed over the reflection again, this time so that its right edge differs from the timing lines between the top and bottom traces by just the amount ΔT. Part *C* shows this placing of the triangle. In the present example, ΔT is -0.005 of a second, so we place the top of the right edge of the triangle opposite the top point on the timing line, as shown. The bottom of this edge opposite the bottom point is 0.005 of a second to the right of the timing line. Thus the triangle is tilted so as to measure the ΔT of this reflection. With this placing of the triangle, we can remark the spots on the template as indicated. The remarked template is then ready to be used for measuring the ΔT's of the remaining reflections. It is readily, easily, and quickly used for this purpose. The determination of ΔT is, however, only a secondary purpose of the template.

It is sometimes maintained that a template is not needed because the reflections can easily be seen without it. But it must be admitted that, with almost any record that can be studied, if we pick more and more reflections we eventually get to the point where the good ones are all picked and only the very weak ones remain. It is these very weak reflections that can be seen more readily with a template.

Some seismologists take great pride in having a photographic memory; they remember how each trace is shifted. Although this may be possible for a few individuals, it is not possible for most people and it is utterly impossible when the reflections are very weak and of very poor quality. When, in addition, we realize that sometimes the weakest reflections are very important in solving an economic problem, we realize that a template has its usefulness.

However, a template can be misused. Accordingly we wish to consider some rules for using one.

11.4.2. Rules for Using a Template

Any instrument is capable of being misused, especially if it is not understood. There are a few simple rules for using a template which are based on the physical principles involved in the transmission of seismic pulses. One of the simplest rules was given in the directions for making the template, namely, to pick the time and ΔT on the *reliable* reflections so that the template itself will be reliable to a certain extent.

When picking any reflections, reliable or not, we must remember that when we move to a different reflection the effect of move-out variation due to the varying depth of the reflector is continuously variable. It occurs smoothly, not abruptly. It is therefore important to use the template without abrupt changes in the way we pass across a reflection. The writer has seen a template used halfway across a record along a series of troughs, for example, and then abruptly rotated through a fair-sized angle to get the remainder of the reflection! This is impossible. Each reflection curves smoothly and gradually across the record. It may go a certain distance and disappear, to be followed across the record by an entirely different reflection, as was the case in the example just mentioned. But it would be a serious mistake, in principle, to call these two separate reflections one single reflection. Thus, one of the primary rules for using a template is that it must be fitted to the reflections in a smooth continuous manner. The variation of move-out should be indicated on the template itself; this can easily be done, as is shown by the pair of short segments at the top and bottom of part *C* in Figure 11.10. At the top is a segment that is marked 2.62. A similar segment appears at the bottom. These correspond to the positions which that point on the template would have had if we had used a particular reflection at 2.62 seconds. The situation is similar for a shallower reflection at 0.42 second, where the move-out is much larger. These little segments at the top and bottom are intended to give some idea of the effect of move-out, so that in using the template on reflections at different times we can have reasonable limits to the move-out variation, and thus know how smoothly to vary the angle of the template as we follow across a reflection.

If this rule of continuity of slope of the reflection is followed faithfully—as it certainly should be, because it is based on sound physical principles that will be discussed in greater detail a little later—the template will be found to be a very useful and valuable tool in the more accurate selection of reflections.

Accuracy is increased in more than one way. For example, the determination of ΔT is made to depend, not on the top and bottom traces alone, but on the averages of several traces near the top and several near the bottom. It is usually found that this procedure gives rise to a smoother dip cross section, suggesting that the ΔT's have been determined with a little greater accuracy by this process.

Once the template has been made, the ΔT's on the good reflections can be determined very quickly. It may be noticed, in passing, that the ΔT's on the partial reflections can also be determined. It is important in determining these ΔT's to grade them down, since they are considerably less reliable than those on reflections that carry reliably all the way across the record. However, these partial reflections often contribute information of vital importance to the investigation.

11.5. "POOR REFLECTIONS"

Poor reflections are often of such major importance that they are given particular attention in this section.

11.5.1. Accidental Line-Ups

Although any individual place on a record is very unlikely to have an accidental line-up of troughs or peaks, there are so many opportunities for these accidental line-ups that they do occasionally occur. This situation can be illustrated by considering the accidental line-ups of spaces between words on the page of a book. For example, at the lower right corner of page 371 of Volume 1 of the *Geophysical Case Histories,* there is a space between the words "the" and "west." This space goes upward at approximately 7° to the left for 10 lines, with a few exceptions. If the reader will hold the book on a slant and follow the spaces upward, he will see an accidental line-up such as might, on a reflection record, be taken for a reflection. Similar examples can be found on many pages of the Case Histories volume. Although each one is highly improbable, so many opportunities are presented on the many pages that such alignments do occur fairly often.

In the same way, there are so many opportunities for accidental line-ups on a series of seismic records that although each such line-up is extremely improbable, they do occasionally occur, and we must be on guard not to interpret them as reliable reflections. The object in calling attention to these accidental line-ups is not necessarily to cast serious doubts on every weak reflection, but only to make clear that accidental line-ups without any geological significance occasionally occur. Perhaps the interpreter can be alert for them without necessarily concluding that all weak-appearing reflections are accidental line-ups and are to be ignored for that reason.

11.5.2. Weak Reflections

In the geological section there are often interfaces between different stratigraphic horizons which give rise to very weak reflections. Sometimes these weak reflections are of great economic importance. In this case it is important that they be given serious attention.

The principal difficulty associated with a weak reflection is that the background noise usually tends to overshadow it, making it extremely difficult to see. There is, of course, nothing to be gained by straining the imagination to such an extent that reflections are seen where none exist. Such a ridiculous procedure should be avoided. Though it is ridiculous it is quite natural, especially in those cases where a certain zone of weak reflections is known to be of major importance. It may be assumed that the time and ΔT of a really weak reflection will be very poorly determined. Thus, in addition to doubts about the existence of the reflection at all, there is a large range of indeterminacy for the time and ΔT. All these characteristics point to the necessity or desirability of plotting the weak reflections so they can hardly be seen. When plotted in this way they disturb the rest of the cross section to a minimum extent but are on hand in case they are required for the interpretation.

11.5.3. Partial Reflections

The seismologist who has studied a good many reflection records will recall many cases of partial reflections. Some of these, especially in the deeper part of the section or at later times on the record, will be very strong on part of the record and disappear altogether on the remainder. In other cases a reflection may pass over a part of the traces, disappear on another part, and reappear later.

Partial reflections may be due to a number of causes. One cause is the actual variation in the strength of the reflection; thus on some of the traces the reflection appears strong, whereas on others it does not override the background noise. Another cause of partial reflections is interference with other reflections. In any case, partial reflections have their use. It is the writer's opinion that they should be plotted on the cross section with lowered grading, indicating their incompleteness. Since they do not go all the way across the record, their ΔT's are not very reliable, at least not as reliable as those of reflections that are complete. The ΔT's for partial reflections can be easily determined by the template procedure outlined in Section 11.4.

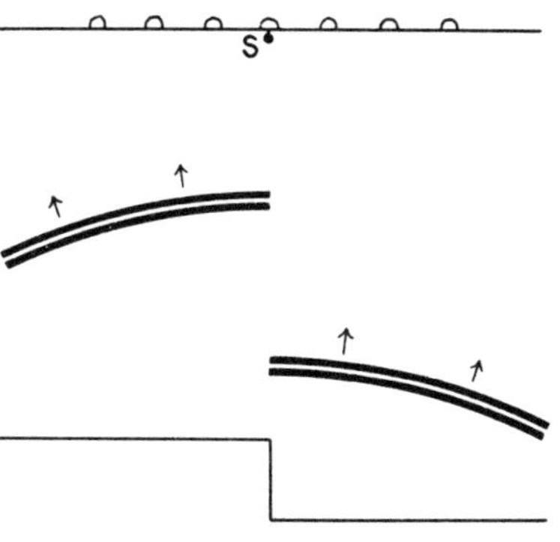

FIG. 11.11. Oversimplified picture of a reflection by a fault.

11.5.4. DIFFRACTIONS. FAULTS

What we mean by a diffraction can be explained in terms of what happens in a simple fault. For this purpose refer to Figure 11.11, which shows a shot point and a row of geophones at the surface, a fault below, and two reflected parts of the pulse corresponding to the high and low sides of the fault.

In several publications, in exhibiting records showing faults, the assumption has been made that the situation shown in this figure corresponds roughly to the facts. This, however, is a great oversimplication, for the pulses do not stop abruptly at their *shadow edges,* which correspond to a line between the fault and the shot point in the figure. Instead of stopping abruptly at these edges, the energy is *diffracted.*

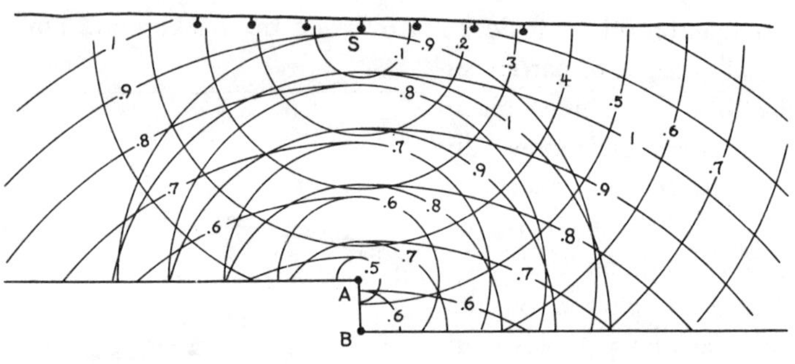

FIG. 11.12. Production of diffractions by a fault.

The diffraction process is approximately illustrated in Figure 11.12, which shows the shot point S, the fault *AB*, and image points of the shot, S′ in the upper reflector and S″ in the lower reflector. A pulse is sent out from the shot and is followed every tenth of a second. During the first 0.4 of a second the pulse does not reach either reflector, but at 0.5 of a second the pulse has reached the upper reflector and is beginning to be reflected in it. The reflected pulse in the upper reflector is a circular arc with its center at S'. This travels upward as indicated, so that it reaches a point just below point S in a little over 0.9 of a second. In addition to this regular reflected pulse there is another pulse corresponding to a circular arc with its center at A. It is this latter pulse which we call the diffraction. The front of this diffracted pulse is drawn for every tenth of a second. What is shown here is that point A behaves as if it were a *new source* of energy traveling outward in all directions from the point.

That this phenomenon occurs has been demonstrated many times in connection with sound waves in fluids. The same process occurs in solids. An article* by Frank Rieber contains pictures of diffracted pulses like those in Figure 11.12.

It may be noted that the reflection from the lower interface also travels upward and that there is a diffraction pattern of circular arcs with centers at point B. This gives another set of diffracted pulses which can also be observed.

If we now consider the main features of the diffraction pattern by reference to Figure 11.13 and contrast the upper part of this figure with Figure 11.11, we will see how the situation has become more complicated. The pulses have been divided into reflected pulses A and B, and diffracted pulses C, D, and E. All these pulses are likely to appear on the record in varying degrees of strength. If they all appear we plot the cross section as in the lower part of Figure 11.13, the letters referring to the portions of the pulse in the upper part. The fault is represented by segments A and B. The secondary parts arising from the fault by diffraction effects are indicated by C, D, and E. A study of many cross sections that are plotted with proper offsets will reveal, in the neighborhood of faults, plotted secondary segments such as those in Figure 11.13. It should be observed, however, that this complicated pattern arises primarily because of the fact that the reflectors have a serious

* *Geophysics*, 1:196 (1936).

irregularity, that is, the edge of the fault. Another type of irregularity would also give rise to a diffraction pattern, not exactly the same as the present one, but a diffraction pattern nevertheless. Since a complete diffraction pattern can seldom be observed because it is so weak, it is very difficult to be sure that the presence of a given diffraction pattern indicates a fault. It might indicate a *sudden change of dip*, as illustrated in Figure 11.14.

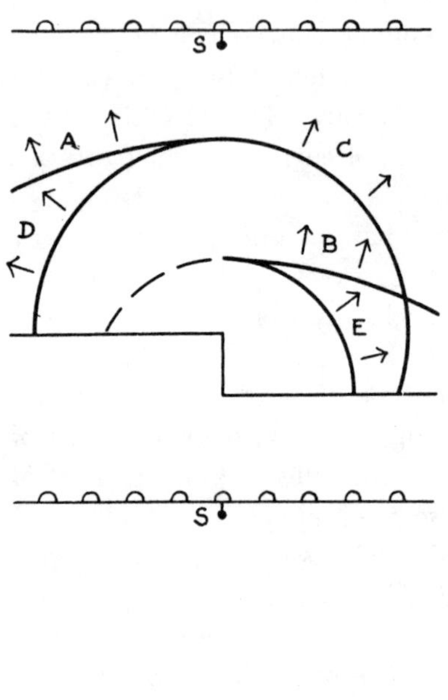

FIG. 11.13. Fault diffractions and reflections and corresponding plots.

Figure 11.14 shows the shot point S, and the ground surface. Beneath these is shown a dipping interface with a sudden change of dip at the point *K*. Image points *S′* and *S″* have been constructed as shown in the figure and in relations 1 and 2. Relation 3, which follows from the construction, is important because it shows that the distance from *K* to *S′* is the same as the distance from *K* to *S″*. This relationship secures the continuity of the reflected pulse front, *AB* and *CD*, with the diffracted pulse front, *EBCF*. Thus *ABCD* forms a *continuous* pulse front, that is, a pulse front with a continuously turning tangent (but with discontinuously changing curvature). The diffracted pulse will add to the reflected pulse to the left of B and alter the pulse shape or character there as compared, say, with the character of the reflected pulse to the left of S.

Thus a sudden change of dip gives rise to diffractions which form a pattern that is somewhat different from the pattern formed in the case of a fault. It may happen that, because of

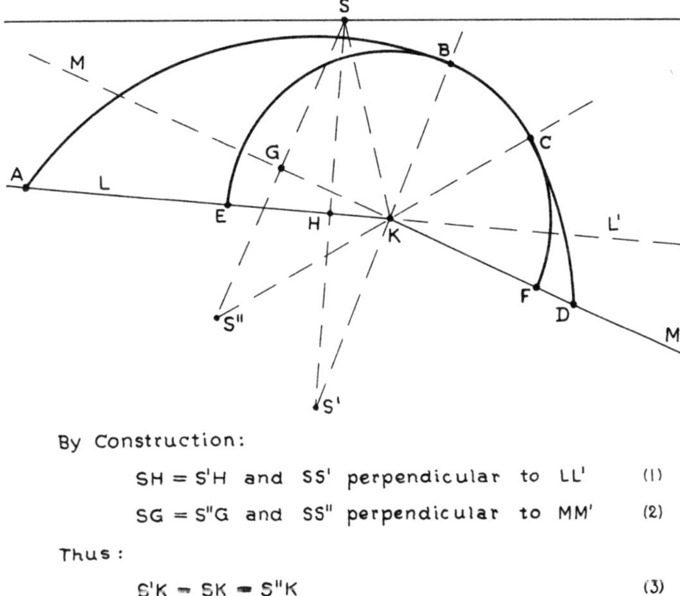

Fig. 11.14. Diffractions caused by sudden change of dip.

the fragmentary nature of the evidence, a distinction cannot be made. Any irregularity in the physical properties of the material that forms the section can produce diffraction effects. Thus a buried river channel will give diffraction effects, as will also an escarpment or the edge of a reef.

When there are diffraction effects, like those in Figures 11.13 and 11.14, we run into the danger of identifying as reflections pulses that are not reflections. The only true reflections in Figure 11.13 are represented by segments *A* and *B*. The other segments represent diffraction effects. It would be more proper to plot these diffraction effects as if they were points instead of reflectors, thus emphasizing the fact that our information tells approximately where the energy is coming from but tells nothing about the dip of any reflector.

It is sometimes very difficult to distinguish between a true reflection and a diffraction, but in many cases it can be done. One of the distinguishing properties of a diffraction, as compared with a reflection, is that the time move-out of a diffraction is much larger than that of a reflection, because the origin of the diffraction is much shallower than the apparent origin, or image point, of the reflection. Another feature of a diffraction which can occasionally be used for its identification is that as the observation position is moved along the surface, the point of origin of the diffraction does not move because it is fixed in the subsurface material, whereas the apparent point of origin of reflections does move. Though sharp in theory, this distinction is rarely usable in practice because the diffractions are so poorly observable. They often appear to branch off the regular reflections; and since they seem to originate at the rear of the corresponding reflection, they are usually more seriously disturbed by the reflection itself than it is disturbed by them. Therefore, in most cases great accuracy in making this distinction is impossible.

In spite of the difficulty of identifying diffractions and locating faults or reefs by this procedure, it can frequently be used with considerable success. The results, however, should be regarded as just another kind of evidence and they should be bolstered by further evidence wherever possible. Usually the diffractions provide a bit of secondary evidence of faulting; but in most cases their presence should be looked at as simply an anomaly, in the dictionary sense of the term, and this anomaly should be further investigated in greater detail by seismic or other methods.

We have listed diffractions under poor reflections, and this is how they usually appear on the records. But it should not be supposed that diffractions are reflections in the ordinary sense. We limit the term reflection to that type of event which yields information regarding the dip of an interface. Diffractions yield no such information but only information regarding the position of an irregularity in the section. The writer has seen cases of faulting where numerous diffraction events could be picked from several shooting locations, thus providing information in some detail regarding the hade of the fault. If these events had been picked as true reflections, they would have been plotted as inconsistent reflections and have given rise to a good many questions. If they are interpreted as diffractions, they may be understood in their proper setting and the dips plotted for them may be understood as meaningless. This is one of the strong reasons for plotting weak reflections, which may turn out to be diffractions, as very short segments so that the dip is not easy to see. This makes possible a sort of compromise type of plotting which allows a weak reflection to be interpreted as either a true reflection or a diffraction.

Any discussion of diffractions from faults would be seriously deficient if it failed to refer to the pioneer study by Curtis H. Johnson.* There can be little doubt that the Geo-Sonograph was admirably suited to studies of this type.

11.5.5. Surface and Shot Hole Noise

Extraneous noise may in some cases weaken reflections, especially if it is somewhat localized. It may happen that a reflection will appear well on one part of the record and not on another. Careful inspection of the record will generally show that noise is present and in some cases it may be attributable

* *Ibid.*, 3:273 (1938).

to local situations. For example, a single tree in the wind may give rise to a local distribution of noise at a certain time on the record that spoils one or two otherwise usable reflections.

Another type of noise is the shot hole noise generated by a disturbance in the hole. This frequently disturbs the places nearest the mouth of the hole and has often led to offsetting the geophones away from it. In some cases this offsetting is good practice. In others it is desirable to leave the geophones close to the shot hole and, when the noise comes in, use the outer traces and a template made on the basis of earlier reflections to obtain the times for the center traces where the shot hole noise is disturbing. (See Figure 6.9.) Covering the charge with concrete will reduce the hole noise to negligible proportions in most cases.

11.5.6. Curvature Effects

As was mentioned earlier, poor reflections sometimes result from curvature effects. (See Section 11.3.4 for a fuller discussion of this.) All seismologists who have had a reasonable amount of experience will recall that the deeper reflections have a great tendency to go only part way across the record. These reflections may be very strong over a few traces and disappear entirely over the remainder. Such reflections the writer would call poor reflections, because their ΔT is very poorly determined. The probable existence of large phase shifts along the readable part of such a reflection, due to curvature effects, makes the determination of ΔT most unreliable. They should be severely graded down, even though they are very strong on several traces.

This is an ever-present difficulty in deep reflections. A reflection may appear to go all the way across the record and yet be subjected to severe relative phase shifts on one side of the record that will make the determination of ΔT seriously in error. One cause of phase shifts in the deeper reflections is the presence of *buried foci*, which we shall discuss in a later section in detail. This is an extremely important effect which distorts many seismic cross sections. It is difficult to take it into account. The presence of buried foci may be suspected in accounting for queer irregular effects on some of the deeper cross sections.

11.6. SYNTHESIS. INTERPRETATION

In the preceding sections we have discussed ways of picking reflections and considered the character and variability of the reflection. We also considered variations across the record due to velocity variations, thickness variations in the weathering, etc. In order to handle these variations across the record, we described a template which has been found, in practice, to serve as a useful crutch, enabling one to pick many more reflections on a record than are usually the concern of the interpreter. This is important, since often the very weak reflections are those of greatest economic importance. Let us therefore summarize by describing the way we would go about handling an interpretation problem.

In the first place, individual reflections may be picked, computed, and plotted on regular cross sections if the dips are relatively small. But if they become too high, the reflections should be plotted on vertical cross sections in order to get a closer approximation of the true situation.

Up to this point we have regarded each individual reflection, or diffraction, as *independent* of all the others. This is a good policy because it avoids any prejudice regarding the information. Thus all the information that can be found is put on a first cross section, either vertical or regular. We say first cross section, because subsequent interpretations may show that we have put on this cross section a great deal of material that is actually noise, especially if it can be definitely interpreted in terms other than as simple reflections from geological interfaces.

After this first cross section is made, we are ready to consider further details. We may look at them from several viewpoints. From one, we examine the weaker reflections in an effort to determine which of them help to make a consistent geological picture. In this way we can grade down isolated, inconsistent, weak reflections. These reflections we may suspect of being diffractions of some sort, or accidental line-ups. When we have a collection of very weak reflections all pointing toward the same structural possibility, we of course have much more confidence in each individual reflection thus included.

Another viewpoint from which we may look at the reflections is correlation. Up to this point we have studied individual reflections without regard to their relationship to other reflections, either on the same record or on different records. We now consider the interrelationship of the reflections. The cross sections and records should be studied together wherever possible, in order to bring out this interrelationship.

Correlations between reflections should be indicated on the cross section. A dotted line joining the centers of the correlated reflections may be used, or some other device such as using the same letter for all the intercorrelating reflections. As was explained in an earlier section, the problem of correlation is exceedingly difficult and it becomes more difficult in the complicated cases in which it is more important. This unfortunate fact makes the work of the seismologist difficult indeed. But this difficulty by no means implies any lack of importance. Whenever time is available, the correlations should be studied.

Another viewpoint that we may take in studying the first cross section is the removal of noise. Some of the reflections that are plotted may correspond to simple noises. Some of them may be multiple reflections; others may be diffractions that should not be plotted as reflections. The records and cross section should be carefully studied to find as much noise of this type as possible so that it can be removed.

Some seismologists consider the removal of any information from cross sections as an unscientific and almost dishonest procedure. The writer agrees with this as far as it relates to the first cross section. This should have plotted on it all the information, including the noise, that can be extracted from the records. The plotting should be done with as much accuracy as possible so that consistency and inconsistency will have real meaning. Obviously, if inaccurate cross sections are plotted, any inconsistency on a cross section may be simply a by-product of the inaccuracy. If, instead of limiting our interpretation to the first cross section, which contains all the possible information, we make an *interpreted cross section,* from which inconsistent, weak reflections and multiple reflections are removed, and on which diffractions are shown by very short segments or points, and correlations are also shown, we shall obtain a product that will be our best effort at interpreting our data. This may be submitted, together with the first cross section containing all the plotted data, so that a clear understanding will be gained concerning what has been left out. But when only one cross section is made, and certain reflections are excluded from it, the procedure is subject to severe, legitimate criticism and should, of course, be avoided.

A large part of the seismologist's work is concerned with finding the individual reflections. This is analogous to observing the trees in a forest. He should not fail to consider the forest, because he has so much difficulty in finding the individual trees. After all, the individual reflections are not the primary purpose of his search. His real purpose is to find geological structure, and he must accordingly carry his interpretation beyond the stage of considering individual separate reflections.

Thus far we have covered most of what may be regarded as the practical part of reflection seismology. Many details have been omitted to keep the size of the book within reasonable limits. Our most important practical omission is refraction prospecting, and this will be discussed in the next chapter.

11.7. COMMENTS TO THE SECOND EDITION

Now, we "stack" coherently. Then, we learned to "see" reflections under the noise. We did our stacking behind our eye balls! I am thankful that this aspect of the work is over.

Part IV

REFRACTION PROSPECTING

CHAPTER 12

Line Refraction Work

This chapter is written primarily for members of reflection seismic parties who may have to do an occasional refraction job. It is assumed that the background is the same as that for reflection work. However, the instruments are somewhat different, because most refraction work requires a somewhat lower-frequency response in the instruments. When first breaks are to be obtained at long distances, it is essential to have a detecting, amplifying, and recording system that has good response down to 10 cycles per second. In some cases it may help considerably if the response carries into lower frequencies. Means should be provided for reducing wind noise by filtering out the higher frequencies, when necessary.

12.1. NO-DIP CASE

We consider in this section the case when no dip is involved. We suppose that we have to deal with two layers, one of which is the *overburden* and the other the *undersection* of higher velocity. In refraction work it is impossible to work with the reverse condition in which the undersection has lower velocity. In Figure 12.1 is shown a time-distance graph in which the travel time, T_x, is plotted against the distance, x, from the shot point to the geophone. Below this graph is a cross section with the shot point, S, from which energy travels down along the ray to the refracting interface, thence along this interface, and finally up along the ray to the geophone.

We want to interpret the graph in terms of the cross section. Actually our practical interest concerns the reverse; we wish to use the graph to find the cross section. Suppose for a moment we have the cross section as indicated; then the travel time, T_x, is that given in relation 1. The first term on the right-hand side of this relation represents the *slant travel time down* to the refracting interface from the shot *and up* to the geophone from the refracting interface. The second term represents *travel time along the high-velocity layer* between the points of refraction at each end. Collecting terms involving the depth, z, we may rewrite this as in relation 2. We then use relation 3, which is an expression of Snell's law. This is a particular case, because the angle of refraction is 90° and the sine of 90° is 1, as indicated in this

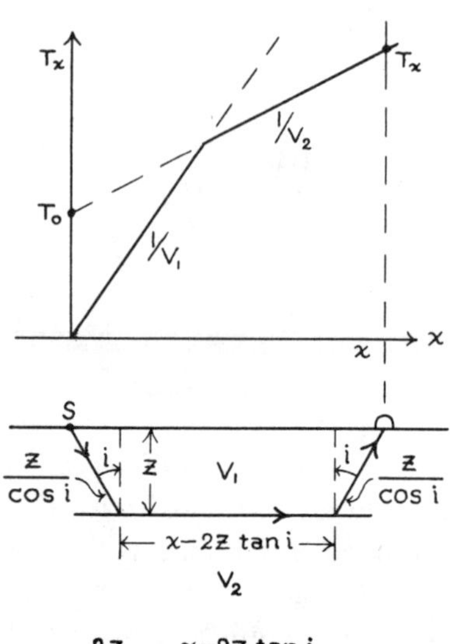

$$T_x = \frac{2z}{V_1 \cos i} + \frac{x - 2z \tan i}{V_2} \qquad (1)$$

$$= z\left(\frac{2}{V_1 \cos i} - \frac{2 \sin i}{V_2 \cos i}\right) + \frac{x}{V_2} \qquad (2)$$

$$\frac{\sin i}{V_1} = \frac{\sin 90^\circ}{V_2} = \frac{1}{V_2} \qquad (3)$$

$$T_x = \frac{2z}{V_1} \frac{(1 - V_1^2/V_2^2)}{(1 - V_1^2/V_2^2)^{1/2}} + \frac{x}{V_2} \qquad (4)$$

$$T_x = (1/V_2)x + (2z/V_1)(1 - V_1^2/V_2^2)^{1/2} \qquad (5)$$

$$T_0 = (2z/V_1)(1 - V_1^2/V_2^2)^{1/2} \qquad (6)$$

$$z = (T_0 V_1/2)/(1 - V_1^2/V_2^2)^{1/2} \qquad (7)$$

FIG. 12.1. Refraction calculation of depth.

equation. Thus, when the velocities are given, the sine of the angle of incidence can be determined and hence the angle of incidence can be found. We substitute in relation 2 the value for i found in relation 3 and obtain relation 4. In doing this we express the cosine of i as $\sqrt{1-\sin^2 i}$ and replace the sine of i by the ratio of V_1 to V_2. Relation 4 may be rewritten as in equation 5. We may separate the part of this equation that is independent of x and T_x. This is the intercept time, T_0, and is given by relation 6. By solving relation 6 for the depth, z, we obtain relation 7. Thus the depth is given in terms of intercept time, T_0, the velocity in the upper layer, V_1, and the velocity in the lower layer, V_2. Equation 7 is therefore a definite expression by which the depth can be computed from the information shown in the graph. We obtain the velocities by measuring the reciprocal slopes of the time-distance graph as indicated. (Slope here is the ratio of increase of T_x to increase in x.) The slope of the first part is the reciprocal of the shallower velocity. The slope of the second part is the reciprocal of the higher velocity.

This illustrates the case in which the layers are of *constant velocity* with *no dip*.

Let us now consider the case where the overburden has a velocity that *increases uniformly with increasing depth*. We have to consider first the behavior of the *direct wave* in such a medium. The upper part of Figure 12.2 is a graph with a curve representing the travel time for such a case. Evidently, the tangent to this curve at the origin has a slope that is the reciprocal of the datum velocity, as shown. Let us consider the relationship between the time and the distance by referring to the diagram below the graph. The upper dashed line represents the line of centers of ray circular arcs. Two such arcs are shown with their centers. We have supposed that the center of the second arc is on the line of centers halfway between the shot point, S, and the geophone, as indicated. If we refer to our consideration of this case (see Figure 8.13), we find that the R value corresponding to the half travel time is the same as the R value corresponding to a dip of 90° at the deepest point on the path, as indicated. This we may substitute in our formula for R and obtain relation 1. This relation may be solved for T_x, as indicated in relation 2. This in turn may be solved for T_x in a slightly different form* in relation 3. Either equation 2 or 3 may be used in plotting the graph in Figure 12.2.

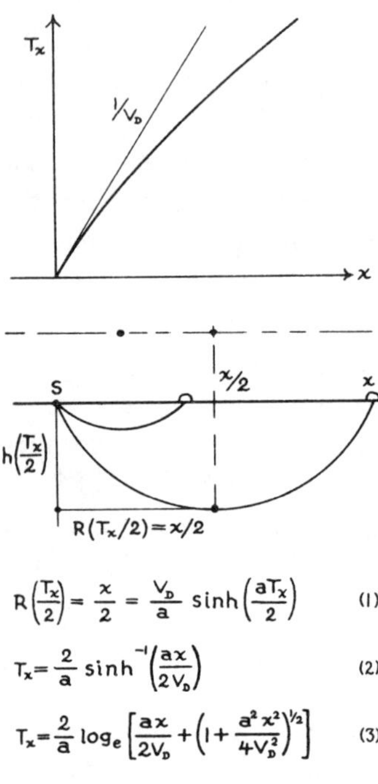

FIG. 12.2. Determination of a and V_D from refraction data.

If we have a collection of data taken from an area where the overburden velocity does increase uniformly with increasing depth, as we have assumed in Figure 12.2, we have a means of determining the rate of this increase. Suppose the situation is like that illustrated in Figure 12.3. The early part of the graph shows the curved section, and we have indicated the curved extension of this. Usually when the velocity of the overburden increases uniformly with increasing depth, the directly received pulse is a strong one and consequently can be picked as a secondary pulse, beyond the point where it appears as a first break. This accounts for the dashed extension upward and to the right in the figure. Taking the first breaks and their extension together, we have a curve of considerable length which can be used for finding the

* See Peirce, *A Short Table of Integrals,* No. 679 (1910).

datum velocity, V_D, and the a value. This can be done by plotting on transparent paper a series of curves with different a values and different datum velocities, using formula 2 or 3 of Figure 12.2. These curves are then compared with the curve in Figure 12.3. Actually, the curves will not be like that in the figure; they will be only a series of plotted data points. We then have to select the combination of a and V_D that gives the best eye fit to the data. With these particular values of a and V_D, we are ready to proceed with the computation shown in Figure 12.3.

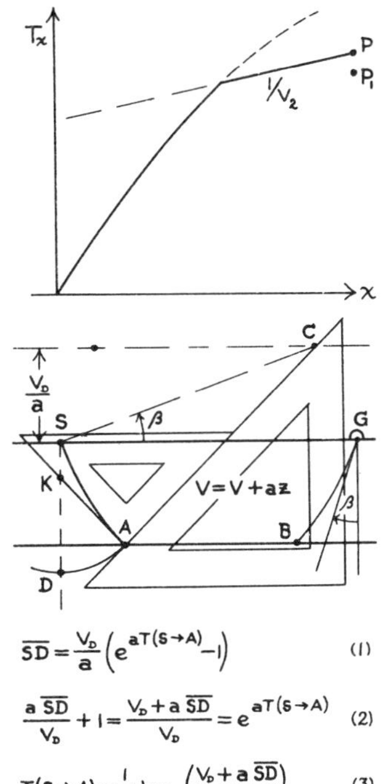

$$\overline{SD} = \frac{V_D}{a}\left(e^{aT(S\rightarrow A)} - 1\right) \qquad (1)$$

$$\frac{a\,\overline{SD}}{V_D} + 1 = \frac{V_D + a\,\overline{SD}}{V_D} = e^{aT(S\rightarrow A)} \qquad (2)$$

$$T(S\rightarrow A) = \frac{1}{a}\log_e\left(\frac{V_D + a\,\overline{SD}}{V_D}\right) \qquad (3)$$

$$T(S\rightarrow A) = \frac{1}{a\log_{10}e}\log_{10}\left(\frac{V_D + a\,\overline{SD}}{V_D}\right) \qquad (4)$$

$$\frac{\sin\beta}{V_D} = \frac{\sin 90^\circ}{V_2} = \frac{1}{V_2} \qquad (5)$$

$$\beta = \sin^{-1}(V_D/V_2) \qquad (6)$$

FIG. 12.3. Calculation of refraction travel time for linear overburden.

It is possible to compute the depth of the high-velocity interface, but the formula is exceedingly complicated and difficult to use. We therefore recommend an alternative procedure which consists of making a *guess* regarding the depth. We may obtain this guess by a calculation, if we wish. For example, an average velocity for the overburden can be estimated from the data. Then, with this average velocity, the depth z, can be computed according to formula 7 in Figure 12.1. When this first approximation is determined, either by computation or by guess, we draw a line such as the one which passes through A and B at this depth. We are then ready to calculate the travel time corresponding to this particular estimate. To do this we have to measure the time from S to A, from A to B, and from B to G. We determine the time from S to A, which in this case we suppose equals the time from B to G, by finding the time from S to D. It is assumed that this time is computed as if the velocity were uniformly increasing down to the level of D. This is a fiction which we introduce solely for the purposes of this computation, for we have already assumed that the velocity is V_2 below the line through A and B. The reason for introducing this is that it is much easier to determine the time from S to D, vertically below it, than from S to A. We could actually find the time from S to A by using the dip slide rule referred to in Section 8.4.3, but this computation would not be sufficiently accurate for our purpose because the slide rule is not particularly sensitive to changes in time.

Let us consider how we calculate the time from S to D. The depth from S to D is given by equation 1, which was derived in Figure 8.20. This equation can be rewritten in the form of relation 2 by multiplying through by a/V_D and adding 1 to both sides of the equation. We thus have an expression that is equal to the exponential of aT, where T is the time from the shot to point A. By the definition of the logarithm of a number as the power to which e must be raised to get the number, we immediately get relation 3. This relation represents the time in terms of the natural logarithm to the base e. For many purposes it is preferable to have the time given in terms of the logarithm to the base 10, as is shown in relation 4. It should be remembered that the logarithm of e to the base 10 is 0.4343. We can thus determine the time from S to A if we have some means of knowing S and A. We know the angle β between the up-coming ray at the surface point, G, and the vertical, as illustrated. This we find from relation 5, which is an expression of Snell's law for this case. Solving this relation for β, we obtain relation 6, giving β in terms of the datum velocity, V_D, and V_2, both of which it is supposed that we have measured. Through point S we then draw a line making an angle β with the horizontal and obtain the point C, which is the center of the ray circular arc. With this center we draw a circular arc through S down to the line AB. This determines the point A. We rest the long edge of a triangle against A and C as shown, and then place a second triangle on the edge CA. Thus we determine the point, K, on the vertical through point S, as indicated. With K as the center, we strike an arc through A to find point D. Having found this point, we use relation 4 to find the time. In this way we can compute the time from S to A, which we assume to be equal to the time from B to G. To get the total time we have only to add to this the time from A to B, traveled at a rate V_2.

These times may be computed, and added, and compared with the graphed information. Suppose, for example, that the observed point is the point P in the graph and that our computed point is the point P_1. P_1 represents the *first approximation.*

We now have to consider the problem of adjusting the depth to get a closer approximation. The adjustment, in the present case, has to be in the direction of increasing the depth. We might guess the adjustment, but since it can be computed with considerable ease we shall explain how to do so.

Figure 12.4 illustrates the situation approximately. The shot point, S, and the geophone, G, are shown. We suppose that we have computed the travel time from S to A to B to G and that it is too small. We have to find a deeper interface, JI, which will give a closer approximation. The velocity below this interface is V_2. We assume that the velocity in a section between the level of AB and the level of JI is V_a, where V_a is either the velocity at the level A, obtained from $V = V_D + az$, where z is the depth of A, or a slightly larger velocity to take account of the fact that the velocity increases as we go deeper. In any case, we assume the section between *AB* and *JI* to have a constant velocity for the purposes of our approximate computation. We then note that the new travel path will be from S to *A* to *J* to *I* to *B* to *G*. We immediately observe that the travel paths from S to *A* and from *B* to *G* are common to the old paths, so

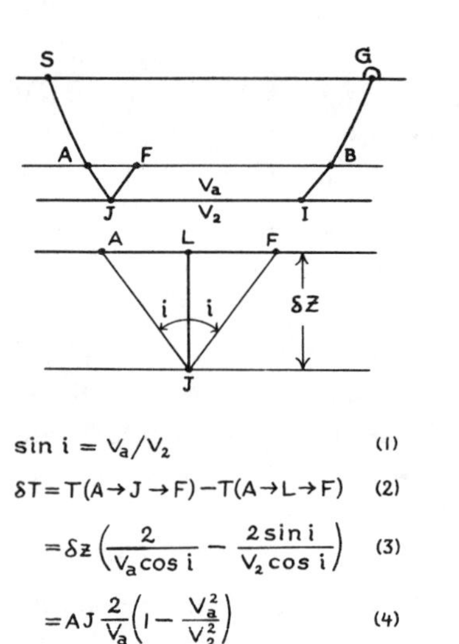

FIG. 12.4. Adjustment of estimated depth.

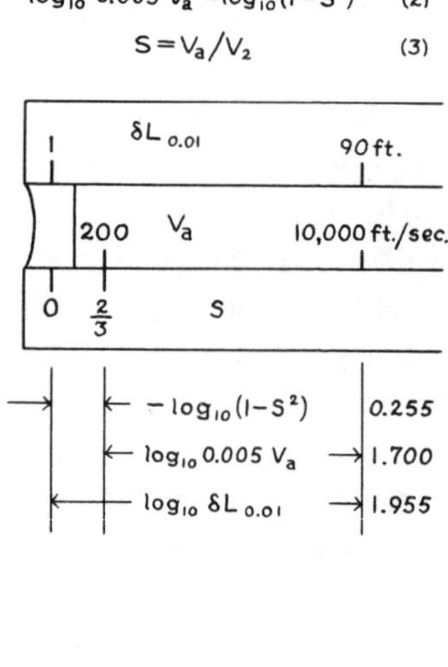

FIG. 12.5. Slide rule for calculation of refraction depth.

we need no longer consider these parts of the path. We also observe that the travel path from *J* to *I* has the same travel time as the travel path from *F* to *B* in the original case. Therefore these paths may be omitted from consideration. Consequently the *change in travel time* corresponds to the difference in the travel time from *A* to *F* at the high velocity, V_2, and the travel time from *A* to *J* to *F* at the lower velocity, V_a. This we can readily compute. We see that the sine of i is the ratio of the two velocities given by relation 1, which comes directly from Snell's law, the angle of refraction being 90°. Thus the change of travel time is defined by relation 2, which we may write in the form of relation 3 if we calculate the individual times.

This relation 3 with the help of relation 1 can be written in the form of relation 4. Relation 4 may be solved for the change in ray length, δL or AJ, to get relation 5. Since the denominator of the right-hand side of relation 5 is $\cos^2 i$ and $\delta z = \delta L \cos i$, we can immediately derive relation 6. I prefer 5 to 6 in use, because 5 requires the construction of the cross section to scale. Relation 5 can be easily used if we suppose δT to be 0.01 second. Then relation 5 gives the change in depth along the ray, per hundredth of a second, between P and P_1 in Figure 12.3.

In order to carry out the calculation we refer to Figure 12.5. In the top line of this figure is the relation for $\delta T = 0.01$ second. Relation 2 is obtained merely by taking the logarithm of both sides of 1. Relation 3 defines the abbreviation S. Relation 2 is the basis for a very simple type of slide rule. The bottom or S scale is *measured* from the zero point according to $-k \log_{10} (1 - S^2)$, k being a constant that determines the size of the slide rule. The bottom scale is *labeled* according to S. The slider scale is measured according to $k \log_{10} 0.005\, V_a$ and labeled according to V_a. As $\log_{10} 1$ is zero, the zero point for the measure corresponds to $V_a = 200$ ft./sec. The top scale is measured according to $k \log_{10} \delta L_{0.01}$ and labeled according to $\delta L_{0.01}$. In the figure the scales are set for the case where the ratio of velocities is $S = 2/3$, the velocity V_a is 10,000 ft./sec., and $\delta L_{0.01}$ is 90 feet. Thus if we missed by 0.01 sec. on our first estimate we would continue the ray deeper by measuring 90 feet along the ray to the interface.

An alternative method is to make a chart plotting V_a against V_2 and graphing equations 5 or 6 in Figure 12.4 for fixed

δL or fixed z.* This has the advantage of avoiding the calcula- tion of S.

This method may seem somewhat cumbersome, but in practice it is both quick and easy to use. It usually gives a result, within the range of observational error, at the second approximation. With a little care, one can become proficient in estimating the first depth so that this first estimation will be close enough. "Close enough" in this case usually means something like ± 0.005 sec. Actually the refraction data are seldom more accurate than this.

Incidentally, in making the first approximation the adjustment slide rule shown in Figure 12.5 can be used to good advantage. We merely extend the high-velocity line back to find the intercept time, T_0. This intercept time contains a certain number of hundredths of seconds, and this number is used with the slide rule to obtain the first estimated depth.

Furthermore, if the upper section is assumed to have a constant velocity, the slide rule can be used for determining the depth. This greatly reduces the amount of computing required.

12.2. CALCULATION OF DIP

In this section we calculate the dip on the basis of a reversed refraction profile. In order to explain the principles of this computation, we present Figure 12.6, which shows time-distance graphs plotted in opposite directions from shot point S to geophone G. The graph at the left represents the case when the energy travels down dip along the refracting interface. We assume that the shallower velocity is V_1 and, in order to simplify the relationships a little, that the depth of the high-velocity layer at the shot point is zero, as indicated. We can then see that the straight-line time-distance relationship in the graph is that given as relation 1. This is so because the angle between the up-coming ray to the geophone and the vertical is α plus θ, where α is the critical angle of refraction and θ is the

* C. H. Dix, *Geophysics*, 6:378-396 (1941). See Fig. 5.

dip angle. This relationship is easily verified from the geometry in the figure. Another way of writing relation 1 is shown in relation 2, in which V_{2d} represents the apparent velocity with

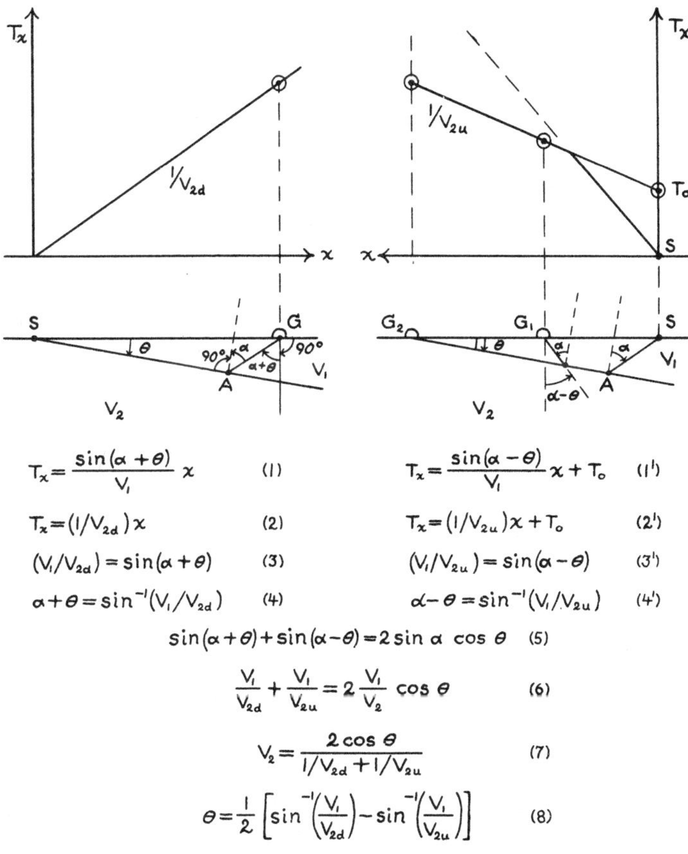

FIG. 12.6. Calculation of velocity and dip.

which the refracted pulse sweeps along the horizontal surface. Since equations 1 and 2 are identical, we have a relationship between the apparent and the upper velocity and the critical and the dip angle that is given in relation 3. This we may solve for the sum of the angles, as in relation 4.

In the graph at the upper right of Figure 12.6, we have *reversed* the positions of shot point and geophone; S is now on the right and G on the left. We still have the same critical angle, α, and the same dip angle, θ. But now instead of the rays coming up to the geophone with an angle α plus θ, they come up with an angle α minus θ. This gives rise to relation 1′. It will be noted that this relation differs slightly from relation 1 by the inclusion of an intercept time, T_0. That an intercept time must be included in this case is evident from the fact that the shot point, S, is no longer at the high-velocity level. We write equation 2′, which again is equivalent to equation 1′. Equation 3′ is similar to equation 3, and can be solved by equation 4′.

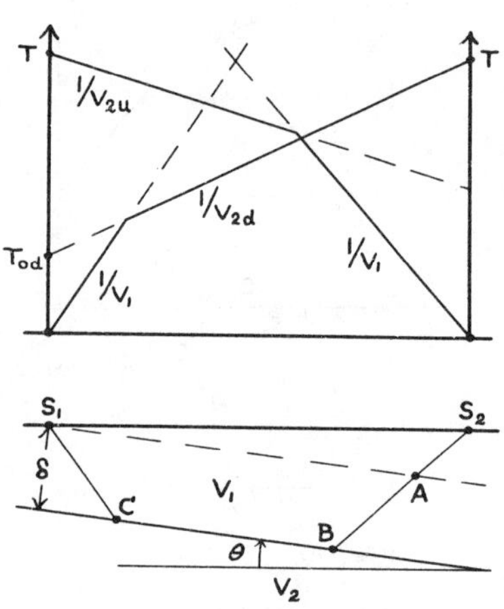

FIG. 12.7. Calculation of depth and dip.

We thus have two relationships, equations 4 and 4′, giving respectively the sum and difference of the critical and dip angles. We use relation 5 with relations 3 and 3′ to obtain relation 6. It is of some interest to note that equation 6 involves the *actual high velocity* V_2 in only one place. This equation can be solved for this velocity, giving relation 7. Relation 7 thus states the true high velocity in terms of the dip angle and the two apparent velocities, one shooting down dip, V_{2d}, and the other shooting up dip, V_{2u}. It may be verified that if θ is close to zero, V_2 is close to the average of V_{2d} and V_{2u}. In many cases this average is sufficiently close to the true value given by equation 7 that the equation does not have to be used.

We may now calculate the dip angle by taking the difference of relations 4 and 4′ and dividing by 2. This gives relation 8. Thus the dip is stated in terms of the measured velocities. We then substitute this dip in relation 7 and we have the complete relationship from which to find velocity V_2. When V_2 and θ have been found, we have all the information we need in order to calculate depths. We now consider the calculation of depth.

At the top of Figure 12.7 is shown the time-distance graph for a reversed refraction profile of the simple type. We may compare the lower part of this figure with the situation in Figure 12.6. It becomes evident that the distance δ in Figure 12.7 can be computed in terms of the intercept time, T_{od}, which is shown on the vertical axis at the left. We have only to use the adjustment slide rule illustrated in Figure 12.5, compute the number of feet per hundredth second, find the number of hundreds in T_{od}, and multiply this number by this adjustment footage to find δ. Thus depth can be calculated in a single step with the adjustment slide rule.

We now consider the calculation when the velocity of the overburden uniformly increases with increasing depth. This is illustrated in Figure 12.8. We again have the apparent velocities, V_{2u} shooting up dip and V_{2d} shooting down dip. We shall suppose, in this case, that V_2 is given with sufficient approximation by relation 2. We can calculate the direction of the upcoming and down-going rays under shot points S_1 and S_2 by means of formulas 1 and 1′ respectively. Also by these formulas angles β_1 and β_2 can be computed; they are then plotted with vertices at S_1 and S_2 respectively. This gives lines for the determination of C_1 and C_2 respectively. In this way we determine the centers of the ray circles on the line of centers, which is the dashed line the distance V_D/a above the datum. We draw in the ray arcs and make an *estimate* of the high-velocity inter-

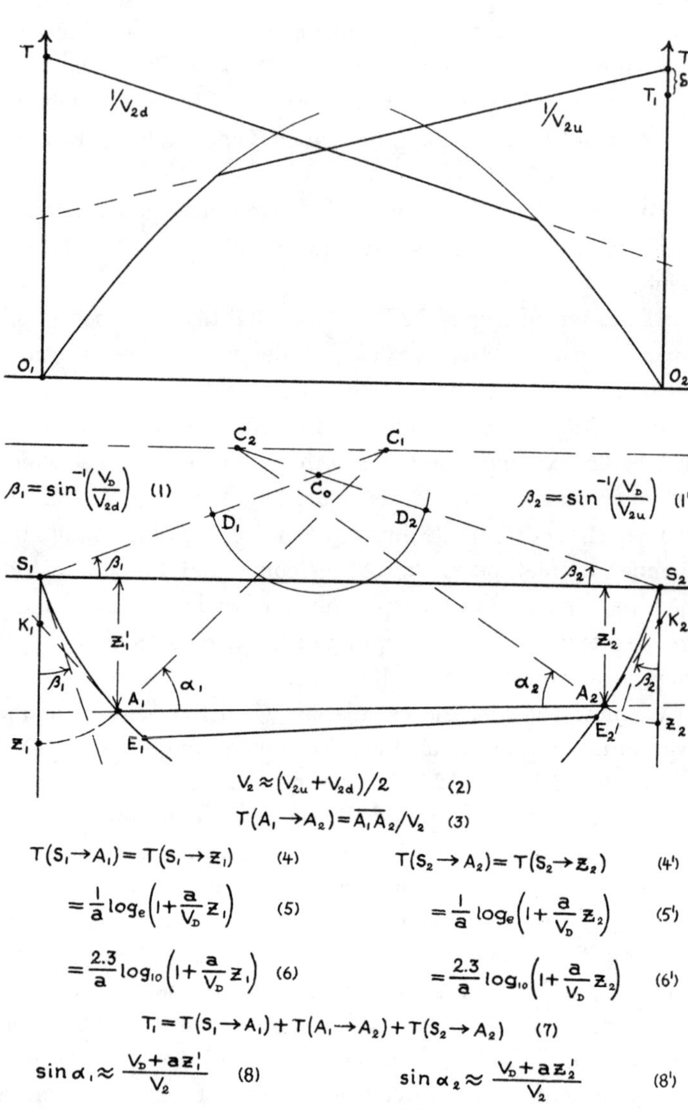

FIG. 12.8. Adjustment of depth and dip for linear overburden.

face; this is A_1, A_2. We calculate the time from A_1 to A_2 as indicated in formula 3. We wish now to calculate the time from S_1 to A_1 and from S_2 to A_2 as indicated respectively in formulas 4 and 4′. As we have seen, this may be done by means of formulas 5 and 5′ or 6 and 6′. Having substituted the numerical values, we add the times and attain a first approximation, T_1. This is the value given in relation 7 and is plotted at the upper right of the graph.

We find that the agreement in this case is not perfect, so we have to make a *second approximation.* We realize that we have estimated the dip angle as well as the depth. In order to determine which way the dip has to be altered we do the two calculations illustrated in relations 8 and 8′. These are shown as approximate relationships. If our first approximation to the dip were correct, these relationships should be correct too. The direction in which they fail to be correct indicates how we must alter the dip. If we compute the dip on the assumption that the velocity does not increase with depth, we find always that the dip has to be increased in the second approximation. We have attempted to indicate this roughly by showing E_1, E_2, with slightly increased dip, as a second approximation. The process by which we determined the time for A_1A_2 must be repeated for E_1, E_2 down through a relation equivalent to equation (7). We also should recompute relations 8 and 8′ for points E_1 and E_2 to check the fact that we have used the correct dip.

It may be mentioned that the dip angle is only roughly determined because the velocity increases with increasing depth. As we have indicated, if the high-velocity section is a plane, it will not give a strictly straight-line time-distance graph. If we wish, we can take some account of this curvature that necessarily exists if one of the time-distance graphs appears perfectly straight, by using relations 8 and 8′. For this we use a fudging procedure that is not particularly difficult, but involves successive approximations. In this case two or three approximations may be required before the correct result is obtained. This difficulty is not too serious, however, unless the dip is rather steep. The reason for this is that refraction data are seldom accurate enough to justify making a sharp distinction between the plane interface and the curved interface.

12.3. UNREVERSED PROFILES

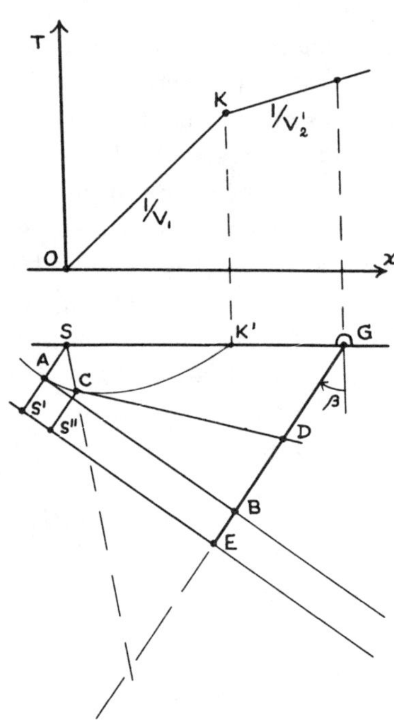

Fig. 12.9. Interpretation of one-way refraction data.

We consider in this section the situation in which a profile is shot in only one direction. In such a case we have no information regarding the dip unless we know the true-velocity value V_2, which we usually do not. Our purpose here is twofold: to show how the data can be used to determine all possible plane interfaces consistent with this particular information, and to show how to arrange a field setup so that a reversal can be computed.

We consider our first aim by reference to Figure 12.9. The time-distance graph in this figure consists of two parts, one showing the low-velocity value V_1 and the other an apparent high velocity V'_2. Since we know nothing about the dip we cannot start with this idea. We do, however, have a certain amount of information from the apparent velocities themselves. For example, we know how to compute angle β between line *GE* and the vertical. This is the line along which the ray comes up to geophone *G*. The sine of angle β is the ratio of V_1 to V'_2. Having computed this sine we can find the angle and draw the down-going reversed ray which has the direction *GE*. Since we do not know the dip we do not know how far to extend this line. On the other hand, we know that if we had an infinitely high second velocity the angle of incidence would have to be zero, which means that the interface would have to be perpendicular to *GE*. We determine the interface on this assumption and it proves to be *AB*. We understand now that the travel time from *S* to *A* plus the travel time from *B* to *G* is equal to the observed time shown on the graph. The interface *AB* represents an extreme situation. Of course it is far beyond anything obtained in practice; as it stands, it is a purely mathematical fiction. Let us construct a line *S′E* parallel to *AB* and the same distance from *AB* as the shot point *S* is. Thus *AS′* equals *AS*. We shall find that this line, *S′E*, represents the *directrix* and the shot point, *S*, represents the *focus* of the parabola through points *A*, *C*, and *K′*. We now consider the construction of this parabola in somewhat greater detail by reference to Figure 12.10.

Figure 12.10 is very similar to Figure 12.9. The line *CF* is drawn perpendicular to *CM*. Angle β is the angle whose sine is V_1/V'_2. The length *NM* is V_1T_1, where T_1 is the observed travel time from the shot point to the geophone at *N*. In this way we determine point *M*. We draw the directrix *CM* through *M*, perpendicular to *MN*. Then we drop a perpendicular from the focus *F*, the shot point, to the directrix and find point *C*. We then draw a line through *CF* as indicated. With an origin of measurement at *C*, we locate a series of points along line *CF* and its dashed extension, and through these points we draw perpendiculars which are shown as light dashed lines. Using the distances from C to these lines as radii and F as center, we draw a series of arcs intersecting these straight lines at a series of points which determine the parabola. The parabola

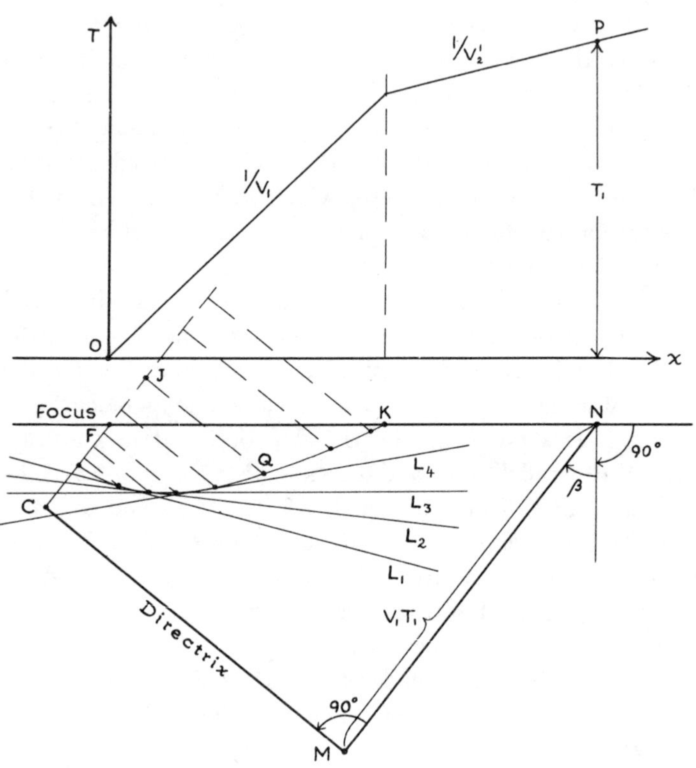

FIG. 12.10. Construction of parabola to which plane interface must be tangent.

through these points is shown in the graph. It will be noted that point Q is on the parabola, as is also K. It is instructive to prove that K must lie under the intersection point of the two straight lines forming the time-distance graph. In addition to the parabola, to which *all possible refracting interfaces must be tangent,* we have drawn four interfaces, L_1, L_2, L_3, and L_4, as samples. It is of course necessary to use different velocities for each sample.

When we have constructed the parabola, we do not know whether we should use one of the lines L_1 through L_4, or some other one, for the refracting interface. All we know is that whatever this interface, if it is a plane corresponding to constant velocity it will necessarily be tangent to the parabola as constructed. We also know that the point of tangency is the point of refraction, so that if we draw a particular tangent line to the parabola we take the point of tangency as the point to which to draw the ray from the focus or shot point, F.

By this construction we utilize to the utmost the rather small amount of data that we have at hand.

We now consider a means of shooting refraction profiles in the field in only one direction, so that the reversal information can be computed. For this purpose refer to Figure 12.11, which shows a time-distance graph and, below it, a refraction profile of several shot points S_1, S_2, and S_3. Suppose first we shoot from shot point S_1. The ray travels downward to the interface and meets it at point A_1. It travels along the interface and is refracted upward to the various surface geophones, giving rise to the high-velocity part of the information on the time-distance graph from shot point S_1. We suppose that the last geophone location from shot point S_1 is G_1. Keeping this position G_1 fixed, we move the shot point up to position S_2 and shoot another refraction profile, thus obtaining times along the line between P_2 and P_3 in the graph. We have already obtained point P_1 shooting from shot point S_1. It is always possible to exchange geophone and shot point locations, and this we do for G_1 and S_1 and S_2. We obtain points Q_1 and Q_2 in this way. Both Q_1 and Q_2 are computed points, but they are computed on the basis of a very sound assumption, namely, that the travel time over the exactly reversed path is the same as the directly measured travel time. Having these two points, Q_1 and Q_2, we may draw a reversed time-distance graph between them; this is indicated by long dashes on the graph. We then shoot from S_2 until we finally reach geophone position G_2, whereupon we move the shot point up to S_3. We shoot another refraction profile, keeping the geophone at G_2, and

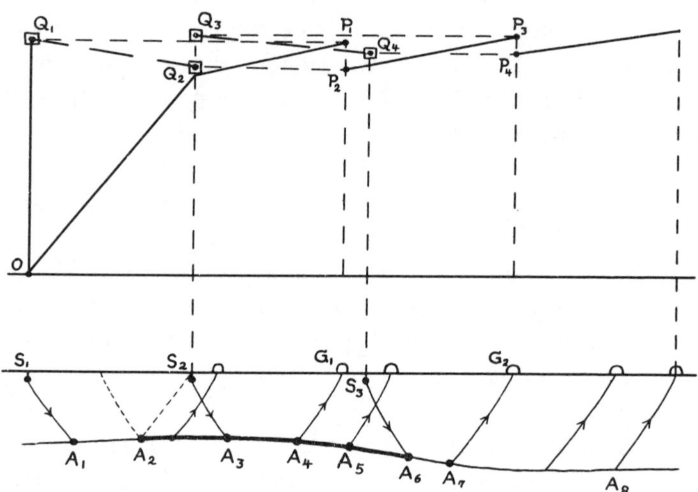

FIG. 12.11. Field arrangement to avoid shooting reversals in refraction profiling.

we obtain the time P_4. We can then proceed with a second reversal by exchanging geophone G_2 with shot locations S_3 and S_2. Thus P_3 corresponds to Q_3 and P_4 corresponds to Q_4. Then we construct another small reversal between Q_3 and Q_4.

We may move along this refraction line in this manner as far as we wish, constructing the reversals as we go along. This process is legitimate, as long as the structure is reasonably simple and without serious faulting.

Now let us see what kind of information we have obtained. Our travel path extends in the diagram all the way from A_1

to A_8. It should be observed, however, that the *reversed travel path* is much more limited. It may be seen directly that this path extends only from A_2 to A_6 (indicated by the heavy line), because the remainder of the subsurface is shot in only one way and consequently does not give a reversal. It is important to note, by reference to Figure 12.11, that a *continuous* set of reversals can be obtained on the interior of a refraction profile line. But this reversal is of limited extent and a good deal more line shooting has to be done on the surface than the amount of reversed subsurface coverage we can obtain. We could of course have stopped shooting at P_4 or G_2. Then our coverage would have been from A_1 to A_7, with complete reversal from A_2 to A_6. This would have saved a little shooting.

The above procedure gives somewhat better detail than is usually obtained by less careful planning, even when reversals are shot. The reason is that the reversals were usually shot in such a way that they gave reversals only on the surface, there was little reversal in the subsurface interface. Thus in much refraction shooting the data are of such a character that the value of V_2 cannot be properly computed. It is therefore necessary in such cases to resort to a procedure like that illustrated in Figures 12.9 and 12.10. Such a procedure proves sufficient in some cases. But sometimes true subsurface reversals are needed in order to determine the required subsurface structure. The method illustrated in Figure 12.11 is the subject of patent No. 2267868.

12.4 FIRST BREAKS vs. SECONDARIES

The refraction computing described above was limited to so-called *first breaks*, that is, the first energy to reach the geophones. This limitation is not necessary, for it is possible to pick secondary refraction breaks on records in many important areas. It should be remembered, however, that secondaries cannot be picked at the very beginning. They have to be picked much the same way as reflections are picked. Thus we have to pick a first strong trough or peak in order to follow along the refracted pulse with any degree of regularity. Since we usually cannot pick the first beginning of the secondaries, the process of reversing shot point and geophone is impossible in many cases. Although the reversal is strictly true for first breaks, it is not necessarily true for first troughs or first peaks. The reason is that the position of the first trough or the first peak depends to a great extent upon the *character* of the refracted pulse. This in turn depends on a great many factors; it is only occasionally the same when we shoot in two different directions. It is therefore necessary, when using secondary refracted pulses behind the first breaks, to shoot the reversals completely. After this shooting, we must be ready to admit that the reversed times may not match. All that we can obtain is a pair of apparent velocities and an *approximate* travel time. With these velocities and this travel time, we can use the methods outlined in Section 12.2 to determine the depth and dip. It is unfortunately true that the variation in character of the refracted pulse may be affected by structure so seriously as to make an interpretation of secondaries extremely inaccurate. The person who has to use secondaries for such an interpretation is working under very difficult circumstances and may not be able to produce a reasonably true picture of the geological structure. Unless circumstances are exceptional, an operation of this sort using secondaries is considered a last resort, to be used when all other types of prospecting have failed.

12.5. STRUCTURE

The geological structure of the refracting interface has a serious influence on the character of the refracted pulse and a gross influence upon its amplitude. This can be easily illustrated if we try to picture the physical process involved in the formation of the refracted pulse, as is done in Figure 12.12. The shot point S is shown in the upper drawing. The position

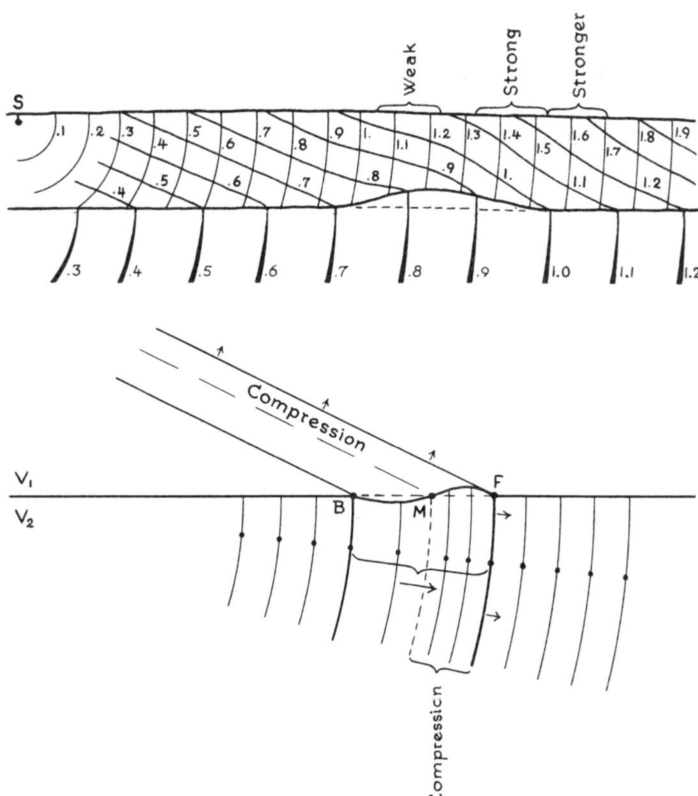

FIG. 12.12. Generation of a refracted pulse.

of the outgoing pulse is shown at various times one-tenth of a second apart. It will be observed that at 0.3 second the critical point is reached on the refractor. At 0.4 second the direct wave is separated from the refracted wave and we have started to generate the wave which is refracted upward and gives rise to the data recorded at the surface. At 0.5 second this separation between the direct wave and the wave in the lower medium is wider. The generated refracted wave is the slanting pulse front that is shown.

Without proceeding further with the above process, we may perhaps understand it a little better by reference to the enlarged cross section in the lower part of the figure. Here the interface separates the two media, of velocity V_1 and V_2. The pulse that travels in the lower medium may be considered to be well separated from the pulse that travels directly from the shot in the upper medium. The front of the pulse at the interface is the point F and the back of it is the point B. A radial line is drawn perpendicular to the front and back of the pulse, and a series of points is laid out on this line. These points represent particles. Behind and in front of the pulse these particles are equally spaced. We suppose that the particles in the pulse were similarly spaced equally before it arrived, but that its presence has displaced them from their equilibrium positions, so that the front particles are compressed closer together and the rear ones are farther apart. Thus the front of the pulse consists of a compression and the rear of it consists of a rarefaction. The point at which the pulse is not under either compression or tension is the point M. Now if such a pulse is compressed between M and F, it will not only displace the particles forward along the radial line, but will also push them up toward the upper medium as indicated by the upward bulge between M and F. The rarefaction between M and B will produce a downward bulge into the lower medium. The upward bulge between M and F will give rise to a compression in the upper medium. This compression is complicated by the fact that there is also a crowding together of particles horizontally along the interface. The important thing is the compression in the upper medium near the interface between M and F and the rarefaction between M and B

in this medium. Thus we see that a source of compression and rarefaction travels with the wave along the interface and that this source gives rise to the refracted wave which is shown in the upper medium with a plane intersection corresponding to a straight-line pulse front and pulse rear.

The above description is greatly oversimplified, for many details have been left out. But it does give a gross picture of the generation of the refracted wave used in refraction prospecting, and gross as it may be, it also indicates the variations to be expected in the intensity of this refracted wave. For example, if energy is transmitted into the upper medium it must be withdrawn from the lower medium, and this withdrawal must give rise to a thinning of the pulse as it travels in the lower medium along the pulse front up toward the interface. An attempt to indicate this thinning has been made by making the line representing the pulse front thinner as it approaches the interface from below.

We can also see another property of the refracted wave when the structure is like that in the upper part of the figure. No effects of this structure upon the refracted wave are visible until 0.7 of a second after leaving the shot, when the traveling wave reaches the edge of the structure. At 0.8 of a second we see that the structure has formed a sort of shadow indicated by the dotted line that extends the interface straight across. It is evidently necessary that the energy be diffracted into this shadow zone and then sent across the interface into the upper medium. This may be expected to introduce an additional weakening corresponding to the surface interval indicated above. However, after going nearly across the structure, the diffraction into the shadow zone is likely to be sufficiently strong so that as the diffracted pulse begins to strike the interface at less of a glancing angle it starts generating a stronger refracted wave. This stronger wave is followed by one that is still stronger. The increased strength is due to the convergence of rays toward the surface, resulting from the fact that the interface is concave upward and thus gives a semifocusing action to the rays. Then a certain distance beyond the structure the refracted energy resumes its normal variation of intensity.

We thus see, from Figure 12.12, that we can to some extent diagnose a high structure in terms of the amplitudes of the refracted pulses alone. The first part will show a weak refracted pulse of low amplitude, followed by a strong pulse, this in turn followed by an even stronger pulse. Of course there may be no stronger pulse; and this will be the case when there is no concave upward section on the other side of the structure away from the shot point S.

We thus have an additional tool for interpreting refraction results that is not based entirely on timing. This tool has often been very useful to the writer. He recalls one case where the presence of a structure so influenced the amplitude of the refracted wave as to make it disappear altogether, so that no matter how large a shot was fired no energy could be detected in this region although it could be easily detected just beyond this region. It is fairly evident that if the line is reversed in such a case the pattern is also reversed, and the surface position at which the weak and the strong and stronger refractions are picked up is usually also shifted. Thus verification of the presence of structure can be obtained by reversing the profile.

12.6. DETAILING

It often happens that the time-distance data, as plotted, do not line up and form a perfectly straight line. Sample situations are shown in Figures 12.13 and 12.14. Figure 12.13 represents a small high and 12.14 represents a small fault. The correction required in such an instance can be computed simply. Since the corrections are similar, we present this material together.

The situation shown at the top of Figure 12.13 need not be described in detail, for it is a type of situation that occurs fairly often. Point H is the point that is off the line. We want to compute how much of a rise in interface is required to fit

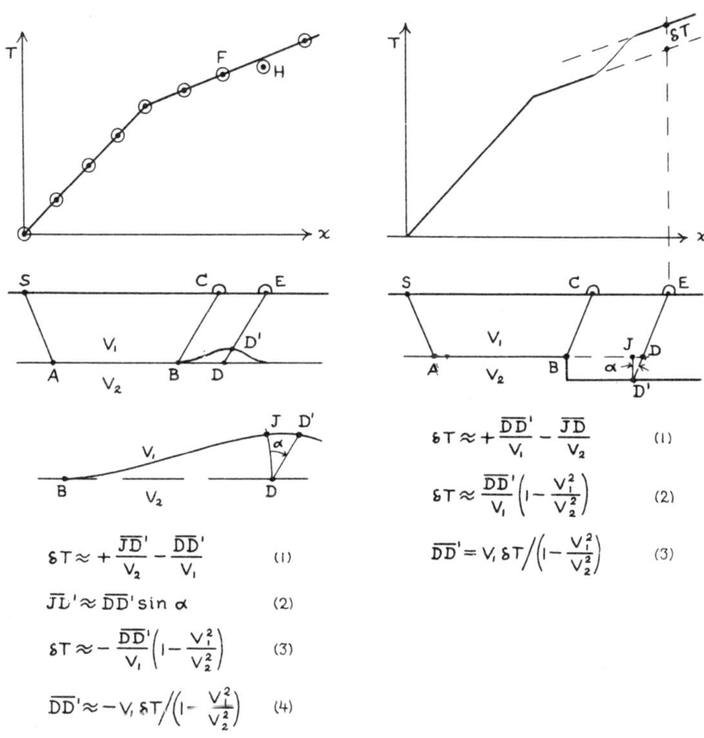

FIG. 12.13. Adjusting an interpretation for a small high.

FIG. 12.14. Calculating the throw of the fault.

the information observed; that is, we wish to compute the distance DD'. This portion BDD' is enlarged in the bottom drawing. We find that the difference between H and the point on the line is due to the difference between the travel time over path JD' at velocity V_2 and the travel time over distance DD' at velocity V_1. This relationship is given in formula 1.

Formula 2 gives the approximate relationship between JD' and DD'; this may be substituted in formula 1 to give formula 3. If formula 3 is solved for DD', formula 4 results. Thus, knowing V_1, V_2, and δT, we can compute DD' with formula 4. We suppose that δT is negative if we wish to make DD' positive. We can then measure DD' and sketch in the little high required to displace the point to its observed position, H, off the line.

We now discuss the small fault shown in Figure 12.14. Here, due to the fault, there is a displacement of δT in the time-distance graph for the high-velocity material. It is easily seen that δT can be computed on the same basis as before, and, as before, distance DD' is given by formula 3, which is very similar to formula 4 above, except that here we assume that δT is positive, representing an increase in time corresponding to the positive value of DD'. Thus the throw of a fault may be computed from the shift of time, δT, when the velocities V_1 and V_2 are known.

With the methods just described in connection with Figures 12.13 and 12.14, we can give a more detailed interpretation of a refracting interface. In this way we can fit every individual point if we wish.

Such careful fitting is not necessarily desirable. The reason is that some of the variation of the points from a straight-line graph may be due to variations of weathering. Weathering variations are usually disregarded in refraction shooting, the refraction work being too inaccurate to justify the added expense. Some may think that this argument is slightly circular, because the added expense would make the refraction work more accurate. This is indeed true, but only to a limited extent, for the additional accuracy obtained by measuring the weathering velocity and making a better correction to a level datum would not be sufficient to compensate for a more serious kind of inaccuracy arising from the general character of the refraction method.

12.7. ACCURACY

The accuracy of refraction line profiling depends to a great extent upon two simple factors. When the contrast between velocities V_1 and V_2 is small, the amount of correction per 0.01-second difference in travel time becomes very large. Thus, in such a case, a small variation in structure will not give rise to a measurable time difference and hence the accuracy must be very low. Another factor that reduces the accuracy of the work is increase in depth. This is so partly because of the tendency to make the velocity contrast smaller and smaller, and partly because of the introduction of many uncontrolled variables in the intermediate section.

A depth may be computed on the basis of refraction shooting and be much too deep. For example, in part of west Texas there is a salt section that is rather thick, roughly between a top of 1700 feet and a base of 2700 feet. Below this, at a depth of around 4000 feet over part of the area, is the top of the Permian lime. Refraction methods were used to determine the top of the Permian lime in earlier refraction work. If the salt is assumed to extend down to the top of the lime, the refraction depth for the lime will be far too large. The reason is that below the base of the salt and above the top of the lime there is a layer whose velocity is lower than that of the salt, and this cannot be detected by refraction methods. Failure to detect this low-velocity deeper layer makes the top of the lime appear much deeper than it is, because it increases the travel time observed for the refraction at the lime.

The error may take the reverse direction. Again in certain parts of west Texas, if attempts are made to measure the depth of the top of the salt, the depth determined by refraction methods is too shallow. The reason apparently is that over part of the area there is an overlayer of anhydrite whose velocity is not quite as high as that of the salt. This overlayer, however, is too thin to be detected by a refraction line. It therefore does not appear in the refraction data and the computation proceeds as if it did not exist. The presence of this high-velocity section, which is neglected, shortens the travel time and makes the top of the salt appear shallower than it really is.

These two types of effects, one making the depth too great and the other making it too small, are much more likely to play their part in a *deep* cross section. In this case, not all of the detailed velocity changes throughout the whole section can be detected. Under such conditions the depth error may be very large.

Since the error in depth may be large for a deep refraction horizon, and since the cost of the exploration increases very rapidly with increasing depth because of the increase in the size of charge required, we have a situation in which the information decreases while the cost increases. It is therefore necessary to cut off operations at a certain level in order to have an economical exploration program.

It is sometimes considered desirable in certain areas to make a detailed study of the randomly scattered errors in the data. This has been done in some detail by the author and has also been discussed in a published paper.*

A least-squares discussion of the errors is very useful in this connection. The errors could be discussed in a substantially equivalent alternative way as follows. If several hundred apparent velocity determinations are made for a given profile, they can be grouped in a frequency diagram which will show an approximately normal distribution of variation of both velocity and intercept. From the normal distribution of velocity and intercept time can be obtained a measure of the range of variation or of the standard error corresponding to the normal distribution. The process just outlined is laborious. The standard least-squares procedure involves less labor. For this

* E. C. Bullard and others, *Philosophical Transactions of the Royal Society of London*, 239:24-94 (1940).

procedure, the reader is referred to the paper by Bullard and colleagues.*

The use of the method of least squares in geophysical problems has sometimes been objected to on the grounds that the least-squares best fit is not really any better than any other fit. This is true, but the real point is that the method of least squares has been incompletely and improperly used when only the best fit has been obtained. A major purpose of this method is to obtain the *range of variation* about the best fit. This latter procedure is often not carried out in calculations in which the method of least squares is used in applied geophysics. There are, of course, some notable exceptions, but for the most part the method has been used only for finding the best fit, which, if given without the range of variation, is often misunderstood as a claim of great accuracy. The procedure involving finding the range of variation cannot be easily done directly because it requires the determination of the slope and intercept a great many times by different individuals and this entails a great deal of work. We might better use the method of least squares, which will arrive at essentially the same result, to find the range of variation of slope and intercept for the given set of data.

This method can also be used to determine the velocity required for the elevation correction of refraction data. This is done by introducing the velocity of an upper layer as an unknown that is to be calculated with the intercept time and the apparent velocity. Thus, we have three unknowns altogether, each to be determined by the method of least squares; this gives a least-squares best fit and a range of variation. The author has used this procedure in a number of cases and has found that sometimes it is satisfactory. When it has been unsatisfactory, this has been due to the fact that a layer of

* *Ibid.* A more detailed and more fundamental discussion will be found in H. Jeffreys, *The Theory of Probability*, pp. 121-131 (1939).

uniform velocity with a level base must be assumed to account for all the elevation variations observed. Since this is not necessarily the case, this method of approach is not always valid. When, however, the base of the weathering is the water table which is almost level, such a base may be used with considerable confidence, especially if the weathering velocity does not vary too much.

12.8. SEVERAL LAYERS OF CONSTANT VELOCITY

A major and very complicated part of any discussion of refraction line prospecting usually concerns the case when there are several layers of constant velocity. The number of layers that can be handled is very small. The total change in velocity in the whole section is, at most, near 15,000 ft./sec. Assuming at least three or four points for each interface on the time-distance plot of data, the maximum possible number of layers, in practice, is around three or four. If statistical tests of significance are applied to the data, it is quickly found that the selection of short segments is difficult and no more valid than a smooth fit such as is obtained by a linear increase of velocity with depth. Furthermore, the correlation of interfaces on neighboring lines becomes impossible when there are too many layers in the interpretation. Although this does not show the invalidity of the multiple-layer interpretation, it serves to indicate the hopeless difficulties it presents in most cases.

If a multi-layer interpretation must be made, it may be done straightforwardly by the process outlined below.

1. The first layer is calculated by procedures discussed in foregoing sections. The velocities V_1 and V_2 of the first and second layers will thus be found. Also the location of the first buried interface will be determined; this is plotted graphically using equal vertical and horizontal scales.

2. The plotted data give apparent velocities V_{3a}, V_{3b}, V_{3c}, etc., from the third layer. Using these and V_1, angles β_{3a}, β_{3b},

β_{3c}, etc., can be found from sin $\beta_{3a} = V_1/V_{3a}$, etc. Here β_{3a} is the angle between the up-coming ray in the upper medium and the vertical. Thus all the up-coming rays can be drawn down to the first buried interface. On this basis the down-going ray from the shot to the first interface can be computed, or at least estimated. This ray can be used to construct a new shot point at the first interface for use in finding the second interface. This is found by subtracting from the observed travel time the time from the actual shot down to the new constructed shot location on the first buried interface. Constructed geophone locations are likewise found on the backward retraced paths of the up-coming rays where they meet the first buried interface. The corresponding time reduction is evident. Thus we scrape off the first layer and have new reduced times and new reduced distances. We plot these reduced times against the reduced distances and obtain a time-distance graph very similar to the one we would have obtained, had the shot and the geophones been at the first buried refraction interface.

3. With the reduced time-distance graph we compute the second layer by methods discussed in the preceding sections. This gives us V_3 and the location of the second buried interface.

4. We next scrape off the second layer and get a new reduced time-distance graph in which the shot and geophones are located on the second interface.

5. We then compute the third layer, etc.

With the above procedure, every step of which is straightforward, any multilayer problem can be solved that can be legitimately solved at all. No formulas are required, except those based on the single-layer case. Quite aside from the greater ease of making the interpretation, mistakes are less likely to be unnoticed, because of the closer contact required of the interpreter by the graphic method, in which each layer is plotted to the selected scale of the cross section.

In summary, line refraction profiling is easily used in the field in order to find faults, and in some cases, to find structure. The shallower the structure the more accurately can the interpretation be made. Furthermore, the greater the velocity contrast the greater the degree of accuracy that is possible. Usually the refraction method is used as the last resort when reflection methods have failed. In a few cases the refraction method can give better results, results with more reliability, than can be obtained by reflection shooting. This is especially true when we are looking for faults in a relatively shallow basement.

We have now covered most of the so-called practical aspects of seismic prospecting for oil. But this practical work would never have been done at all, had not several imaginative people pictured some of the possibilities. Furthermore, the solution of the more difficult interpretative problems of the present day requires people who can apply to their problems reasonably sound views of how the part of the world they are investigating should behave. We now proceed to investigate the *background picture* the competent seismologist must have for his work.

12.9. COMMENTS TO THE SECOND EDITION

Usually, prospecting "refraction" lines are shot when reflection surveys are in trouble. Then reflection crews and reflection instruments are likely to be used. A transfer to a new routine has to be established—a very risky change. Not only is efficiency likely to suffer but also "refraction" signals may not be obtained. For long offsets, very good low-frequency uniform sensitivity is needed and is not likely to be obtainable through the nondigital parts of reflection receivers. Extending geophone (and geophone environment) capabilities far beyond their normal frequency range may lead to low-frequency amplitude and to phase characteristics that vary too much from receiver to receiver to give correlatable "refraction" signals.

In such a case, try filtering out some of the low-frequency signal and see if this helps the correlation. If this filtering is helpful, probably you are not looking at a "refracted" signal at all but at a long offset reflection from a deep reflector, short of the critical distance for that reflector to produce a "refraction."

The above remarks do not apply as seriously for very shallow "refractions."

Part V

THE PHYSICAL PROCESSES INVOLVED

Part V considers in some detail the physical background of the processes involved. Our purpose is not merely to acquaint the reader with physical principles, which he could probably get better elsewhere, but to gather together the ideas that help the seismologist and geophysicist in understanding the practical aspects of their work.

It is a matter of considerable practical importance to consider the physical picture the seismologist carries to his work. This picture is his mental background. It contains his ideas regarding the way his world should behave.

The reason this picture is of so much importance is that it influences the seismologist's *selective action* when he is making interpretations. It is this influence upon his selective action that should be kept as clear of errors as possible.

One might suppose that the best way to do this is to omit absolutely everything that is not completely founded on fact. To a certain extent this is true and may well be applied to the case at hand. Sometimes this attitude is carried so far that the seismologist is supposed to be better off without the prejudices of the physicist. The latter is supposed to have an altogether different attitude toward his science from that required of a seismologist. This is to some extent true, for in many of its branches physics has become a very mature science. However, it is certainly a mistake to call it an exact science. On the other hand, exploration seismology is largely a matter of getting as much information as possible from rather scanty data. Often the investigator must be satisfied with making good judicious guesses. He cannot seek the extreme accuracy that was sought, say, by the classical physicist.

But there is a body of knowledge in classical physics which forms the background of seismic interpretation and this body of knowledge must be clearly understood by any seismologist who pretends to carry his art and his science beyond the strictly routine grooves into which much of it has fallen.

The questions involved here are of no small economic importance. The individuals who have almost no background of mechanics and physics tend to settle into a single routine which may be quite suitable for one part of the world, and then they adhere to this routine under all possible circumstances. The fact that so many geophysicists came from the Gulf Coast, where dips are very low, has had a decided influence on the character of exploration seismology all over the world. This has not always been fortunate. Gulf Coast methods are not as widely applicable as they have been applied. Thus millions of dollars have been spent to secure seismic information which, in large part, has been secured by methods developed in the Gulf Coast states and applicable mainly to very small dips. It has sometimes been suspected that the methods were not adequate and could be improved. But any consideration of improvement leads to thoughts about reworking great masses of data. This often stops any idea of improvement because the reworking of data is very costly; hence in many cases the data continue to be inadequately presented and interpreted, just in order to keep the system uniform. The writer feels that this type of procedure must reflect, to a certain extent, a misunderstanding of the great importance of adequate use of information. No matter how much the information costs, it isn't worth very much unless it can be properly used, and it cannot be properly used if understanding it falls too far short of what is possible.

It is therefore our intention in the next three chapters to discuss, in some detail, some of the physical processes involved in seismic exploration for oil.

CHAPTER 13

Basic Physics Assumed

This chapter, as with the other chapters of Part V, is written primarily for seismologists interested in the interpretation of seismic results. The present chapter outlines the main ideas used in Chapter 14.

13.1. CLASSICAL MECHANICS

By classical mechanics we mean the Newtonian mechanics involving principally and fundamentally the three Newtonian relations between *force, mass, acceleration, time, and momentum,* both for the linear or straight-line type of motion and for rotary motion. For rotary motion, *moment of inertia* must be used in place of mass, *angular acceleration* in place of linear acceleration, and *moment of force,* or *torque,* in place of force.

In addition to the fundamental Newtonian relations, usually called Newton's laws of motion, the reader is expected to be familiar with some of the consequences of these laws, namely, the principle of *conservation of energy* and the principle of *conversation of momentum.* These laws are valid in all our considerations. There are very few facts of a general nature in this world that have such a high degree of reliability as the principles of conservation of energy and of momentum. Any interpretation which leads to the conclusion that energy is not conserved or that momentum is not conserved is erroneous in some respect and should always be avoided.

13.2. MOTION OF AN APPROXIMATELY CONTINUOUS MASS DISTRIBUTION

This is the main section of this chapter; it is broken up into subsections below. We may remark, in passing, that we assume that the distributions with which we shall deal are continuous distributions of mass. We know at the outset that this assumption is not precisely correct because all the rocks with which we are concerned have a certain granular aspect. It will appear later, however, that the seismic view of our material is necessarily very blurred, since fine granularity, or even the coarse granularity of a conglomerate, need concern us only as a cause of annoyance. The granularity has no serious influence on any of the deeper observations. Its only effect, though this is sometimes serious, is its influence on the source or the shot neighborhood. Sad experience has proved that shooting among a lot of boulders is not likely to yield good results. If, on the other hand, the conglomerate, however coarse, is buried a few hundred feet, it appears to our blurred seismic view as though it were practically a continuous distribution of matter.

13.2.1. Deformation and Strain

We now consider the purely geometrical problem of describing a deformation of matter. Figure 13.1 shows, at the upper left, a region, R, bounded by a surface, S_R, inside of which is a point, C, that is the center of a sphere, S. This is an *initial shape*, as indicated in the figure. Now we introduce a deformation of this shape. The *deformed shape* is shown at the upper right of the figure with the initial shape dashed in the background. The original boundary, S_R, is deformed into a new boundary, and point C has been displaced to point K. Point

P_1, originally on the spherical surface, has been moved to P_1' and the diametrically opposed point, P_2, has been moved to P_2'. In an attempt to show how the sphere has been deformed, the deformed surface has been drawn passing through P_1' and P_2'. It will be observed that the mass within the sphere has been slightly rotated in a counterclockwise direction and

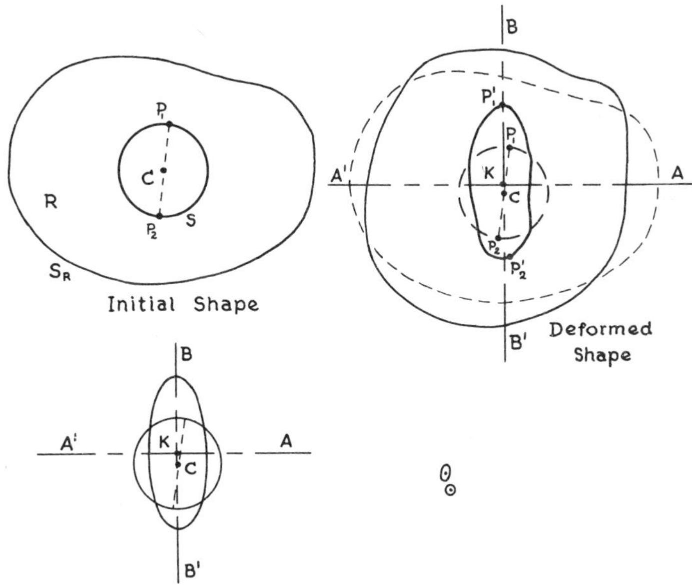

FIG. 13.1. Descriptions of deformations.

considerably elongated in the direction BB' and shortened in the direction AA'. An *approximate description* of this situation is shown at the lower left of Figure 13.1, where the sphere is compared with an ellipse. Of course, in the three-dimensional case the sphere would have to be compared with an ellipsoid but we are considering only the cross section for simplicity.

The deformation of a sphere which is only approximately

represented by the translated and rotated ellipsoid may or may not yield a sufficiently accurate description of the true deformation. A more accurate one involves a great deal of detail.

For our purposes, the ellipsoidal description of deformation is quite adequate, for we are primarily interested in *very small deformations* of very small spheres. In the lower right portion of the figure is shown a sphere and its deformed ellipsoid. Here, however, it should be pointed out that, although the sphere is small, the deformation is not small because the difference in the axes of the ellipsoid is large. Our interest is confined to the case in which a sphere is deformed into an ellipsoid that is almost a sphere. Nevertheless, we must illustrate the principles involved by using large deformations, because the small deformations, the ones that really interest us, are too small to be seen.

With this understanding, we proceed to further consideration of deformations. The upper portion of Figure 13.2 shows two *strains*. That on the left is a *pure contraction*; that on the right is a *lengthening* and *shortening*. The middle drawings show two *rigid displacements*, the one on the left being a *rotation*, the one on the right a *translation*. *Rigid displacements are not strains*. The three lower drawings show what might be called a *shear* strain. In this strain, the side AC is rotated through the angle α, so that the square cross section $ABDC$ is deformed to the parallelogram, $ABFE$.

Let us consider how we can synthesize the shear strain shown at the lower left of the figure. The center drawing shows the original figure $ABDC$. We then introduce a strain which shortens CB to $C'B'$ and lengthens AD to $A'D'$. This is followed by a translation of A' to A and D' to D''. To complete the procedure we have to refer to the drawing at the right, which shows a rotation through an angle, $\alpha/2$. We thus see that the original shear strain shown in the lower left drawing can be synthesized by a lengthening, a shortening, a translation, and a rotation, as indicated.

We now consider a more precise description of strain. For this purpose refer to Figure 13.3, which shows the cross section of a cube $ABDC$ prior to deformation. Suppose that after de-

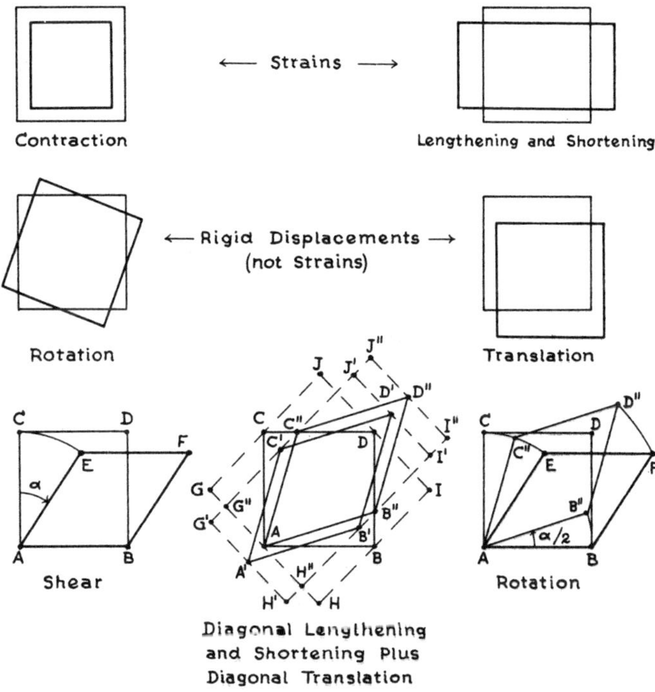

FIG. 13.2. Strains.

formation this cube becomes $A_1B_1D_1C_1$, as indicated, and that the deformation does not change the center point. Then point G is moved to G_1 and point F to F_1. Suppose that G has the coördinate x, and that F has the coördinate $x+dx$. These are the initial coördinates of points G and F. After the deformation, G_1 has the coördinate x_1, which is $x+u$, where u is the displacement in the x direction. F_1 has the coördinate x_1+dx_1, which is $x+dx+u+du$, as indicated diagrammatically. We then form the ratio of the difference between F_1G_1 and FG to FG as indicated in relation 3. This ratio we define as the

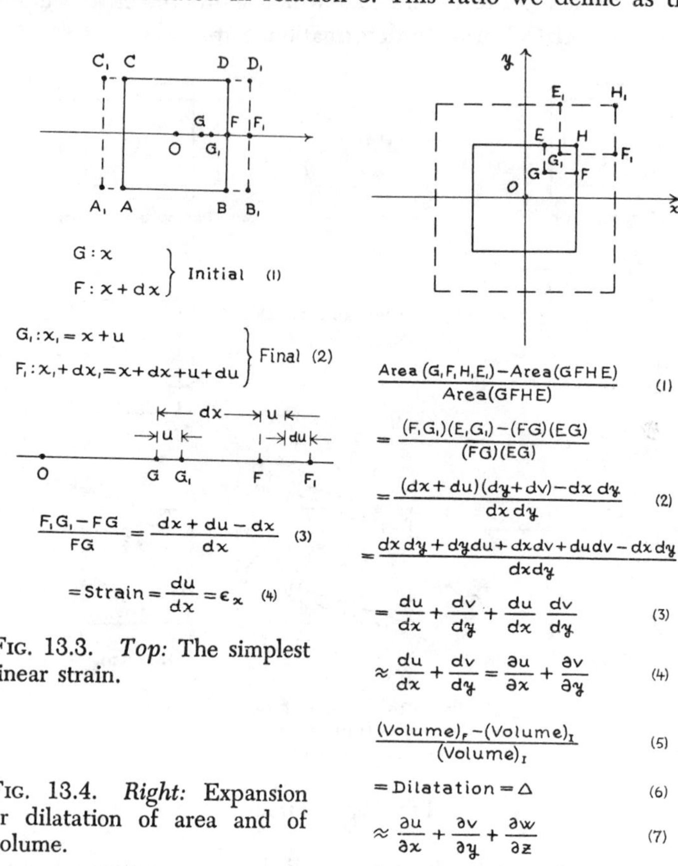

FIG. 13.3. *Top:* The simplest linear strain.

FIG. 13.4. *Right:* Expansion or dilatation of area and of volume.

strain. We readily see that the strain in this case is the rate of change of displacement with respect to distance.

The next most complicated situation is illustrated in Figure 13.4, where the cube is deformed into a rectangular parallelepipedon so that the little portion, $EGFH$, is deformed into

$E_1G_1F_1H_1$. The *strain of the area* is represented by the ratio in the first expression. A simple calculation based on a straightforward extension of the method of Figure 13.3 leads to expression 2, which may be rewritten as expression 3. Since in all our work we shall suppose that the strains are always very small, we can neglect the product of two of them in comparison with their sum, which we do in going from expression 3 to expression 4. It will be observed that in equation 4 we pass from ordinary derivatives to partial derivatives. This transition may or may not be made. It need not be made when the displacement, u, depends only upon the x direction, and the displacement, v, depends only upon the y coördinate. If, on the other hand, displacement u depends partly upon the y coördinate, we wish to indicate that, in calculating this part of the derivative, we consider only the dependence upon the x coördinate. Similar remarks apply to the other partial derivative.

We shall now consider *volume strain*. For this purpose refer to the last three expressions in Figure 13.4. The volume strain is defined by the ratio in expression 5. This is called the *dilatation*. It is positive if the final volume is greater than the initial volume. The dilatation is usually expressed by Δ, as indicated in the sixth relation. The approximate form of the dilatation is merely an extension of relation 4 and is shown as relation 7. We shall use this dilatation relationship a great deal in what follows because the principal waves in seismic prospecting are waves of dilatation.

We now wish to describe more precisely *shear strain* and *rotation*. For this purpose refer to Figure 13.5. Remembering that all the quantities are much smaller than illustrated (for instance, both the angles α and r are far smaller than those shown), we find that the angle $\alpha/2+r$ is approximately the rate of change of the x displacement with respect to y, as indicated in relation 1. Similarly the angle between the edge and the x axis is given by relation 2. If we add these two equations we get for the angle α, approximately the *sum* of $\partial u/\partial y$ and $\partial v/\partial x$, as shown in relation 3. If, on the other hand, we subtract the second from the first relation, we get an expression for the angle of rotation, r, in terms of the *difference* of the derivatives, as shown in relation 4. These relationships, which are shown as approximate, become closer and closer as the

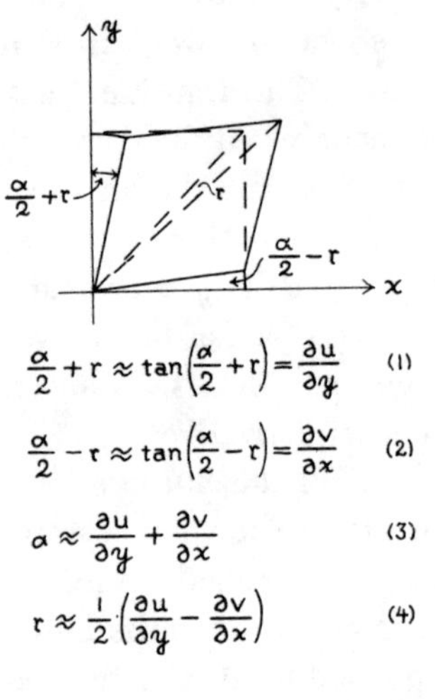

Fig. 13.5. Rotation and shear strain.

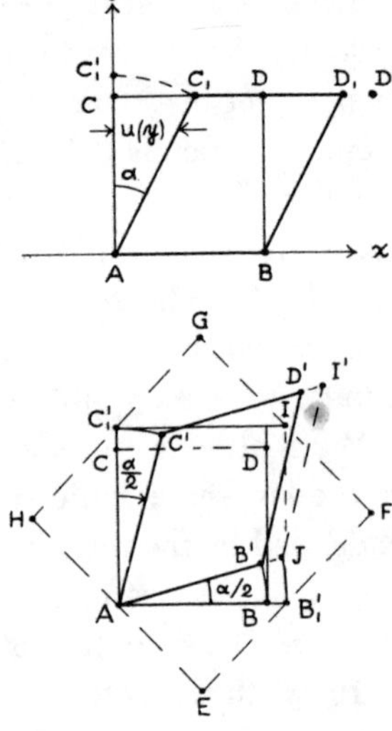

Fig. 13.6. The rotation involved in transverse waves.

strains and rotations grow smaller and smaller. Accordingly we shall regard them henceforth as exact relationships and not indicate that they are only approximate.

There is a particular form of strain that plays an important part in connection with transverse waves; it is illustrated in Figure 13.6. In this case, the original cube *ABDC* is deformed into a parallelogram ABD_1C_1 that has the same cross-sectional area as the original cube. The measure of shear strain is the angle α. Using the process described in Figure 13.2, we can generate this shear strain by means of simple lengthenings and shortenings, together with rotation, as illustrated in the lower part of Figure 13.6. This particular strain is important because it represents the type involved in transverse waves that have no dilatation.

13.2.2. Stress

We now consider the way forces may be described in a continuous mass distribution such as a solid or a fluid. The solid may be a plastic or elastic solid. Nothing is said about its detailed properties. At the upper left of Figure 13.7 is shown a three-dimensional region, R, with a boundary, B. On the boundary surface is a point, P; also shown is a little patch of the boundary, with area, A. We may suppose that the *total force* exerted on the area A by some agency is given by the vector, $\overline{F}$, issuing from point P. At the upper right of the figure we see that the main change is that there is a smaller element of area, ΔA. The corresponding total force is a smaller element of force, $\Delta\overline{F}$. (A bar over a letter indicates a *vector* quantity.) We now consider the definition shown in the figure; it refers to the drawing above it. The *stress at P across the surface B* is defined to mean the *limit of the ratio of the element of force to the element of area as the element of area approaches zero*. Thus stress is a force per unit area.

A familiar example of a stress is a *pressure,* which again is a force per unit area. We shall always understand stress to be the force exerted *on* the body, and pressure to be the pressure exerted *on* the body. Note that we have drawn our arrow outward as if we were applying a tension. We shall find that consistency requires that we always assign ordinary hydrostatic pressure a negative sign.

At the lower left of Figure 13.7 is shown the stress $\overline{S}(P,B)$. The stress is thus a vector quantity; its direction is parallel to the force vector and it depends upon the point P and the surface B.

It will be noted that the stress can be broken down into two components, a *normal component* in a direction perpendicular

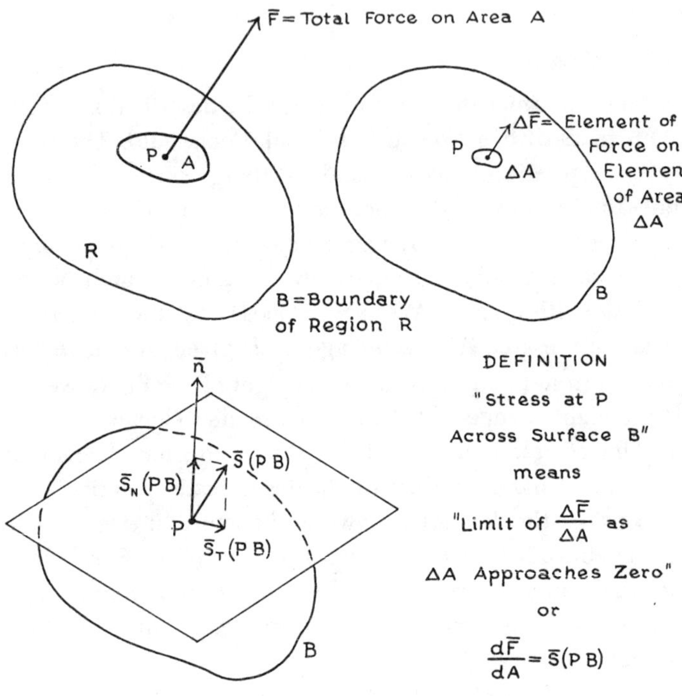

Fig. 13.7. Stress on the surface of a solid.

to the surface B at point P, and a *tangential component* lying within the tangent plane.

We have discussed in Figure 13.7 the most fundamental aspects of the application of force to a bounded continuous distribution of mass. No ambiguity is involved, because the boundary surface is available for the application of force. However, we need to consider the *internal* forces as well. Figure 13.8 shows an internal point, P, and an internal surface. The $\bar{n}$ is the normal to this internal surface at P. We may suppose that the stress at the right of P is the stress imposed by the material at the right on the material at the left. The equal and opposite stress, or reactive stress, $\bar{S}_R$, is shown on the left.

We can describe not only the vector of stress at internal points but also the components of this vector. Consider, for

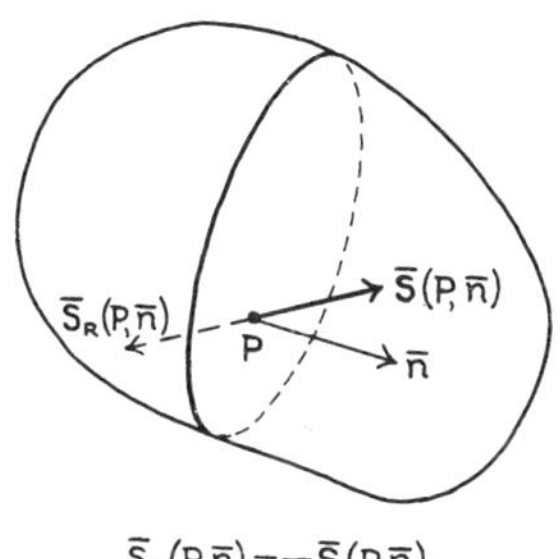

FIG. 13.8. *Top:* Internal stress in a solid.

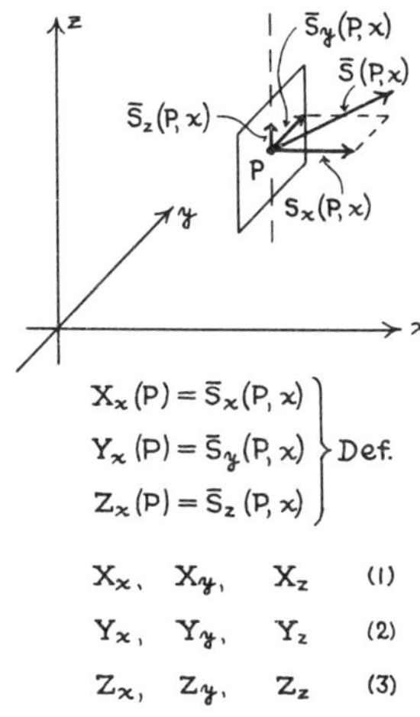

FIG. 13.9. *Right:* X, Y, and Z components of stress in and on a solid.

example, the system of coördinates x, y, and z, together with the small element of a plane perpendicular to the x axis, shown in Figure 13.9. At point P in this plane there may be a stress, $\bar{S}(P,x)$. We indicate in this way that this stress depends upon the location of point P, and upon the direction of the normal to the surface to which the stress is referred. This stress may be broken down into three components: an x, a y, and a z component. When this breakdown is made in terms of x, y, and z components, the stress components may be stated as indicated in the first three definition equations. It is thus seen that the x subscript refers to the direction of the plane across which the stress is determined. X, Y, and Z refer to the components of stress across this plane, and P, of course, refers to the coördinates of the point at which the stress is determined. Thus the three components of stress across the x plane are indicated as the first expressions in relations 1, 2, and 3. The middle column, X_y, Y_y, and Z_y, gives the three components of stress across the plane perpendicular to the y axis. X_z, Y_z, and Z_z give the three stress components across the plane perpendicular to the z axis. Thus there are nine components of stress at each point, P, within the mass.

We now consider some properties of the stress components. Figure 13.10 shows a set of shear stresses which would tend to deform the cube with center at point P. We see from the figure that one pair of the shear stresses is larger than the other pair. It is evident that this difference in the size of these stresses will introduce a rotation into the cube. We wish to prove that this is impossible if the cube is arbitrarily small, that is to say, that, instead of being different, the shear stresses are equal. We prove this by considering the torque, or moment of force, in relationship to the moment of inertia and the angular acceleration of the cube. The moment of force, M_F, is given in relation 2. Equation 2 can be reduced to the approximate equation 3. The transition from equation 2 to 3 may be made exactly if we assume that the stresses are homogeneous, that is, that they do not vary from point to point. Since most of the stresses that we shall deal with are roughly of this character, we may assume that the approximation is a very good one. To get the force, it is necessary to multiply the stress by the area of the face over which the stress is applied. The area of the face for the Y_x stress is $2\ dy\ 2\ dz$ and the lever arm is dx. This accounts for the factor, $4\ dy\ dz\ dx$. The other factors are accounted for similarly. It can be easily established by means of integral calculus or by reference to a handbook* that the moment of

* *Handbook of Chemistry and Physics*, 30th ed., p. 2475 (1947).

inertia of a small cube is the expression in relation 4. Therefore the angular acceleration is given by expression 5. Thus, unless the shear stresses are equal, the angular acceleration of this little cube will approach infinity as the size of the cube ap-

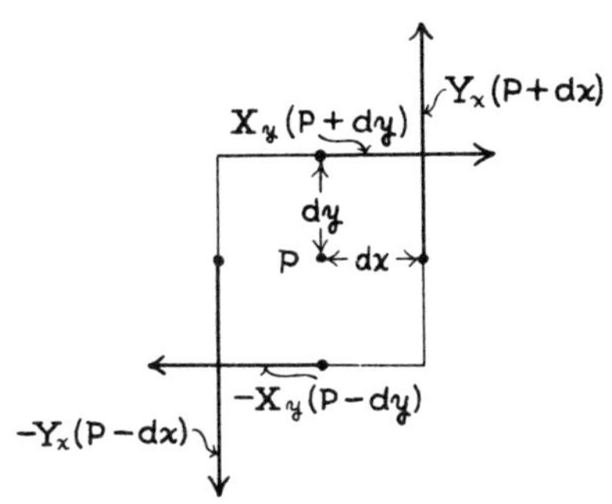

Proof that $X_y(P) = Y_x(P)$

Torque = Moment of Inertia times Angular Acceleration

$$M_F = Ia \tag{1}$$

$$\begin{aligned} M_F = {} & Y_x(P+dx)\,4\,dy\,dz\,dx - X_y(P+dy)\,4\,dx\,dz\,dy \\ & + Y_x(P-dx)\,4\,dy\,dz\,dx - X_y(P-dy)\,4\,dx\,dz\,dy \end{aligned} \tag{2}$$

$$M_F \approx 8\,[Y_x(P) - X_y(P)]\,dx\,dy\,dz \tag{3}$$

$$I = 8\,dx\,dy\,dz\;\rho(dx^2+dy^2)/3 \tag{4}$$

$$a \approx \frac{3[Y_x(P) - X_y(P)]}{\rho(dx^2+dy^2)} \tag{5}$$

Unless $Y_x(P) = X_y(P)$, as $dx = dy \to 0$, $a \to \infty$

FIG. 13.10. Symmetry of shear stress in a solid.

proaches zero. Since we assume that matter does not behave this way, with many unbounded angular accelerations distributed all through it, we must conclude that the x component of shear stress on the face perpendicular to the y axis is equal to the y component of shear stress on the face perpendicular to the x axis. We have three such relationships; hence, instead of having nine independent components of stress as in Figure 13.9, we have only six independent components.

Another property of stress which we use is illustrated in Figure 13.11. Here we consider that the stress is referred to planes perpendicular to the x axis. We show three little plane segments that are three little squares. The middle square has at its center the point P. We wish to show that the stress referred to these planes cannot be discontinuous at P. Since this point is any point in the medium, the stress must be continuous at all points for motion perpendicular to the planes across which the stress is measured. Let us suppose, if possible, that the stress is actually discontinuous at point P. Then the stress that points to the right is different from the stress that points to the left. This being the case, we can apply Newton's relationship and obtain equation 1. This may be rewritten in the form shown in equation 2, and this in turn may be rewritten as relation 3, a form that is slightly better for our purposes. We see immediately that if the numerical value of the stresses on the two sides is not equal, the acceleration approaches infinity as δx approaches zero. Since infinite accelerations do not exist

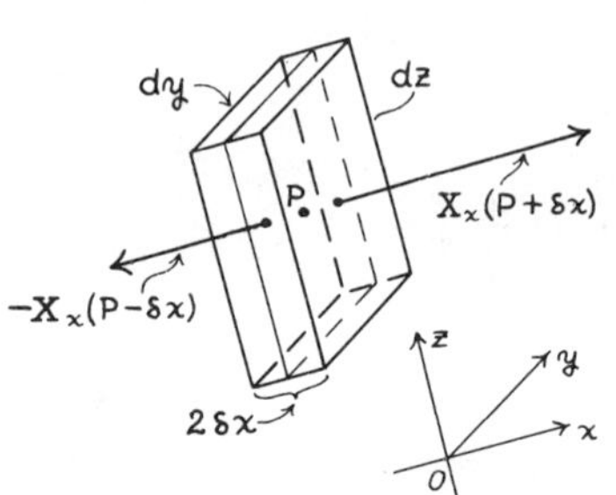

Proof that $X_x(P)$ is continuous in the X Direction

$$\text{Mass } (d^2x/dt^2) = (\text{Force})_x \qquad (1)$$

$$2\delta x \, dy \, dz \, \rho (d^2x/dt^2) = X_x(P+\delta x) dy \, dz - X_x(P-\delta x) \, dy \, dz \qquad (2)$$

$$\frac{d^2x}{dt^2} = \frac{X_x(P+\delta x) - X_x(P-\delta x)}{2\rho \, \delta x} \qquad (3)$$

Unless $X_x(P+\delta x) - X_x(P-\delta x)$

approaches zero as

δx approaches zero

(d^2x/dt^2) approaches ∞

FIG. 13.11. Continuity of normal stress in a solid.

within this medium, we must assume that the stress across the interface at P has to be continuous.

It may be observed that, in this proof, we have said nothing whatever about the kind of medium. Therefore, even though there is a discontinuity, say a transition from sandstone to shale, at point P, in which the interface is perpendicular to the x axis, there is no discontinuity in stress. This is one of the fundamental conditions that we shall use many times in subsequent considerations.

Although we have said that the stress is continuous as the point moves in the x direction, we have said nothing at all about the stress or its continuity as the point moves in the x plane. A simple example will show that the stress need not be continuous for such motion. Thus, if we place a steel rod of square cross section and a rubber rod of equal square cross section side by side and apply stresses at the ends of both rods so as to lengthen them by the same amount, these stresses will be very different because of the difference in the material. Thus, in passing from the rubber to the steel, there will be a severe discontinuity in the stress across the plane perpendicular to the axes of the rod.

This description of stress was not limited as our description of strain was. The stress can be as large or as small as we please and the description will still be valid.

13.2.3. Hooke's Relations

Hooke's Relations, usually called Hooke's law, constitute relationships between stress and strain. Since we consider only small strains, we need consider only small stresses. If the strain is small enough, we may confuse the initial state with the final state in our description without causing any trouble. This vagueness could not be tolerated if we were dealing with large strains and stresses. But if we did not permit this vagueness, we would find ourselves with a very difficult problem on our hands and we could not present the discussion in any simple terms. Apparently the most accurate discussion of this latter case thus far is the one by F. D. Murgnahan.* Brief inspection of his article will show clearly that the treatment is difficult. It may be remarked in passing that this difficulty explains to a great extent why geologists have never found a theory of elasticity very useful. Geologists are usually interested in large strains and stresses, and the part of the theory of elasticity that can be used with any reasonable degree of ease is automatically excluded by this fact. For our purposes, however, the small-strain theory is almost adequate—almost, because of the exceptional case involved in the study of conditions in the immediate neighborhood of the shot. In this neighborhood the material does not behave elastically, and so this also should probably be excluded as long as we are considering elastic problems.

Let us consider the simplest Hooke's relation by reference to Figure 13.12, which shows a cube that is elongated by X_x only. Hooke's relation says that the stress, X_x, is proportional to the strain ϵ_x. This is given as relation 1. The constant of proportionality, E, is defined below this; it is generally known as *Young's modulus* for the material. It is assumed, in determining this modulus or constant of proportionality, that the strain is very small. This is the simplest case.

Let us now consider the first generalization of this case. Figure 13.13 shows a cube that has been stretched by the stress X_x. It suffers the natural lateral contraction indicated by dv. Hooke's relations are expressed by relationships 1, 2, and 3. The first one we have already encountered in Figure 13.12. The second one introduces the effect of the lateral contraction. This lateral contraction is assumed to be proportional to the quantity σ, which is called *Poisson's ratio* and is defined by the last line in Figure 13.13. Thus Poisson's ratio is the ratio

* *American Journal of Mathematics*, 59:235-260 (1937).

of the lateral contraction to the extension in length, as shown in the figure. It may be observed that if there is no lateral contraction, as in Figure 13.12, Poisson's ratio is zero. On the other hand it may be shown that if there is no change of volume the lateral contraction is such that Poisson's ratio is 0.5. For actual materials, this ratio is more likely to be around 0.25. For a good many rocks it is of the order of 0.3, although for blocks of granite at atmospheric pressure it is more nearly

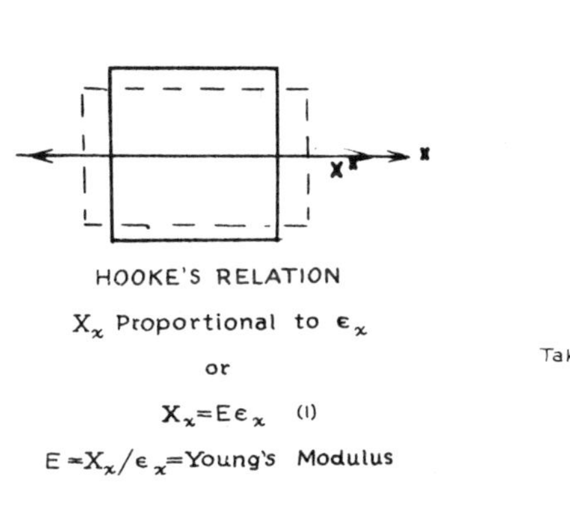

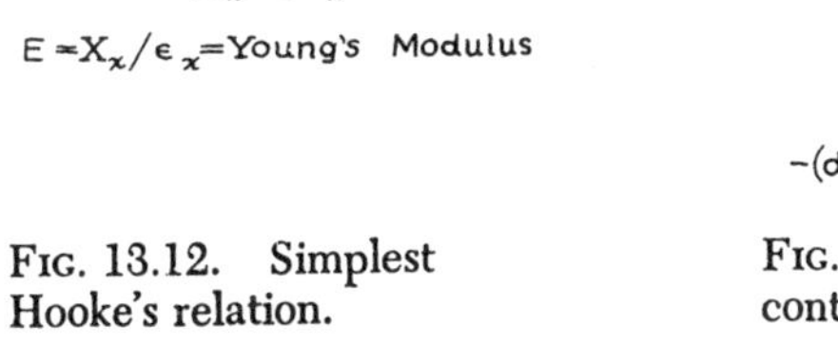

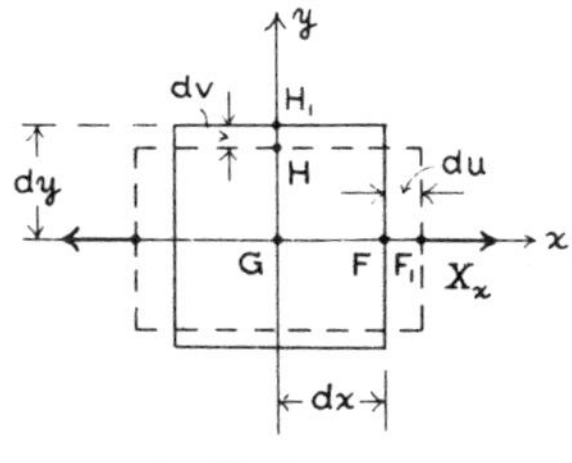

HOOKE'S RELATION

Taking Account of Lateral Contraction

$$E(du/dx) = X_x \quad (1)$$

$$E(dv/dy) = -\sigma X_x \quad (2)$$

$$E(dw/dz) = -\sigma X_x \quad (3)$$

$$-(dv/du) = \sigma = \text{Poisson's Ratio}$$

FIG. 13.12. Simplest Hooke's relation.

FIG. 13.13. Effect of lateral contraction. Poisson's ratio.

0.1. This may perhaps be explained in terms of the probable structure of granite blocks. They are probably made up of layers which appear to be very solidly put together but which are actually put together relatively loosely. These correspond to the cleavage planes. When the granite cube is extended, the lateral contraction is small because the extension is mostly taken up in terms of a slight opening of the cleavage planes. Needless to state, Poisson's ratio, σ, is an exceedingly important elastic constant.

We now proceed to a generalization of what was just described. Refer to the three cubes in Figure 13.14. The first one is subjected to a stress in the x direction; the second is stressed in the y direction, and the third is stressed in the z direction.

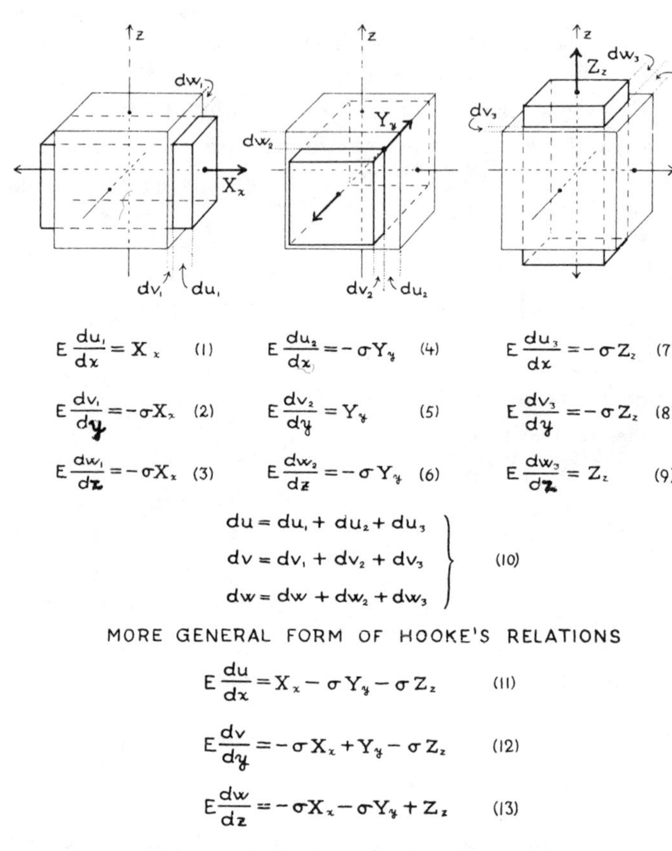

FIG. 13.14. Hooke's relations in a three-dimensional case.

These three effects are described by means of the three relations directly below each cube. We now suppose that the displacements, which are indicated by subscripts 1, 2, and 3, are added as indicated in relation 10. This gives the superposition of the effect of all three sets of stresses. We thus obtain, by adding the relations above, relationships 11, 12, and 13. Equation 11, for example, is obtained by adding equations 1, 4, and 7 and using the first line of equation 10. Equations 12 and 13 are obtained similarly.

The last three equations in the figure may represent the most general stress-strain relationships providing no shear stresses or shear strains are involved. As we have seen, it is possible to orient the axes in such a way that no such strains are involved. Thus these three equations are, in this sense, perfectly general relationships. Unfortunately, however, in any specific case, we need to know about shear strains also in order to decide how to orient the axes. The orientation of axes, which eliminates all shear stresses and strains, gives what are called *principal axes.* The principal axes of stress and strain are the same in the case under consideration. It may be well, at this point, to say that we have been dealing with an *isotropic* elastic medium. An isotropic elastic medium is one for which the orientation of the principal strain axes is always the same as the orientation of the principal stress axes. The isotropy at a point is defined as elastic uniformity with respect to directions around the point. As we look out in all directions from this point we find no preferred directions. The elastic constants are the same in all directions. Crystals of course are not iso-

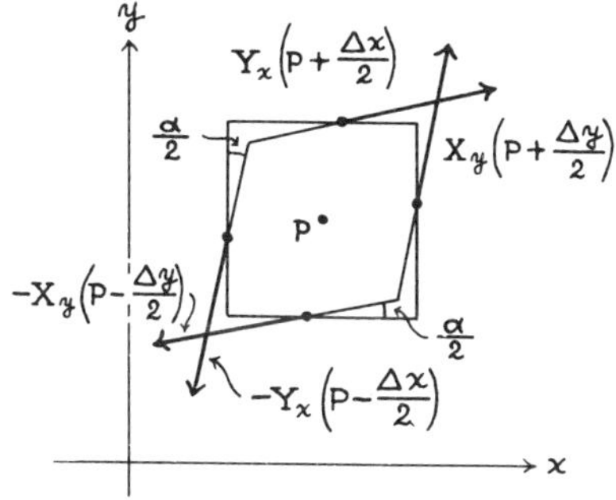

FIG. 13.15. Hooke's relation for shear.

tropic. Usually an isotropic medium has to be a glass. But even a glass under stress is not an isotropic medium. However, we shall make the assumption of isotropy, because it enables us to get certain important results that appear to be approximately correct.

In order further to generalize the relations in Figure 13.14, we now consider Hooke's relation for shear, shown in Figure 13.15. In this case we consider only the shear stresses. Hooke's relation assumes that the shear stress is proportional to the shear strain. The factor of proportionality, μ, is called the *modulus of rigidity*.

We have already introduced Young's modulus, Poisson's ratio, and the modulus of rigidity. We now show that the modulus of rigidity is related to Young's modulus and Poisson's ratio and can be immediately computed from them. For this purpose refer to Figure 13.16. The author should perhaps apologize for the complexity of this deduction. This deduction is rarely made. It usually appears as a simple mathematical identification that has no close relationship to the physical picture. It is our purpose, however, to base the relationship on physical considerations already discussed. At the upper left of the figure there is an inner square and an outer square. The inner square is oriented perpendicularly to the x,y coördinates below. The outer square is oriented with respect to the x',y' coördinates, also shown below; it makes an angle of 45° with the x,y coördinates. We imagine now that tension is applied to the face upon which points C, F, and B lie. Contractions or pressures may be needed on the face upon which points A, E, and C lie. The tension and compression are applied in such a way that the volume of the material is unchanged. The z direction is not shown, but we shall assume that it is the same as the z' direction.

We can easily find out how to apply the tension and pressure so that the volume is unchanged. We add relationships 11, 12,

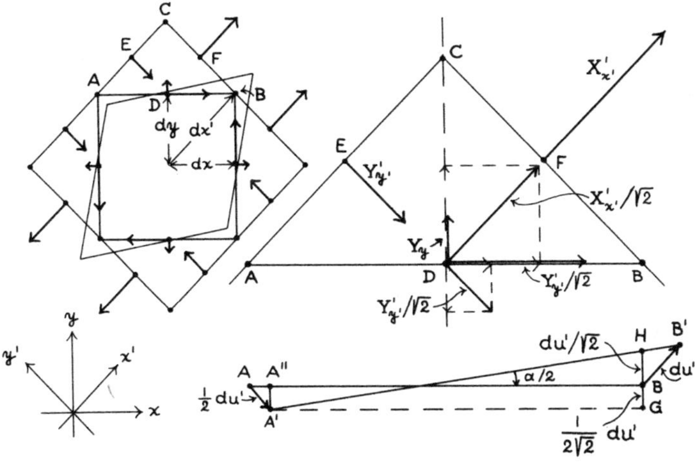

Relation of μ to E and σ

$$(1-2\sigma)\left(X'_{x'}+Y'_{y'}+Z'_{z'}\right)=E\left(\frac{du'}{dx'}+\frac{dv'}{dy'}+\frac{dw'}{dz'}\right) \tag{1}$$

If $Y'_{y'}=Z'_{z'}$ and there is no volume change, then

$$Y'_{y'}=-\frac{1}{2}X'_{x'} \tag{2}$$

$$E\frac{du'}{dx'}=X'_{x'}-\sigma\left(Y'_{y'}+Z'_{z'}\right)=(1+\sigma)X'_{x'} \tag{3}$$

$$\mu\left(\frac{\partial u_1}{\partial y}+\frac{\partial v_1}{\partial x}\right)=X_y=\frac{3}{4}X'_{x'} \tag{4}$$

$$\frac{\partial u_1}{\partial y}=\frac{\partial v_1}{\partial x}=\tan\frac{\alpha}{2}\approx\frac{\alpha}{2}\approx\frac{\frac{du'}{\sqrt{2}}+\frac{du'}{2\sqrt{2}}}{2dx}=\frac{\frac{3}{2\sqrt{2}}du'}{\frac{2dx'}{\sqrt{2}}} \tag{5}$$

$$=\frac{3}{4}\frac{du'}{dx'} \tag{6}$$

$$\mu\frac{4}{3}\,2\,\frac{3}{4}\frac{du'}{dx'}=X'_{x'}=\frac{E}{1+\sigma}\frac{du'}{dx'} \tag{7}$$

$$2\mu=\frac{E}{1+\sigma} \tag{8}$$

FIG. 13.16. Relation between the shear modulus, Young's modulus, and Poisson's ratio.

and 13 in Figure 13.14 and, after some collecting of terms, we get relation 1 in Figure 13.16. We can then show that the proposition given as relation 2 in the figure is true, because the expression at the right of relation 1 measures the change of volume and is zero if there is no change of volume. Thus we find that the lateral compression, $Y'_{y'}$, is one-half the tension stress, $X'_{x'}$. This being the case, we can rewrite equation 11 in Figure 13.14 as indicated in relation 3 in the present figure.

We now consider another description of the process involving the rigidity modulus, μ. This is given in relation 4, the first part of which may be considered as merely a definition of μ. The second part gives the shear stress as three-quarters the tension stress, and requires a detailed inspection of the upper right drawing in Figure 13.16. This shows an expanded view of the triangular cross section *ABC* taken from the left part of the figure. We wish to consider the stresses on face *ADB* which will be equivalent to the stresses already considered on faces *CFB* and *AEC*. Consider, for example, the stress that is equivalent to $X'_{x'}$ acting at point *F*. This is the stress acting at point *D* which is parallel to $X'_{x'}$. It will be noted, however, that we have reduced the length of this stress vector by dividing by $\sqrt{2}$. This reduction is required in order to make the corresponding *forces* the same. For instance, the stress $X'_{x'}$ acting at *F* acts on face *CB*, whereas the stress acting at *D* acts on the larger face *AB*. Since the force is the stress times the face area, the division by $\sqrt{2}$ becomes necessary because the area of face *AB* is equal to $\sqrt{2}$ times the area of face *CB*. A similar remark applies to stress $Y'_{y'}$ applied to *E* and its corresponding value applied to *D*. Thus each of the stresses applied at *E* and *F* is now reduced to two stresses applied at *D*. These two stresses can be analyzed into tangential and normal components, as indicated. We are primarily interested in the total tangential component Y_x or X_y. This component is equal to three-quarters of $X'_{x'}$, as can be seen by measuring and adding the tangential components of stress at D. This accounts for the second part of relation 4. The next step is to relate the displacements introduced by Y_x and X_y to those introduced by $X'_{x'}$ and $Y'_{y'}$. This is done in the fifth relationship. The first part of it results from the fact that the stresses are so applied that there is no rotation, so the two parts of the shear strain are equal. These two parts are each individually equal to the tangent of $\alpha/2$, which is approximately equal to $\alpha/2$, the angle of shear. We now look at the figure below that shows the shear introduced into AB and introduces the angle $\alpha/2$. Point A moves to A'. Point B moves to B'. It can be easily verified that BB' is twice AA'. It may also be noted that approximately HG/AB is the tangent of $\alpha/2$. We see immediately that HG can be written in terms of du', which is the displacement BB'. We do this and, after some arithmetical operations, we get relation 6. Substituting relation 6 in relation 4 gives the first part of relation 7. Using relation 3, we obtain the second part of relation 7. By canceling all the quantities that can be canceled, we get relation 8, which is the one we were seeking. This gives the rigidity modulus, μ, in terms of Young's modulus, E, and Poisson's ratio, σ.

We are now ready to discuss the deduction of a general form of Hooke's relation for small strains. For this purpose refer to Figure 13.17. The first relationship is obtained by multiplying relation 12 in Figure 13.14 by σ, adding the result to relation 11 of that figure, and collecting terms. The second relationship in Figure 13.17 is obtained by multiplying equation 12 of Figure 13.14 by σ and adding this to equation 13 in that figure. As indicated in Figure 13.17, we multiply the first relation by $1-\sigma$ and the second by σ and add to get the third relation. This relation can be rewritten in the form shown in relation 4 by adding to and subtracting $E\sigma\, du/dx$ from the third

GENERAL FORM OF HOOKE'S RELATIONS FOR SMALL STRAINS

$$E\left(\frac{du}{dx}+\sigma\frac{dv}{dy}\right)=(1-\sigma^2)X_x-\sigma(1+\sigma)Z_z \qquad (1)$$

$$E\left(\sigma\frac{dv}{dy}+\frac{dw}{dz}\right)=-\sigma(1+\sigma)X_x+(1-\sigma^2)Z_z \qquad (2)$$

Multiply (1) by $(1-\sigma)$ and (2) by σ and add to get

$$E(1-\sigma)\frac{du}{dx}+E\sigma\frac{dv}{dy}+E\sigma\frac{dw}{dz}=(1+\sigma)(1-2\sigma)X_x \qquad (3)$$

or

$$E(1-2\sigma)\frac{du}{dx}+E\sigma\left(\frac{du}{dx}+\frac{dv}{dy}+\frac{dw}{dz}\right)=(1+\sigma)(1-2\sigma)X_x \qquad (4)$$

or

$$X_x=\frac{E}{1+\sigma}\frac{du}{dx}+\frac{E\sigma}{(1+\sigma)(1-2\sigma)}\left(\frac{du}{dx}+\frac{dv}{dy}+\frac{dw}{dz}\right). \qquad (5)$$

$$\lambda=\frac{E\sigma}{(1+\sigma)(1-2\sigma)} \quad \text{(Def.)} \qquad (6)$$

(Result)

$$X_x=2\mu\frac{\partial u}{\partial x}+\lambda\left(\frac{\partial u}{\partial x}+\frac{\partial v}{\partial y}+\frac{\partial w}{\partial z}\right) \qquad (7)$$

$$Y_y=2\mu\frac{\partial v}{\partial y}+\lambda\left(\frac{\partial u}{\partial x}+\frac{\partial v}{\partial y}+\frac{\partial w}{\partial z}\right) \qquad (8)$$

$$Z_z=2\mu\frac{\partial w}{\partial z}+\lambda\left(\frac{\partial u}{\partial x}+\frac{\partial v}{\partial y}+\frac{\partial w}{\partial z}\right) \qquad (9)$$

$$X_y=Y_x=\mu\left(\frac{\partial u}{\partial y}+\frac{\partial v}{\partial x}\right) \qquad (10)$$

$$Z_x=X_z=\mu\left(\frac{\partial w}{\partial x}+\frac{\partial u}{\partial z}\right) \qquad (11)$$

$$Y_z=Z_y=\mu\left(\frac{\partial v}{\partial z}+\frac{\partial w}{\partial y}\right) \qquad (12)$$

FIG. 13.17. Deduction of Hooke's relations for the general case where only small strains are involved.

relation and combining terms. Relation 5 is solved for X_x, the stress in the x direction. The definition of λ in the sixth relationship enables us to write equation 7, if we use relation 8 in Figure 13.16 giving the relation between the rigidity modulus, Young's modulus, and Poisson's ratio. In exactly similar manner we can deduce equations 8 and 9. Thus equations 7, 8, and 9 give relationships that are equivalent to equations 11, 12, and 13 in Figure 13.14. To these three we add relations 10, 11, and 12 in Figure 13.17, thus securing a relationship between the six independent quantities of stress and the six independent quantities of strain. We assume, of course, that the dilatation which involves the three strain quantities is not counted as an independent quantity but that each of its three component parts is thus counted.

We thus have the relationships between the stress and strain expressed in equations 7 through 12 of Figure 13.17. These are the relationships we shall use in the future.

13.2.4. Equations of Motion

We now consider the equations of motion and we shall do this first by reference only to the stress distribution. At the left of Figure 13.18 is shown a region, R, as before, in which there is a small cube with sides Δx, Δy, Δz. This cube is shown greatly enlarged at the right of the figure. The center point, P, is shown, and also a pair of normal stresses and a pair of tangential stresses. Clearly, this is only a very small part of the stress system associated with the cube. However, to show the complete stress system would require three stress components for each of the six faces and would make the picture exceedingly complicated.

We first discuss the effect of the normal stresses. These are perpendicular to the x axis. It will be seen that the stress to the right of point P is X_x taken at the point, $P+\Delta x/2$. The stress to the left of P on the left face is X_x $(P-\Delta x/2)$. Both

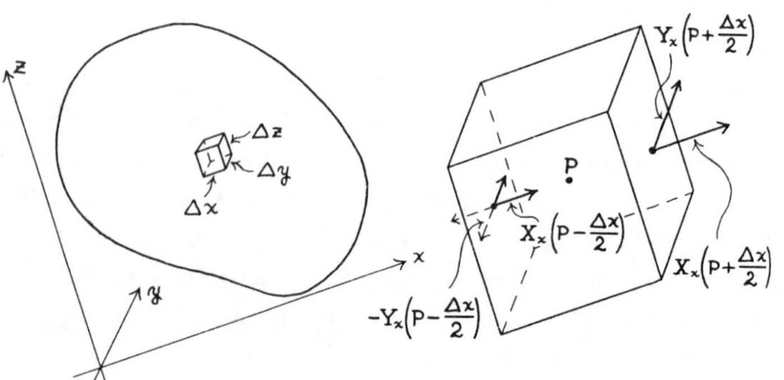

x Component of Force on Faces Normal to x Axis

$$X_x\left(P+\frac{\Delta x}{2}\right)\Delta y\,\Delta z - X_x\left(P-\frac{\Delta x}{2}\right)\Delta y\,\Delta z = \left[X_x\left(P+\frac{\Delta x}{2}\right) - X_x\left(P-\frac{\Delta x}{2}\right)\right]\Delta y\,\Delta z$$

$$= \left\{\frac{X_x[P+(\Delta x/2)] - X_x[P-(\Delta x/2)]}{\Delta x}\right\}\Delta x\,\Delta y\,\Delta z \rightarrow \frac{\partial X_x(P)}{\partial x}\,dx\,dy\,dz$$

y Component of Force on Faces Normal to x Axis

$$Y_x\left(P+\frac{\Delta x}{2}\right)\Delta y\,\Delta z - Y_x\left(P-\frac{\Delta x}{2}\right)\Delta y\,\Delta z = \left[Y_x\left(P+\frac{\Delta x}{2}\right) - Y_x\left(P-\frac{\Delta x}{2}\right)\right]\Delta y\,\Delta z$$

$$= \left\{\frac{Y_x[P+(\Delta x/2)] - Y_x[P-(\Delta x/2)]}{\Delta x}\right\}\Delta x\,\Delta y\,\Delta z \rightarrow \frac{\partial Y_x(P)}{\partial x}\,dx\,dy\,dz$$

x faces y faces z faces

$$\left(\frac{\partial X_x}{\partial x} + \frac{\partial X_y}{\partial y} + \frac{\partial X_z}{\partial z}\right)dx\,dy\,dz = \rho\,dx\,dy\,dz\,\frac{\partial^2 u}{\partial t^2}$$

$$\frac{\partial X_x}{\partial x} + \frac{\partial X_y}{\partial y} + \frac{\partial X_z}{\partial z} = \rho\frac{\partial^2 u}{\partial t^2} \qquad (1)$$

$$\frac{\partial Y_x}{\partial x} + \frac{\partial Y_y}{\partial y} + \frac{\partial Y_z}{\partial z} = \rho\frac{\partial^2 v}{\partial t^2} \qquad (2)$$

$$\frac{\partial Z_x}{\partial x} + \frac{\partial Z_y}{\partial y} + \frac{\partial Z_z}{\partial z} = \rho\frac{\partial^2 w}{\partial t^2} \qquad (3)$$

Fig. 13.18. The dynamic relations—Newton's equations for a small cube.

of these stresses are shown pointing in the same direction, but with a small difference of magnitude. We now consider the x component of force. This is given in the first equation under the figure, where we multiply the stress components, X_x, by the areas of the two faces giving the two forces. We have to take the negative of the stress on the left of P in order to get the force acting on the interior mass of the cube. It is evident that we can rewrite this difference as shown in the second line, dividing by the distance Δx, if we multiply also by this distance. The arrow indicates that the size of the cube is assumed to become arbitrarily small so that the difference quotient approaches the derivative, that is, the partial derivative of the stress, X_x with respect to x multiplied by the volume of the cube. In a similar fashion the y component of the force on this same face is given in the next two equations. A similar procedure is applied to the three components on each of the three pairs of faces. Each pair of faces gives an x component of force on the cube. Thus there are three contributions to the x component of force on the cube, namely, those given by the x faces, the y faces, and the z faces; see the equation immediately above relation 1 in the figure. Each of these contributions has the element of volume, $dx\ dy\ dz$, as a factor. So we write the corresponding element of mass, dm, as the density, ρ, times the volume element, $dx\ dy\ dz$. The volume element may be divided out of both sides of the Newtonian relation to give relation 1. Similarly we obtain the second and third relations. Thus relations 1, 2, and 3 in Figure 13.18 give the force (per unit volume) equations in the medium in terms of the displacements u, v, and w, and the stress components. Hence, if we know the distribution of stress throughout a body at any given time, and also the density, we can compute the accelerations by using these three relations. Unfortunately, these data are usually not given to us. We know almost nothing of the stress distribution of the interior of a medium unless we make some further assumptions.

If we assume that Hooke's relations, namely, equations 7

EQUATIONS OF (SMALL) MOTION

$$\frac{\partial X_x}{\partial x} = 2\mu \frac{\partial^2 u}{\partial x^2} + \lambda \frac{\partial \Delta}{\partial x} = \mu \frac{\partial^2 u}{\partial x^2} + \mu \frac{\partial}{\partial x}\left(\frac{\partial u}{\partial x}\right) + \lambda \frac{\partial \Delta}{\partial x} \qquad (1)$$

$$\frac{\partial X_y}{\partial y} = \mu \frac{\partial^2 u}{\partial y^2} + \mu \frac{\partial^2 v}{\partial x \partial y} = \mu \frac{\partial^2 u}{\partial y^2} + \mu \frac{\partial}{\partial x}\left(\frac{\partial v}{\partial y}\right) \qquad (2)$$

$$\frac{\partial X_z}{\partial z} = \mu \frac{\partial^2 u}{\partial z^2} + \mu \frac{\partial^2 w}{\partial z \partial x} = \mu \frac{\partial^2 u}{\partial y^2} + \mu \frac{\partial}{\partial x}\left(\frac{\partial w}{\partial z}\right) \qquad (3)$$

$$\nabla^2 \equiv \frac{\partial^2}{\partial x^2} + \frac{\partial^2}{\partial y^2} + \frac{\partial}{\partial z^2} \qquad \text{(Def)} \qquad (4)$$

$$\mu \nabla^2 u + (\mu + \lambda) \frac{\partial \Delta}{\partial x} = \rho \frac{\partial^2 u}{\partial t^2} \qquad (5)$$

$$\mu \nabla^2 v + (\mu + \lambda) \frac{\partial \Delta}{\partial y} = \rho \frac{\partial^2 v}{\partial t^2} \qquad (6)$$

$$\mu \nabla^2 w + (\mu + \lambda) \frac{\partial \Delta}{\partial z} = \rho \frac{\partial^2 w}{\partial t^2} \qquad (7)$$

$$(\lambda + 2\mu) \nabla^2 \Delta = \rho \frac{\partial^2 \Delta}{\partial t^2} \qquad \text{(Compressional Waves)} \qquad (8)$$

$$\underline{\text{If}} \quad \Delta \equiv 0 \quad \underline{\text{then}} \quad \mu \nabla^2 u = \rho \frac{\partial^2 u}{\partial t^2} \quad \text{etc.} \qquad (9)$$

$$\text{and} \quad \mu \nabla^2 r = \rho \frac{\partial^2 r}{\partial t^2} \quad \text{etc. (Rotational Waves)} \qquad (10)$$

FIG. 13.19. The dynamic equations taking account of Hooke's law.

through 12 in Figure 13.17, are valid within the medium, we can substitute them in relations 1, 2, and 3 in Figure 13.18, as indicated in Figure 13.19. Adding these together and taking account of the definition in relation 4 of this figure, we obtain relation 5. We obtain relations 6 and 7 similarly.

Relations 5, 6, and 7 involve only the displacements. We then have three equations involving three displacement components which may be regarded as unknowns. The importance of these three equations can hardly be overemphasized.

It may be noticed that if we differentiate equation 5 with respect to x, equation 6 with respect to y, and equation 7 with respect to z and add, we obtain relation 8. This is the equation for *compressional waves,* or *dilatational waves* as they are sometimes called, since Δ is the dilatation of the medium.

We see immediately the proposition shown in relations 9 and 10. This is, *if* the dilatation is identically zero, equation 5 is simplified to the form shown in relation 9. This same simplification occurs in relations 6 and 7 under the hypotheses that the dilatation is identically zero. We may thus expect expressions of the form shown in relation 9 to govern the transmission of transverse waves, those with rotation but with no dilatation. If we write this relation for v as well as for u we can differentiate these, the one for u with respect to y and the one for v with respect to x, and take the difference of the two resulting expressions to give equation 10. This shows that the rotation in a transverse wave is governed by expressions identical in form to equation 9. We have considered only the component of rotation about the z axis. The other two components of rotation satisfy equations just like equation 10.

We have thus completed the deduction of the fundamental equations governing pure compressional waves and pure rotational waves. However, certain surface waves are neither pure rotational waves nor pure compressional waves; hence in such cases it is necessary to use equations 5, 6, and 7 of Figure 13.19 directly.

13.2.5. Conservation of Momentum and Energy

We shall consider in some detail the conservation of momentum. The discussion of the conservation of energy follows

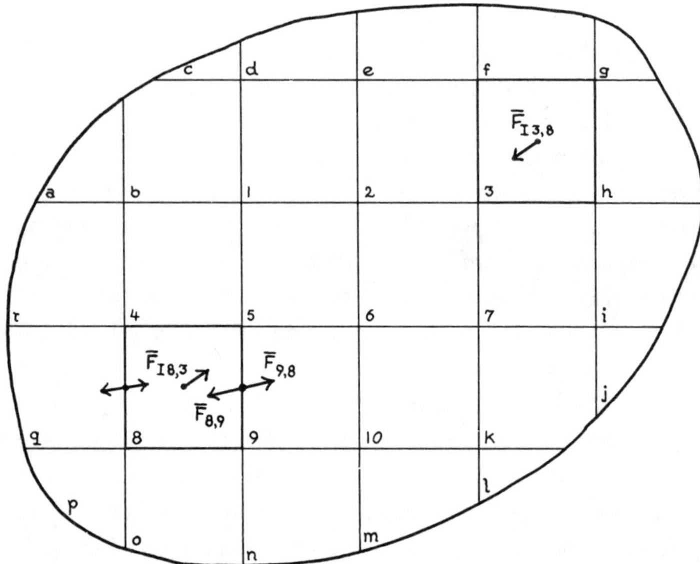

$$\frac{\partial}{\partial t}(m_1 \bar{v}_1) = \bar{F}_{E1} + \sum_{1}^{6} \bar{F}_{1j} + \sum{}' \bar{F}_{I1k} \tag{1}$$

$$\frac{\partial}{\partial t}(m_2 \bar{v}_2) = \bar{F}_{E2} + \sum \bar{F}_{2j} + \sum{}' \bar{F}_{I2k} \tag{2}$$

....

$$\frac{\partial}{\partial t}\left(\sum m_i \bar{v}_i\right) = \sum \bar{F}_{Ei} + \sum \sum_{1}^{6} \bar{F}_{ij} + \sum \sum{}' \bar{F}_{Iik} \tag{3}$$

$\sum \sum{}' \bar{F}_{Iik} = 0 =$ Sum of Interior Body Forces (4)

$\sum \sum_{1}^{6} \bar{F}_{ij} =$ Sum of External Surface Forces (5)

$\sum \bar{F}_{Ei} =$ Sum of Exterior Body Forces (6)

Free Motion Implies $\frac{\partial}{\partial t}\left(\sum m_i \bar{v}_i\right) = 0$ or

$\sum m_i \bar{v}_i =$ Constant (7)

FIG. 13.20. Principle of the conservation of momentum.

similar classical lines and will not be presented in detail in this section.

Figure 13.20 shows a region bounded by a curve and split up into square or cubical pieces which have been numbered or lettered. It will be observed that the numbered pieces are interior pieces. The lettered pieces have part of their boundaries in common with the boundary of the medium. For each of the interior pieces we can compute the time rate of change of momentum in terms of all the forces acting upon the piece. For example, in relation 1 we have computed the rate of change of momentum for piece No. 1. This is made up of three kinds of parts. The first part, $\overline{F}_{E1}$, represents an exterior body force acting upon this piece. This exterior body force may be a gravitational force due to exterior masses not contained within the body itself. The next part consists of the six surface forces acting upon the piece. The piece being a cube, it has six faces, and on each of these six faces we have to consider the action of a force due to the neighboring pieces. These six forces are indicated by the sum Σ from 1 to 6, indicating the summation over the neighboring faces. The third part consists of forces which we may call interior body forces. The Σ' here indicates a sum over all the pieces except the piece for which the time rate of change of momentum is being computed. These forces, for example, are the forces of gravitational attraction between the various pieces of the body. If electrical and magnetic forces exist they also are included under this heading. Thus, for each cubical interior piece we have a relation indicated in equations 1 and 2, etc.

These relations for the interior of the medium may be extended to the exterior parts as well, and we suppose that this extension is made. The only modification would consist of a change in the sum of the number of surface forces; for example, a piece like c at the upper left would perhaps not have as many as six surface forces to be considered, but only five. On the other hand, a piece like b might have more than six surface forces to be considered. Thus in order to include the exterior pieces it is necessary only to make the proper modifications of the second term at the right of equations 1 and 2 so that it will include the correct number of surface forces. This we assume has been done and that all the pieces are accounted for in relations 1, 2, etc. The notation Σ from 1 to 6 is used with the understanding that the 6 will be modified as required for an exterior piece. The dots below relation 2 indicate the continuation of this series.

The lines below the dots indicate that all the equations above are to be added. This addition is done in relation 3. The ordinary Σ indicates the sum over all the pieces of the body. The term at the left of relation 3 is the time rate of change of the total momentum of the medium. It is seen that this time rate of change may be split up into three parts. We shall consider these parts in order from right to left. The first part being the sum of the interior body forces is easily seen to be zero, since for each term there is an equal and opposite term corresponding to the reaction force. Thus in all cases the sum of the interior body forces will be zero. The next term is the sum of the external surface forces. Because of the equal and opposite reactions of the surface forces of adjacent parts, the sum of the interior parts will be zero, and the sum indicated in relation 5 will be the sum of the external surface forces. These forces may, in our case, be assumed to be zero; that is, no external forces will be applied to the outer surface. The third term represents the sum of the exterior body forces. Although in general this sum is not precisely zero we may assume it to be zero, thus neglecting a small unimportant quantity. If we do this, we find that, for *free motion* of the whole medium, the time rate of change of the total momentum is zero, or the total momentum is a constant, as indicated in relation 7.

This constancy of the total momentum of a system, which

of course is a general feature of every mechanical system that is not acted on by exterior forces, is very important, because the momentum is a quantity that can be computed. For example, it can be computed before and after a reflection process, or before and after a refraction process, or before and after an explosion process. Thus we have a means of checking the physical correctness of the solution of such a problem by computing the momentum. This is especially important in certain cases, as, for example, when it is felt to be impossible to follow details of the process, as in an explosion. Any result that we obtain will have to satisfy the conservation of momentum relations; that is, the momentum before the process must be the same as the momentum afterward.

13.3. SPHERICAL COÖRDINATES

In this section we translate relations which we formerly expressed in rectangular coördinates, into spherical polar coördinates. This translation is a step that we must take in order to make possible the convenient discussion of pulses originating in a cavity source, such as a shot hole. The spherical coördinates are the radial distance and the various angular measures which give the orientation. We simplify the discussion of this angular orientation by confining our attention to the case in which there is symmetry about the center point, so the angular orientation does not need to be specifically considered.

13.3.1. Strain

We now consider the expression for the volume strain, or dilatation. For this purpose, refer to Figure 13.21. In connection with this figure, reference should be made to equations 5, 6, and 7 in Figure 13.4, where a similar discussion is presented in terms of rectangular coördinates. In Figure 13.21 the center of the spherical distribution is at point *C*. Going downward

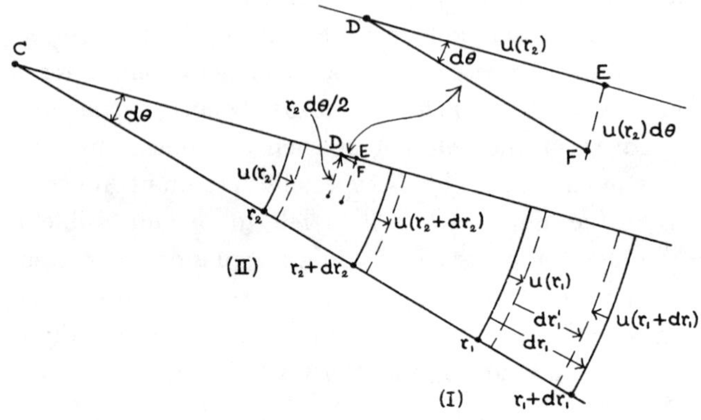

(I) Volume Strain $= \dfrac{dr_1'\, dA_1 - dr_1\, dA_1}{dr_1\, dA_1}$ (1)

$$= \frac{dr_1' - dr_1}{dr_1} = \frac{dr_1 - u(r_1) + u(r_1 + dr_1) - dr_1}{dr_1} \quad (2)$$

$$= \frac{d\,u(r_1)}{dr_1} \quad (3)$$

(II) Volume Strain $= \dfrac{dr_2\, dA_2' - dr_2\, dA_2}{dr_2\, dA_2}$ (1')

$$= \frac{dA_2' - dA_2}{dA_2} = \frac{dA_2 + 2\pi\, r_2 (d\theta/2)\, u(r_2)(d\theta/2) - dA_2}{dA_2} \quad (2')$$

$$= \frac{(\pi/2)\, r_2\, (d\theta)^2\, u(r_2)}{\pi r_2^2 (d\theta)^2/4} = \frac{2\,u(r_2)}{r_2} \quad (3')$$

(III) General Radial Volume Strain $= \dfrac{dV' - dV}{dV}$ (1'')

$$= \frac{du}{dr} + \frac{2u}{r} \quad (3'')$$

FIG. 13.21. Two kinds of radial volume strain.

and to the right of C, we encounter two positions, I and II. These indicate two kinds of volume strain that are possible in the present case. We shall consider these two separately.

In I, the volume strain is defined in relation 1. Here we see that the change of volume does not involve a change in the average area, but only a radial change in which the outer face is displaced inward and the inner face is displaced outward. The displacements are indicated by $u(r_1)$ and $u(r_1+dr_1)$. These displacements are indicated on the drawing by the little arrows and are seen to be approximately equal and opposite. Relation 1, which is equivalent to relations 5 and 6 in Figure 13.4, may be simplified as indicated in relation 2 by canceling the common element of area. This leaves only the changes in radius from the center C. These changes can be expressed as indicated in the second part of equation 2, in which we have expressed dr_1' in terms of the displacements and dr_1. A glance at the figure will verify this substitution. In passing from relation 2 to relation 3, we merely use the definition of the differential of u. Thus, the volume strain, in I, is simply the rate of change of u with respect to r. Since it is assumed that there is no lateral displacement, we see that relation 3 is identical in form with relation 7 in Figure 13.14, in which x corresponds to r, and the two remaining derivatives in relation 7 are zero because of the lack of expansion or contraction in the lateral direction.

We now consider II, the other type of volume strain that is possible. Here the displacements are equal and outward, so we cannot consider that any *change of displacement* is involved. In such a case, the volume strain must be described not in terms of the change of radial displacement, but in terms of the change of area of the cross section of the little element involved. This is expressed in relation 1′. We go from this relation to relation 2′ by canceling dr_2. We then go from the first expression in relation 2′ to the second by calculating the change of area involved. This calculation involves the angle $d\theta$, and the radius. We may then rewrite this relation in the simpler form shown in relation 3′. After cancellation of terms in equation 3′, we find the term at the right, which gives the volume strain of dilatation if the situation is like that illustrated in II. This is the case mentioned before, in which there is no change in the outward displacement. It is seen that an outward displacement necessarily increases the volume and so contributes to the dilatation.

The formulas numbered III illustrate the general radial volume strain, derived by adding cases I and II. Thus in general the radial volume strain is given by expressions 1″ and 3″.

Expression 3″ could be derived by purely mathematical manipulations involving the change of coördinates from rectangular to spherical polar. However, our deduction has illustrated the *significance*—a strictly geometrical significance, of course—of the two terms involved in equation 3″.

13.3.2. Stress

Stress in polar coördinates is defined just as it is defined in rectangular coördinates, by reference to diagrams such as those in Figures 13.7 and 13.8. The curvature of the spherical cap element of surface or of the lateral surface formed by rays from the center has no effect upon the definition, for we ultimately have to consider elements of volume of vanishing dimensions.

In our present work we consider only the radial stress R_r and the lateral stress L. These quantities will be described under Hooke's relations, in the following subsection.

13.3.3. Hooke's Relations

Refer to Figure 13.22, which may be directly compared with Figures 13.14 and 13.17. Figure 13.22 shows two situations. In the one at the left, only radial forces act and consequently the radial stress R_r is the only one that we need consider. We have

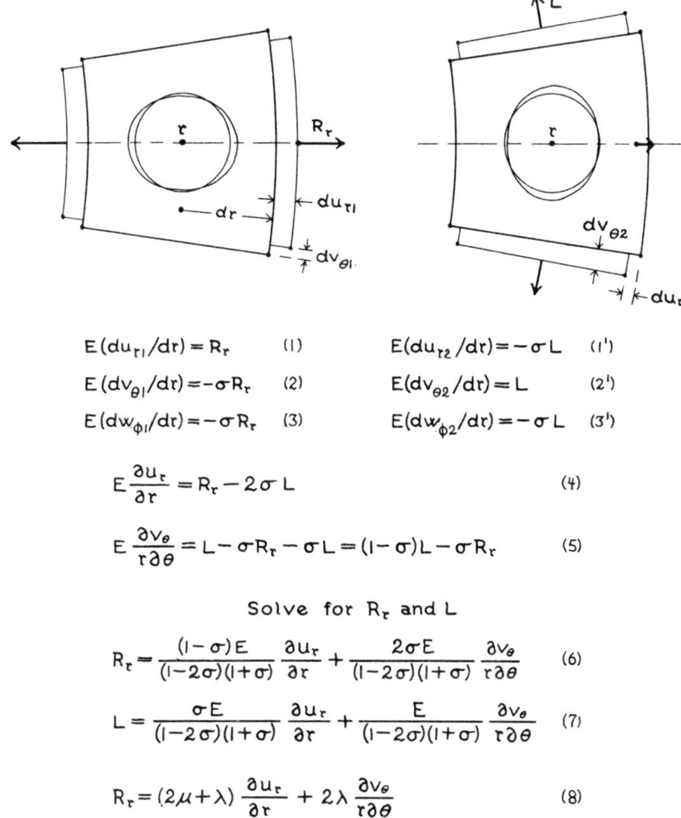

$$E(du_{r1}/dr) = R_r \qquad (1) \qquad\qquad E(du_{r2}/dr) = -\sigma L \qquad (1')$$

$$E(dv_{\theta 1}/dr) = -\sigma R_r \qquad (2) \qquad\qquad E(dv_{\theta 2}/dr) = L \qquad (2')$$

$$E(dw_{\phi 1}/dr) = -\sigma R_r \qquad (3) \qquad\qquad E(dw_{\phi 2}/dr) = -\sigma L \qquad (3')$$

$$E\frac{\partial u_r}{\partial r} = R_r - 2\sigma L \qquad (4)$$

$$E\frac{\partial v_\theta}{r\partial\theta} = L - \sigma R_r - \sigma L = (1-\sigma)L - \sigma R_r \qquad (5)$$

Solve for R_r and L

$$R_r = \frac{(1-\sigma)E}{(1-2\sigma)(1+\sigma)}\frac{\partial u_r}{\partial r} + \frac{2\sigma E}{(1-2\sigma)(1+\sigma)}\frac{\partial v_\theta}{r\partial\theta} \qquad (6)$$

$$L = \frac{\sigma E}{(1-2\sigma)(1+\sigma)}\frac{\partial u_r}{\partial r} + \frac{E}{(1-2\sigma)(1+\sigma)}\frac{\partial v_\theta}{r\partial\theta} \qquad (7)$$

$$R_r = (2\mu+\lambda)\frac{\partial u_r}{\partial r} + 2\lambda\frac{\partial v_\theta}{r\partial\theta} \qquad (8)$$

$$L = \lambda\frac{\partial u_r}{\partial r} + 2(\lambda+\mu)\frac{\partial v_\theta}{r\partial\theta} \qquad (9)$$

FIG. 13.22. Hooke's relations applied to radial stress and strain.

attempted to show how the circular cross section of a sphere is deformed into an ellipse by the action of radial stress. We then apply Hooke's relations taken directly from equations 1, 2, and 3 in Figure 13.14. These correspond exactly to the first three relations in Figure 13.22. The fact that the volume in this case is not precisely a cube does not in any way affect our discussion of the deformation of a sphere, because ultimately we are considering elements of mass of vanishing dimensions. The curvilinear variation in shape in this figure contributes nothing of essential importance to our discussion.

Let us now consider the second diagram in Figure 13.22, in which there is a lateral stress, L. Corresponding to this lateral stress there is a lateral displacement which is given by $dv_{\theta 2}$. The resulting equations 1′, 2′, and 3′ may be regarded as corresponding to equations 4, 5, and 6 of Figure 13.14. We could have written another series of equations, essentially like 1′, 2′, and 3′, to correspond to relations 7, 8, and 9 of that figure; but we are assuming symmetry about every radial axis and hence this third series will not differ from the second series, 1′, 2′, and 3′.

We can now combine the two kinds of stresses and strains as indicated in relations 4 and 5 of Figure 13.22. These two relations correspond to relations 11 and 12 of Figure 13.15. In going from relations 1 and 1′ to 4 we have also used what would correspond to relation 1″, which is just like 1′. This accounts for the factor 2 in relation 4. Likewise, in going from 2 and 2′ to 5 we have used what would have been written as relation 2″. In addition, in relation 5 we have replaced (dv_θ/dr) by $\partial v_\theta/r\partial\theta$. This we can do because we started with a little curvilinear cube, so that $dr = rd\theta$.

It is clear that relations 4 and 5 can be solved for the stress quantities R_r and L, as indicated in relations 6 and 7. These latter relations might well be compared with relations like 5 in Figure 13.17, except that equation 5 must be rewritten so that all the factors of $\partial u/\partial x$ are combined instead of being separated as they are in the present equation. We may go from relations 6 and 7 to relations 8 and 9 merely by substituting for Young's modulus, E, and Poisson's ratio, σ, the shear modulus, μ, and the other elastic modulus, λ. Thus we obtain relations 8 and 9.

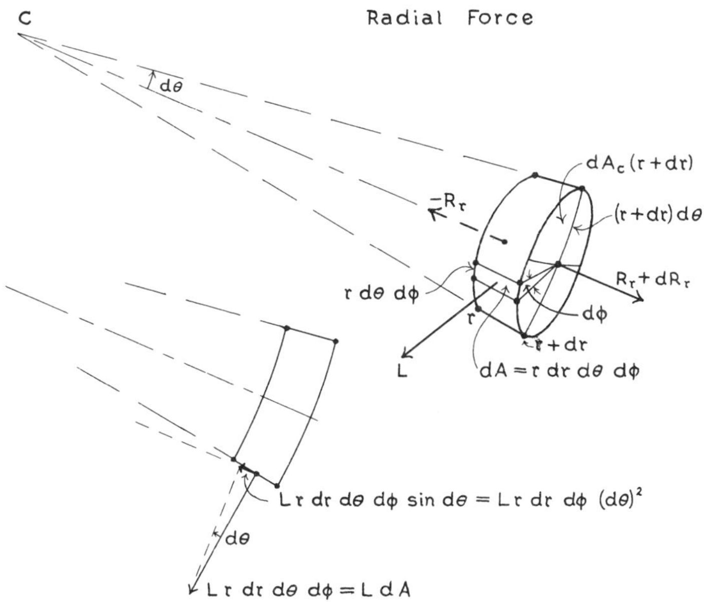

$$(R_r + dR_r)\pi(r + dr)^2 (d\theta)^2 - R_r \pi r^2 (d\theta)^2 - 2\pi L r \, dr (d\theta)^2$$
$$= \rho \pi r^2 d\theta^2 dr (\partial^2 u_r / \partial t^2) \qquad (1)$$

$$r^2 dR_r + 2r R_r dr - 2r L \, dr = \rho r^2 dr (\partial^2 u_r / \partial t^2) \qquad (2)$$

$$\frac{\partial R_r}{\partial r} + \frac{2R_r}{r} - \frac{2L}{r} = \rho \frac{\partial^2 u_r}{\partial t^2} \qquad (3)$$

FIG. 13.23. Dynamic relations, force equations, for radial stresses and strains.

13.3.4. EQUATIONS OF MOTION

We now refer to Figure 13.23, which shows a center point, C, and an element of volume containing a mass whose density is ρ. The stress system on this element of mass is shown, the inner face having a radial stress R_r and the outer a radial stress $R_r + dR_r$. The radial forces on the inner and outer faces are found by multiplying the corresponding stresses by the face areas. Regarding each of the caps as small enough ($d\theta$ small enough) so that they may be taken as plane circles, we have for the inner cap the area $d\ A_c\ (r) = \pi(rd\theta)^2$, and for the outer cap the area $\pi[(r+dr)\ d\theta]^2 = d\ A_c\ (r+dr)$. Thus the radial force on the element due to stresses on the inner and outer caps is given by the first two terms at the left of relation 1.

The lower diagram in Figure 13.23 shows another contribution to the radial force due to lateral stress L. Careful examination of this diagram shows clearly that a lateral stress contributes a radial force, not to the face, but to the volume element. The lateral force on the little element dA is $LdA = Lr\ dr\ d\theta\ d\phi$, and its projection on the central radial direction for the volume is $Lr\ dr\ d\theta\ d\phi\ \sin d\theta = Lr\ dr\ d\phi\ (d\theta)^2$, as shown. To get the total contribution we add all the lateral contributions for $d\phi$. Assuming symmetry around the central axis of the volume element through C, we have only to add the $d\phi$'s. These integrate to 2π.

Thus, except for algebraic sign, we account for the third term on the left side of equation 1. The negative sign of this contribution results from the fact that when the lateral stress is positive, or outward, the resolution along the radial central direction gives a force that tends to decrease r. Since there is this tendency toward decreasing r, and since we have taken forces that tend to increase r as positive, we must use a negative sign for this third term.

The expression at the right of relation 1 is the density times the element of volume times the acceleration in a radial direction. We can then do the indicated multiplications, combine terms, divide all the terms that can be commonly divided through the relationship, keeping only terms of the first order in the elements of distance and of change of stress, and get relation 2. From this relation we can readily get relation 3 by dividing through by r^2dr. We have passed to partial derivatives in expression 3 because R_r depends not only on r but also on t. We assumed, in the first place, that R_r did not depend upon the angle θ; but it does depend on a second variable—the time, t—so that the partial derivative notation should be used. We may regard relation 3 as a direct analogue of relation 1 in Figure 13.18.

It is interesting to derive the equation of motion for the dilatation, that is, an equation in polar coördinates similar to equation 8 in Figure 13.19. For this purpose refer to Figure 13.24. The upper part of this figure presents the deduction of a slight modification of relations 8 and 9 of Figure 13.22. This modification has to be taken into account when we consider that all the motion must take place along radial lines, so that lateral strain terms are somewhat more restricted than those shown in equations 8 and 9 of that figure. We can see relation 1 from the geometry of the diagram; that is, the radial displacement, u_r, multiplied by the angle $d\theta$ must equal the lateral displacement change, dv_θ. We can rewrite relation 1 as relation 2. Then, making use of the latter relation, we can rewrite relations 8 and 9 of Figure 13.22 as relations 3 and 4 of Figure 13.24.

Having made this rearrangement, we are now ready to rewrite the equation of motion, in the case of purely radial strain, for the dilatation. This is shown systematically under the line in Figure 13.24. Relation 1 gives the radial expression for the dilatation Δ_r (see relation 3″ in Figure 13.21). Relation 2 is the derivative of r times this dilatation. Relation 3 gives the expression for the second derivative. We go from this relation to relation 4 by dividing both sides by r. In order to derive relation 5, we substitute relations 3 and 4 from above the line in Figure 13.24 in equation 3 of Figure 13.23. Relation 5 can be differentiated once with respect to r, as indicated in relation 6. To relation 6 we add relation 5 multiplied by $2/r$, and get re-

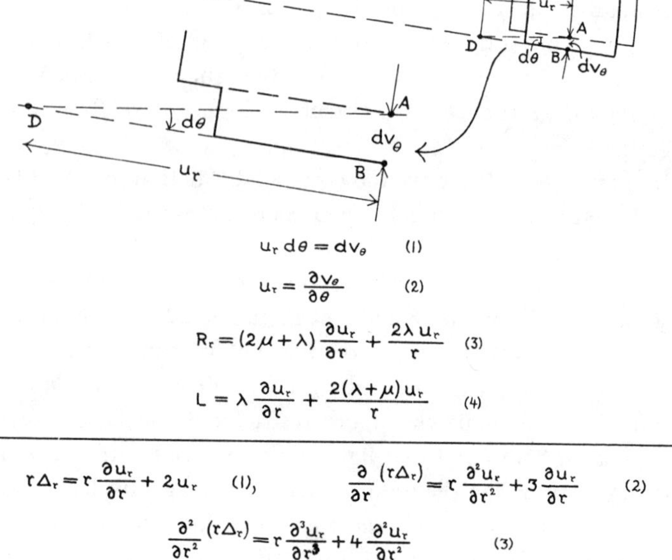

$$u_r\, d\theta = dv_\theta \qquad (1)$$

$$u_r = \frac{\partial v_\theta}{\partial \theta} \qquad (2)$$

$$R_r = (2\mu+\lambda)\frac{\partial u_r}{\partial r} + \frac{2\lambda u_r}{r} \qquad (3)$$

$$L = \lambda \frac{\partial u_r}{\partial r} + \frac{2(\lambda+\mu)u_r}{r} \qquad (4)$$

$$r\Delta_r = r\frac{\partial u_r}{\partial r} + 2u_r \quad (1), \qquad \frac{\partial}{\partial r}(r\Delta_r) = r\frac{\partial^2 u_r}{\partial r^2} + 3\frac{\partial u_r}{\partial r} \quad (2)$$

$$\frac{\partial^2}{\partial r^2}(r\Delta_r) = r\frac{\partial^3 u_r}{\partial r^3} + 4\frac{\partial^2 u_r}{\partial r^2} \qquad (3)$$

$$\frac{1}{r}\frac{\partial^2}{\partial r^2}(r\Delta_r) = \frac{\partial^3 u_r}{\partial r^3} + \frac{4}{r}\frac{\partial^2 u_r}{\partial r^2} \qquad (4)$$

$$(2\mu+\lambda)\left(\frac{\partial^2 u_r}{\partial r^2} + \frac{2}{r}\frac{\partial u_r}{\partial r} - \frac{2u_r}{r^2}\right) = \rho\frac{\partial^2 u_r}{\partial t^2} \qquad (5)$$

$$(2\mu+\lambda)\left(\frac{\partial^3 u_r}{\partial r^3} + \frac{2}{r}\frac{\partial^2 u_r}{\partial r^2} - \frac{4}{r^2}\frac{\partial u_r}{\partial r} + \frac{4u_r}{r^3}\right) = \rho\frac{\partial}{\partial r}\frac{\partial^2 u_r}{\partial t^2} = \rho\frac{\partial^2}{\partial t^2}\frac{\partial u_r}{\partial r} \qquad (6)$$

(6) plus $\frac{2}{r}$ times (5) gives

$$(2\mu+\lambda)\left(\frac{\partial^3 u_r}{\partial r^3} + \frac{4}{r}\frac{\partial^2 u_r}{\partial r^2}\right) = \rho\frac{\partial}{\partial t^2}\Delta_r \qquad (7)$$

$$(2\mu+\lambda)\frac{\partial^2}{\partial r^2}(r\Delta_r) = \rho\frac{\partial^2}{\partial t^2}(r\Delta_r) \qquad (8)$$

Fig. 13.24. Equations of motion using Hooke's law for radial strains.

lation (7). Using relations 7 and 4, we obtain relation 8, which gives the partial differential equation for the dilatation. It will be seen that this relationship is of approximately the same form as the one derived with rectangular coördinates if the y and z coördinates were neglected. This is an important fact because it enables a simple solution of this equation to be given later.

We are now ready to discuss more specific problems connected with the transmission of seismic waves.

This completes our summary of the most important items in the physical picture which we need for our interpretive work. These relationships admittedly involve a very close look at the geological structure—what we might call a microscopic view of the situation. We should never forget, for long, that our primary interest is not in a microscopic view, but in a much larger view. At the present state of knowledge of the subject, however, it is not possible to go directly to the overall view; we must approach it from our presently developed microscopic view. This is somewhat unfortunate, because it tends to place undue emphasis on the microscopic quantities. We should keep this undue emphasis in mind as we proceed.

13.4. COMMENTS TO THE SECOND EDITION

In Figure 13.17, relation (6) is too complicated to give us a clear comprehension of what the λ parameter means. The rigidity, μ, which has the dimensions of stress, is directly and clearly understood from Figure 13.15.

In Figure 13.17, let the displacements u and v be zero. Then equations (7), (8), and (9) become $X_x \equiv \lambda \partial w/\partial z$, $Y_y \equiv \lambda \partial w/\partial z$, and $Z_z \equiv (\lambda + 2\mu) \partial w/\partial z$. If we are standing on the flat level ground and press down everywhere an amount measured by Z_z but do not allow any sideways spreading, that is, $u \equiv 0$ and $v \equiv 0$, then the lateral stress is given by the first two simple equations which say, "the lateral stress is measured by λ."

Furthermore, "the up and down stress Z_z is measured by $(\lambda+2\mu)$." Thus, λ and $(\lambda+2\mu)$ have simple stress meanings that all earth scientists should almost feel directly!

If you look at Figure 13.19 equation (8) and meditate on it for a while, you will comprehend how $\lambda+2\mu$ measures the velocity of compressional waves.

CHAPTER 14

Propagation of Seismic Pulses

This chapter is written entirely for seismologists. The subject matter had to be selected from a very large field. Hence no pretense is made that the subject is treated completely in this chapter. It is, however, hoped that the average reader will obtain a general picture of the present status of the subject.

In spite of a rather large body of theoretical literature on the propagation of seismic waves and pulses, it must be admitted that the subject has not achieved a very mature stage of understanding. The reason appears to be that the subject itself is exceedingly complicated and difficult in its essentials. A secondary reason, however, and one to which the author attaches great importance, is the lack of close ties between theory and observation. There is a twofold reason for this lack. On the one hand, experimental work has not been undertaken until very recently, and on the other, the theories have been oversimplified to a gross extent. This failure to link observation and theory is rapidly being overcome, as a result of the experimental model studies of seismic pulses being vigorously conducted at the present time in various laboratories. It is to be hoped that these model studies may be used to remove much of the ambiguity in the theories.

The incompleteness of the present chapter is due mainly to two factors. One is that the writer has assumed a certain background for the reader and in many cases the theory exceeds the limitations of this background. The second and more important cause is the fact that the major investigations solve the wrong physical problems. A theoretical investigation can solve a certain physical problem that is set by the investigator; but if this problem does not correspond to one that is important in practice, then its solution cannot be of much value to us. This is unfortunately true of a good many studies of a theoretical nature that have been published.

It is our prime purpose in this chapter to supply the seismologist with a background to guide his judgment in interpreting seismic data. This is a difficult task for reasons which should be fairly evident by now. We now proceed to our study of the various parts of this subject.

14.1. PULSE FRONTS. HUYGENS' PRINCIPLE

We wish to study two fundamental aspects of the propagation of seismic pulses. For this purpose refer to Figure 14.1, which shows two kinds of pulses. One is a *left moving pulse* and the other is a *right moving pulse.* In each case four stages of the propagation of the pulse are shown. It is our task to investigate this movement in some detail and determine some of the properties of such a propagation in the simplest cases.

The two simplest cases consist of plane pulses and spherical pulses; both may be considered at the same time. For the plane pulse, refer to equation 8 in Figure 13.19, in which it is assumed that the dilatation is independent of y and z. In other words, the compression or dilatation varies with only one coördinate, x. For the spherical case, refer to relation 8 in Figure 13.24. In this case we have to consider a slightly different function, which is r times the dilatation, but we can apply our theory here just as well as to the plane case. In discussing

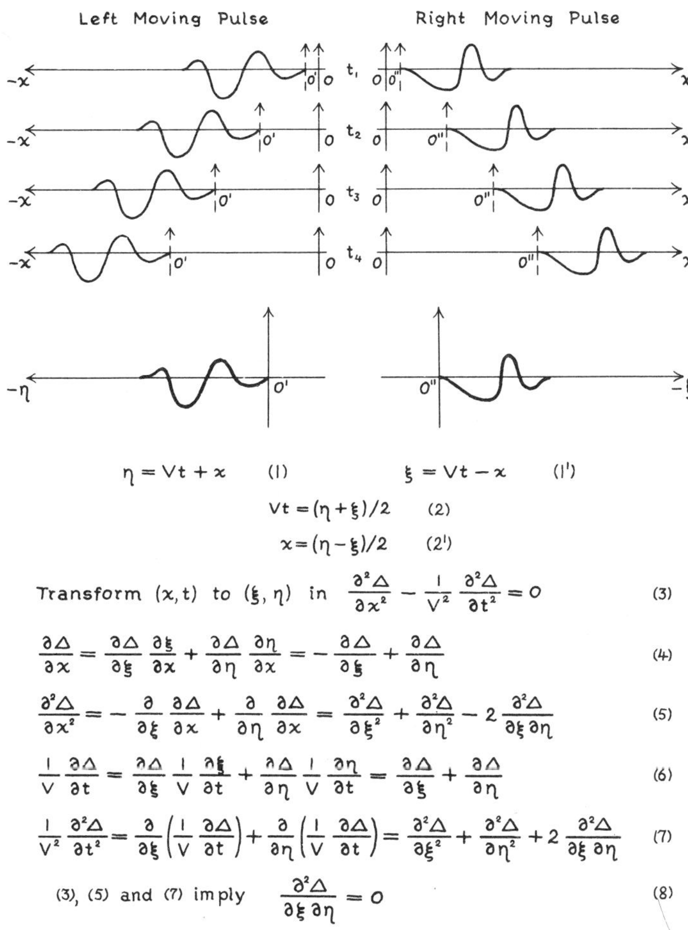

Solution of (8) is $\Delta = f(\xi) + g(\eta) = f(Vt - x) + g(Vt + x)$ (9)

FIG. 14.1. Right and left moving seismic pulses and their description.

Figure 14 1, both the plane and the spherical pulse problems should be kept in mind.

In the case of the left moving pulse, we make a transformation of coördinates. This is a change of our descriptive language and this change is expressed in relation 1. Our new coördinate is η, which is $x+Vt$. In making this transformation we have shifted from a fixed coördinate system to a coördinate system that moves with the pulse. This system moves toward the left with the pulse, so that the shape of the pulse appears fixed in the new coördinate system, as indicated in the drawing at the left, just above relation 1. We can apply the same type of argument to the right moving pulse and introduce a coördinate system that moves with the pulse as indicated by relation 1′. This gives a coördinate system fixed with respect to the pulse, as shown in the drawing at the right, just above that relation. Having expressed two kinds of moving coördinate systems, we can solve for t and x in terms of the moving coördinate variables η and ξ, as indicated in relations 2 and 2′. It is customary to call η and ξ the phase values of the pulse. The various numerical values of η, for instance, correspond to various positions on the pulse, and these, in turn, correspond to various phases of the pulse.

Now consider the equation of motion, which appears as relation 3. This relation is of the same form as relation 8 in Figures 13.19 and 13.24 referred to above. Our problem, as it is stated in the figure, is to transform this equation from x and t variables to η and ξ variables. This transformation is done in relations 4 through 7. We substitute in relation 3 the values found in relations 5 and 7, thus getting relation 8. This relation is equivalent to relation 3 which we started with, but is in a form that admits a very simple solution. It may be verified that the solution of 8 is given in relation 9. In this solution, the functions f and g are arbitrary to a very large extent. We are not interested in mathematical intricacies in this section, so we allow ourselves any degree of continuity for functions f and g that is required in order to make the treatment as simple as possible. A certain amount of continuity is needed so we can carry out the differentiation, say of f with respect to ξ or of g with respect to η. If these differentiations cannot be made, then relation 9 cannot be a solution of relation 8. We therefore assume as much continuity as is required in order for 9 to be a solution of 8.

The solution, 9, of equation 3 is a kind of milepost in the theory of the propagation of waves. It represents an exceedingly important step for all our future considerations. It is of interest to note that the solution has two parts. The part represented by f consists of a pulse traveling toward the right, and the part represented by g consists of a pulse traveling toward the left. If, therefore, we know, from the physical circumstances, that the pulse travels only in one direction, say toward the right, we simply assume that the value of g is zero and our solution is $\Delta = f\ (Vt - x)$. This corresponds to a fairly arbitrary pulse, that is, a pulse of fairly arbitrary shape traveling toward the right.

The application of these results to a plane pulse is direct. Their application to a spherical pulse involves relation 8 in Figure 13.24, where it is seen that $r\Delta_r(r,t)$ is the sum of the two arbitrary functions of phase. This means that Δ_r is the sum of these two arbitrary functions divided by r. Thus the spherical pulse is similar to the plane pulse, except that the amplitude of the pulse diminishes inversely with the distance r from the center.

The above description refers to what we may call the *regular propagation of pulses*. This propagation has a certain simplicity because the form of a pulse does not change with time. This simplicity has given the study of this particular aspect of propagation a central place in the whole theory of the propagation of seismic pulses. But this simplicity should be distrusted because it represents an enormous oversimplification.

It should be understood, therefore, that although in many cases seismic pulse propagation proceeds approximately according to the theory of plane and spherical pulses, these cases usually represent simplifications.

Hence it is important that we consider a more general aspect of seismic pulse propagation, namely, that described in terms of Huygens' principle. For this purpose refer to Figure 14.2, which illustrates an inner region, I, with a boundary B_O. It is assumed that at the time $t=0$, the seismic disturbance was entirely confined within B_O.

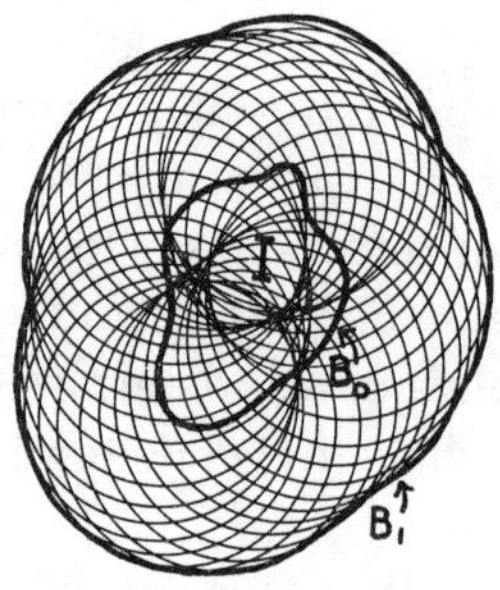

FIG. 14.2. Huygens' principle.

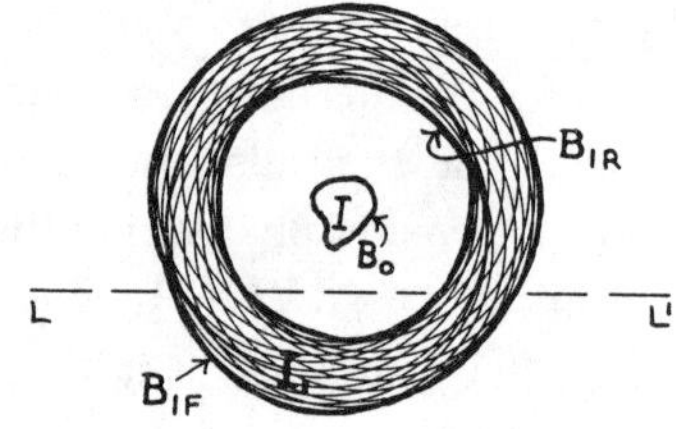

FIG. 14.3. Huygens' principle when a disturbance moves clear away from the source.

We wish to consider what happens to the disturbance after the lapse of a certain time, say until time t_1 after the time $t=0$. In order to find the *region of disturbance*, we multiply V, the velocity with which a plane or spherical pulse would be transmitted, by the time interval t_1-0, or t_1, getting thereby a distance $r_1=Vt_1$. With this distance as a radius, we take every point in the region I as a center and draw a circle (or sphere) of radius r_1. By doing this, we extend into a larger region that is bounded by the outer surface B_1, as illustrated in Figure 14.2. This boundary B_1 includes all the disturbance at time t_1.

Huygens' principle states that the construction described

above enables us to trace the course of a limited disturbance in a qualitative way. We have not described in detail the nature of this disturbance. It is possible, however, to express Huygens' principle in such a form that the details can be described.*

A second feature of Huygens' principle that is valid for three-dimensional pulse and wave processes is illustrated in Figure 14.3. The difference between this and Figure 14.2 is that in Figure 14.3 there is a smaller region of initial disturbance I bounded by B_0. We have allowed more time to elapse so that the disturbance has time to travel farther out in Figure 14.3 than it did in Figure 14.2. The result is that the disturbance travels outward with an outer boundary B_{1F} and an inner boundary B_{1R}, leaving a quiet region behind.

We have assumed, in drawing this indicated propagation of disturbance from I out to I_1 in Figure 14.3, that the material is uniform in all the space that we have to consider. If this assumption is not valid—for example, if the region had a boundary located at LL'—the situation would obviously be entirely different. In part, some energy would be reflected back into the original medium and this reflected energy would have a much more complicated character than we have assumed for the incident energy.

The most up-to-date treatment of the associated problems of Huygens' principle will be found in a long article by Marcel Riesz.† This article also includes a very complete bibliography.

We should point out that in Figure 14.3 we have assumed that the original region of disturbance, I, does not extend very far perpendicular to the plane of the diagram. But if I had been an infinitely long cylinder for example, the figure would

* This quantitative expression of Huygens' principle is beyond the scope of this treatment and we refer the reader for it to the book by Baker and Copson, *The Mathematical Theory of Huygens' Principle*, Oxford University Press (1939).

† *Acta Mathematica*, 81:1-233 (1949).

be entirely different, for the whole interior would then be filled with the disturbance. Thus we have a hint of an essential difference between the propagation of waves in three dimensions and in two dimensions. In the three-dimensional case, when the disturbance starts from a region limited in all three dimensions, after a certain length of time the disturbance will have left this region entirely and spread outward. On the other hand, in the same situation for a two-dimensional region forming an infinitely long cylinder, the quiet interior region is always lacking, no matter how much time we allow to elapse. The writer has shown that this disturbed interior region in the two-dimensional case always exists without an appeal to the three-dimensional situation.*

14.2. SPHERICAL PULSES

We now proceed to a more detailed consideration of spherical pulses, especially their generation. We have already seen that the equation of motion for the dilatation can be solved as indicated in relation 9 of Figure 14.1.

Figure 14.4 shows a spherical cavity on whose wall we have attempted to indicate a uniformly applied pressure, P. The center of the cavity is C and its radius is a. We are interested in the state of motion at a distance r from the center of the cavity. We shall suppose that the pressure is a function of the time, t, as is indicated in relation (1). The second relation gives the solution of the wave equation in a general form. The third relation makes an assertion to the effect that energy is not being transmitted inward. If this were not so, there would have to be some sort of energy source at infinity in order to produce the inward-traveling pulse represented by g. Since we assume no such source at infinity, our solution may be restricted and expressed as at the left of equation 4. We can rewrite the product of the radius and the dilatation as in the third

* See *Geophysics,* 15:447 (1950).

expression of this relation, and this expression can be rewritten as in the fourth expression in this relation. Accordingly, we can write relation 4 as shown in relation 5. This can be integrated from $r = \infty$ back to r, in order to obtain relation 6. We have as-

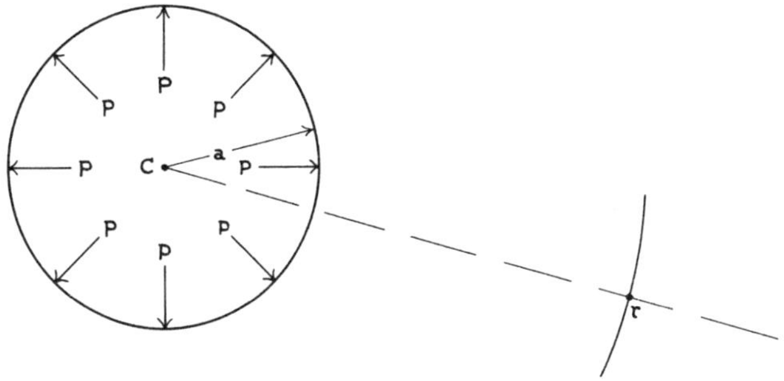

$$P = P(t) \tag{1}$$

$$\Delta(r,t) = f(vt - r)/r + g(vt + r)/r \tag{2}$$

$$g(vt + r) \equiv 0 \tag{3}$$

$$r\Delta(r,t) = f(vt - r) = r\frac{\partial u_r}{\partial r} + 2u_r = \frac{1}{r}\frac{\partial}{\partial r}(r^2 u_r) \tag{4}$$

$$\frac{\partial(r^2 u_r)}{\partial r} = rf(vt - r) \tag{5}$$

$$u_r(r,t) = \frac{1}{r^2}\int_{\infty}^{r} rf(vt - r)\,dr \tag{6}$$

$$R_r(r,t) = \lambda\Delta_r + 2\mu\frac{\partial u_r}{\partial r} = \lambda\frac{f(vt-r)}{r} + 2\mu\frac{f(vt-r)}{r} - \frac{4\mu}{r^3}\int_{\infty}^{r} rf(vt-r)\,dr \tag{7}$$

$$-P(t) = R_r(a,t) = (2\mu + \lambda)f(vt - a)/a - (4\mu/a^3)\int_{\infty}^{a} rf(vt - r)\,dr \tag{8}$$

Fig. 14.4. Relation between cavity pressure, stress, and pulse dilatation.

sumed in carrying out this integration that the displacement, u_r, at any time and at an infinite distance, is exactly zero. This makes the term corresponding to the lower limit of integration zero; hence we directly obtain expression 6. The relation at the left of equation 7 is merely a rearrangement of the first relation 3 in the top part of Figure 13.24, which is the relationship between the stress and strain for a radial problem. We have only to calculate the quantities in terms of the solution given by relations 4 and 6 in order to obtain the remaining part of relation 7. This relation is valid for every radius, r, in the medium for which the wave equation is valid. In particular it is valid at the edge of the cavity at $r=a$. We have set $r=a$ in the eighth relationship and have written $P(t)$ in terms of f.

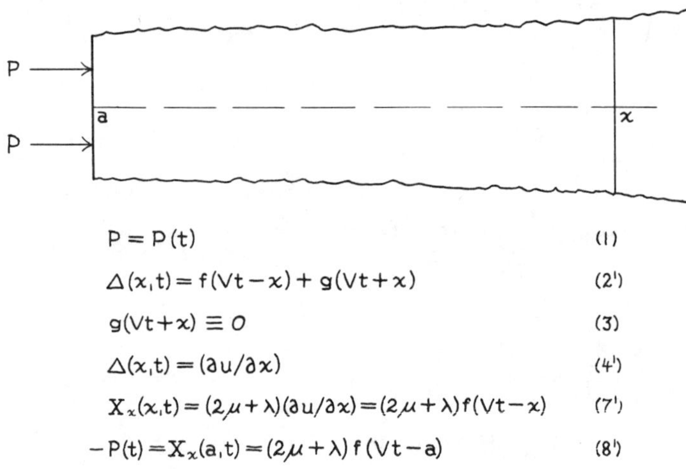

FIG. 14.5. Relation between pressure, stress, and strain when a pressure is applied uniformly to a plane.

Relation 8 represents an integral equation for the determination of the function f, every other function being supposed to be given. Naturally if we could determine f by observation, we would be able to calculate the function $P(t)$ directly from relation 8. Unfortunately the reverse calculation is not nearly so simple and will not be attempted here, although a calculation of this type can be made numerically (see Chapter 15).

Let us now digress a bit and consider the much simpler situation in Figure 14.5, in which plane waves are generated.

This may be considered as a limiting case of an infinitely large spherical cavity, but it is easier to consider the plane case directly, as we have done in the figure. We apply the time variable pressure, $P(t)$. The solution of the wave equation is given in relation 2′, with the part of the expression, g, corresponding to energy traveling toward the left, set equal to zero, as shown in relation 3′. Equation 4′ is the expression for the dilatation. The expression for the relationship between the stress and the strain is given in equation 7′; this corresponds to relation 7 in Figure 14.4. We set $x=a$ as before, and obtain equation 8′, which corresponds to relation 8 in that figure. We see that the two relations would coincide if the integral at the extreme right of relation 8 were equal to zero. This, however, is not usually so; hence the two cases are not identical. One way to weaken the effect of the integral is to make μ small. In other words, if we are dealing with a substance which is almost a fluid, the spherical and the plane wave cases are nearly identical. In this connection it is of some interest to recall that, under certain shot conditions, μ is probably quite small, although not quite zero. For instance, this would be the case if the shot were fired in mushy wet clay because the rigidity modulus for wet clay is very small.

Returning now to relation 8 in Figure 14.4, we see that, in general, the dilatation is not simply related to the pressure variation, P, as it is when the source is a plane.

However, let us consider for a moment under what circumstances the integral part of relation 8 becomes very important. Suppose, for example, that the pressure is applied for an extremely short time and then released so that $P(t)$ differs from zero only during a very short period. We may guess, and it can be proved, that under these circumstances the pulse itself is also extremely short. In such a case, if a is very large, the expression on the right of relation 8 will be very small, for although the integration appears to extend from infinity to a, the function f is identically zero from infinity to the pulse front.

Consequently, the range of integration could well be reduced to the integration from the pulse-front radius back to the cavity radius, a. Thus if the range of integration is very short, as it will be if the pulse is of short duration, the integral will be approximately the average value of rf over the range of integration, divided by a^3 times the difference between the outer and inner range. This is approximately the range divided by a^2, which is quite small compared to the regular term to the left of it. Thus when the pulse is very short, the behavior of a spherical pulse may be regarded as approximately governed by relation 8 without the term on the extreme right. This of course is only a first approximation, but it is mentioned in order to illustrate an important property, namely, that if sharp transmission of a sharp pulse is desired, the pulse should be generated in a large cavity because the argument would break down entirely if a were taken as very small.

The important quantity to be considered in the above argument is the radius, a, in relation to the total length of the pulse that is sent outward. Since in the average shot hole a is very small compared with the pulse, which may be several hundred feet long, the generation of a pulse by a shot in a shot hole must be governed by the complete relation 8, and not by any approximation such as we might consider if we used relation 8′ in Figure 14.5.

It should be emphasized that relation 8 has been derived on the assumption that the wave equation in its simplest form is valid, and that this has been derived on the assumption that Hooke's law is valid. We know, however, that Hooke's law is not valid in the immediate neighborhood of a shot; hence we now discuss this in greater detail.

14.2.1. Spherical Cavity Sources

Refer now to Figure 14.6, in which is shown a cavity surrounded by a region called the potting region. This in turn is surrounded by a region of more than 1 percent failure of recovery which in turn is surrounded by an outer region of less than 1 percent failure of recovery. These regions have been separated in this way primarily for the purpose of this discussion; the separation is arbitrary. A cavity is an open region in which the explosive is located. The explosive is surrounded by a fluid which transmits the pressure to the outer walls of the cavity. The potting region is the region which is broken up by the explosion. Thus after explosion this region can no longer be considered as even approximately similar to the solid it was

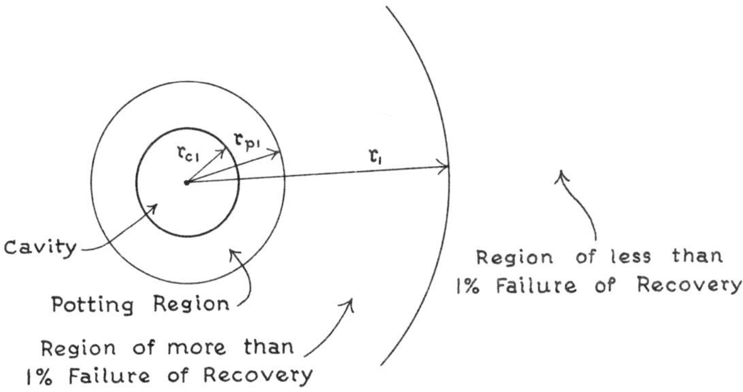

FIG. 14.6. Departures from Hooke's law near a shot source.

originally. The material in the potting region will fall to the bottom after the explosion; hence a cavity that was once spherical will not be spherical after the explosion. In the outer region, where the recovery fails by less than 1 percent, we may regard Hooke's law as approximately valid. In the inner region, exterior to the potting region, there is a more serious failure of recovery, so that it is not legitimate to consider that Hooke's law is valid in it.

The size of this inner region of partial failure is not very well determined at the present time in any specific case. It may be expected to vary to a considerable extent with the kind of material. This variation is illustrated in Figures 14.7 and 14.8. Figure 14.7 shows the relationship between stress and strain in a silty clay.* The part of the curve marked *I* represents the relationship when the pressure is first applied. When peak pressure, *P*, is reached, point *P* on the graph is also reached, after which the return path *R* is followed down to point *F*, which usually does not coincide with *O*. The distance *OF*, which is

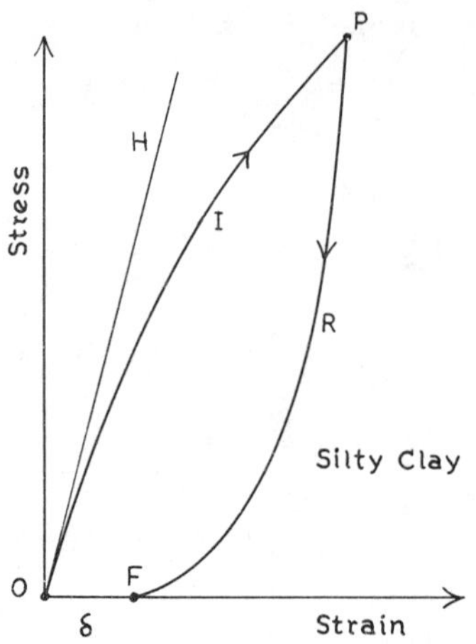

FIG. 14.7. Stress-strain behavior of a silty clay. (From *The Effects of Atomic Weapons*, 1950, p. 410.)

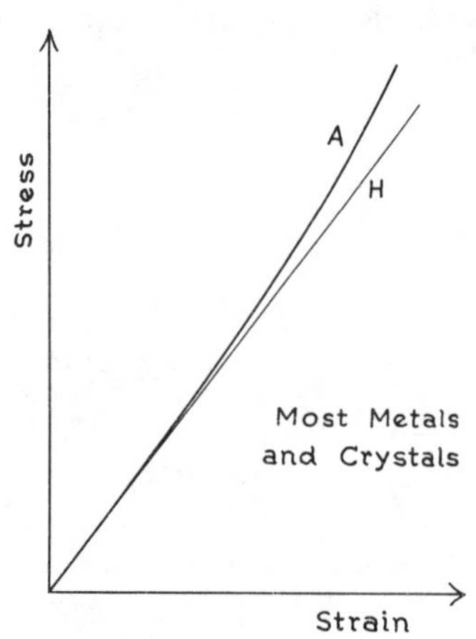

FIG. 14.8. Stress-strain relationships for most metals and crystals.

labeled δ, is what is called the *failure of recovery*. It may be a large percentage of the total strain, as in the figure, or only a small percentage. If it is small, we assume that Hooke's law is approximately valid. If Hooke's law were completely valid in the case illustrated in the figure, the stress-strain curve would

* See *The Effects of Atomic Weapons*, Samuel Glasstone, Executive Editor, U.S. Govt. Pr. Office (1950). See particularly Appendix B, page 410, by C. W. Lampson.

be the straight line *H*, which is drawn tangent to *I* at the origin, *O*.

It is of interest to compare silty clay with certain other common materials such as most metals and crystals. The stress-strain relation for these other materials is indicated in Figure 14.8, which shows a curve, *A*, and a straight line, *H*. The actual curve *A* bends away from *H* in the opposite direction from the direction taken by *I* in Figure 14.7. This difference in character is very important and has great influence on the character of a seismic pulse generated by a shot.

The difference between the situations in these two figures is very important as far as the velocities are concerned. The velocity of transmission of the waves is determined by the slope of the stress-strain graph. If this slope decreases for large strains, as it does in the case of silty clay, the velocity of transmission also decreases. On the other hand, if it increases as in Figure 14.8, the velocity of transmission increases. In other words, in Figure 14.8 the strong waves are transmitted more quickly than the weak ones, whereas in Figure 14.7 the strong waves are transmitted more slowly than the weak ones. For all materials obeying a relation like that illustrated in Figure 14.8, this has the effect of generating what are called shock waves. Shock waves are generated in air and water by explosions, because the stress-strain relations in those substances curve as indicated in Figure 14.8. But when a shot is fired in silty clay and the material satisfies the stress-strain relations illustrated, shock waves cannot be generated, for the weaker pulses travel faster and hence there is a tendency for the pulse to spread out in back of the pulse front instead of building up in front in the form of a shock wave. This illustrates a well-known fact observed by almost every seismologist.

In the light of our discussion of Figures 14.7 and 14.8, it may be well to return to the situation in Figure 14.6 and the discussion of Figure 14.4, especially with respect to relation 8.

The question is, what value are we to take as the radius, *a*, of the source. If we take the shot hole radius itself we will obviously be wrong, for we will have to assume that Hooke's law applies in this neighborhood, which is not the case. It seems more likely that a radius similar to r_1 in Figure 14.6 will be suitable for the relation between the pressure and shape of the pulse. If we use this, we find that we have not actually computed the pressure variation at the boundary of the shot hole. This latter computation requires much more difficult and detailed consideration of the thermo-mechanical properties of the material.

14.2.2. The Shot Hole

In this section we consider in detail another aspect of the source. We suppose that we have an actual shot hole such as is illustrated in Figure 14.9. This figure shows three situations. At the upper left is shown the case in which the tamping fluid, which we assume to be water, has a lower seismic wave velocity than the material surrounding the hole. It is thus evident that the spreading of the pulse may be shown as in the figure, with the pulse in the surrounding material going ahead of the pulse in the water column. This situation is similar to that shown in detail in the lower left drawing. In other words, the shot hole tends to be constricted somewhat in front of the water pulse traveling up the hole. This has the effect of offering resistance for the water pulse that is additional to that offered by the walls of the shot hole itself. We might expect that such a situation would correspond to an *efficient tamping of the charge.*

In the situation illustrated in the middle drawing, the water velocity is much higher than the seismic velocity in the surrounding medium. Hence the pressure wave in the water column precedes the seismic wave in the surrounding medium and the tamping may be expected to be somewhat less efficient. It is difficult to estimate whether this effect will be of major

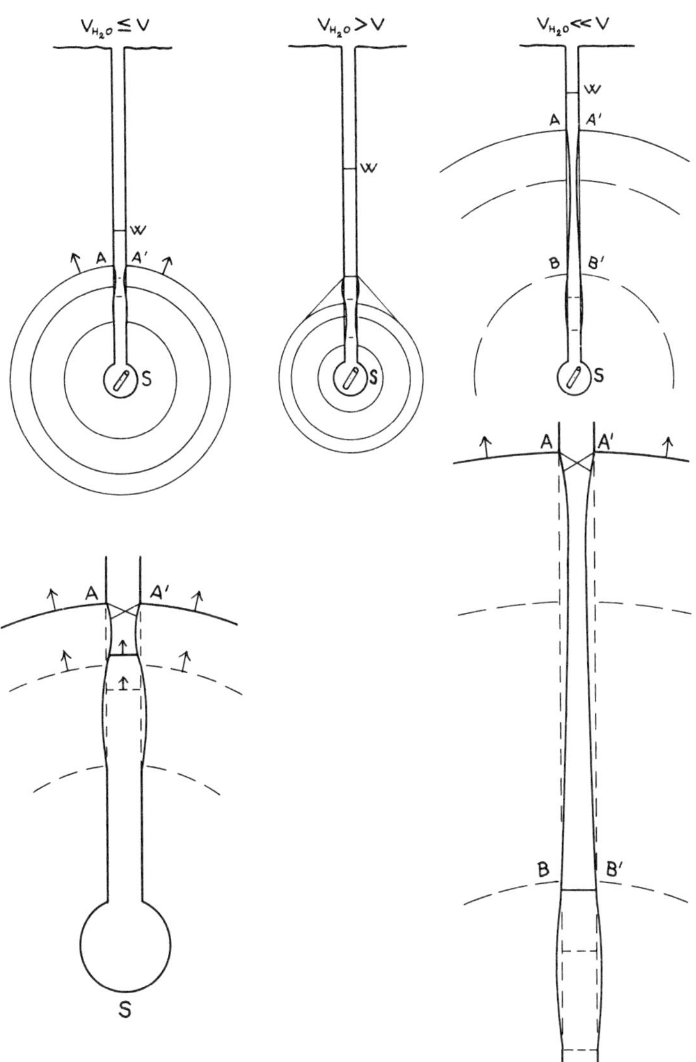

FIG. 14.9. Shot hole tamping relationships.

importance, but a certain amount of inefficiency should be present, as is clearly evident from the diagram.

Another situation sometimes encountered is that in which the seismic water velocity is very much lower than that in the surrounding neighborhood. This is shown in the drawings at the right of the figure. Here the seismic pulse is advancing far beyond the water pulse. The seismic pulse may advance so far beyond that the rarefaction or dilatation in the material behind it will coincide with the compression of the water pulse. Thus the seismic pulse tends to help dissipate the energy in the water column. Such a situation in itself may not be favorable.

Any situation that tends to spread the pulse nonspherically may be undesirable. For example, with vertical jet pulses, when conical charges are used to form long explosive jets, no increase in efficiency is observable. Instead, it sometimes appears as though there were actually a small decrease in efficiency. This is perhaps attributable to the change in the shape of the explosive region so that it no longer has point symmetry.

The difficult source problem has been seriously oversimplified in the above discussion. For another viewpoint, see T. C. Poulter's paper.*

14.2.3. Minimum Oscillatory Character of a Pulse

The writer has discussed the minimum oscillatory character of a pulse in a paper.† We shall not enter into the details of this discussion but will mention only the simplest part of the deduction. Figure 14.10 shows a pulse with a front and a rear. The displacements in the pulse are all forward, as indicated by the little arrows. Since we assume that behind the rear of the pulse the medium has returned to a quiet state without displacement or displacement velocity, the particles must return to their original positions. The displacements shown correspond

* T. C. Poulter, *Geophysics*, 15:181-207 (1950).
† *Ibid.*, 14:17 (1949).

to a compression on the front of the pulse. However, because of the necessity of returning to the original positions, there is a rarefaction on the rear of the pulse. Thus one full oscillation is required.

It may be mentioned that a pulse of this type is actually a

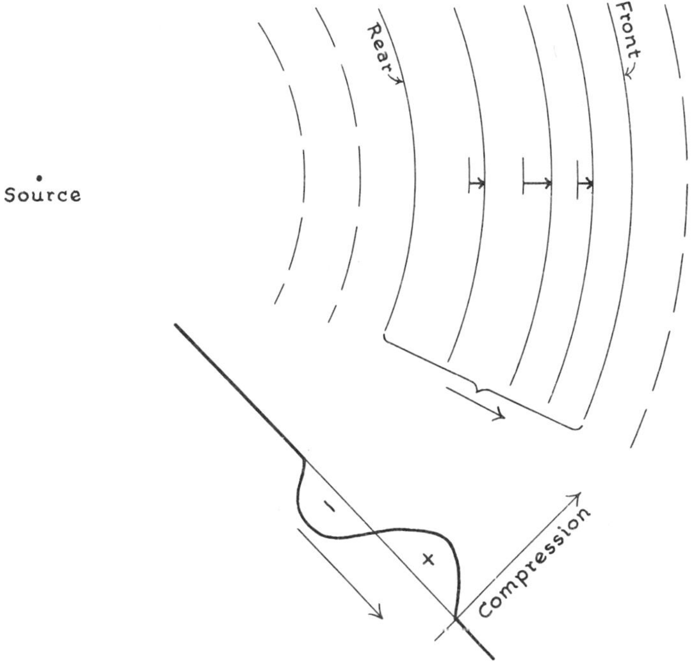

Fig. 14.10. Minimum oscillatory character of a pulse from a point source.

little less oscillatory than is a true spherical pulse in a solid. A little more oscillation is required in such a case, as will be clear from a consideration of relation 8 in Figure 14.4. The integral at the right of this relation must be zero when extended over the entire pulse. As a consequence, the pulse has three half oscillations instead of the two shown in Figure 14.10. The details of this deduction are given in the paper referred to above.

14.3. EFFECT OF A BOUNDARY

We have assumed until now that the medium in which our waves travel is not limited by any boundary. Of course in any actual case the medium must be bounded. This boundary has a serious effect upon all our conclusions regarding the propagation of waves. In the first place the wave equation which we deduced no longer holds. It is true that equations 1 to 3 of Figure 13.18 are still valid, and it is probably true that relations 7 through 12 of Figure 13.17 are also valid. But in this latter case, μ and λ are no longer constant, but vary with position. Because of this, when we differentiate in the equations of motion, we have to differentiate the elastic constants as well as the displacements. If we take this fact into account, we no longer have equations 5, 6, and 7 in Figure 13.19. Consequently we cannot derive equations 8, 9, and 10 in that figure. Our immediate conclusion is that under these more general circumstances the separation of types of elastic wave motion into a purely compressional and a rotational type is not stable. In other words, rotational waves and compressional waves continually interchange. The compressional parts are transferred into rotational parts, and vice versa.

An example of a change of elastic constants is seen when we cross a discontinuity. In such a case, however, we cannot say that the differential equations 5, 6, and 7 in Figure 13.19 are valid at the interface, although they may be valid with their appropriate constants and densities on each of the two sides of the interface. Thus we find that the interface is a place in which waves of one type are transformed partially into waves of another type. For example, when a compressional pulse travels into a homogeneous medium and strikes a boundary, at this boundary occurs a transformation of the waves into new compressional waves and new rotational waves. Furthermore, certain other types of waves associated with the interface may be generated during this process. Unfortunately, this transformation at the boundary is quite complicated and difficult to follow in detail. Nevertheless, we shall try to follow some of its aspects so that we may picture the process of generation to ourselves, at least roughly.

14.3.1. Plane Pulse Reflection. Normal Incidence

In this section we consider the reflection and refraction of a plane pulse at normal incidence, that is, such that the plane front of the pulse is parallel to the interface separating the two homogeneous media under consideration. This situation is illustrated in Figure 14.11, in the upper right of which is shown a medium such that the compressional waves travel with a velocity V_2. The density is ρ_2, and the elastic constants are μ_2 and λ_2. This medium is separated by interface IOI' from the medium on the left, which has a velocity V_1, a density ρ_1, and elastic constants μ_1 and λ_1. A position is shown for the pulse with front F and back B. The pulse travels toward the right. This pulse will strike the interface at a certain time; part will pass through and part will be reflected. At this interface there are certain conditions that must be satisfied. For example, at the interface the value of x is assumed to be zero, as shown in the diagram. But also, at the interface, there must be *continuity of normal stress for all time*, as indicated in the first relation. The second relation gives *continuity of the normal displacement for all time* at the interface. Let us assume that the incident pulse is of the form shown in relation 3, the transmitted pulse is of the form shown in relation 4, and the reflected pulse is of the form shown in relation 5. Clearly relations 3 and 4 represent pulses traveling toward the right, whereas relation 5 represents a pulse traveling toward the left, as it must if it is to represent a reflected pulse. We can use relations 1, 3, 4, and 5 to obtain relation 6. This relation is also obtained by using equation 7 in Figure 14.5, which relates the stress to the solution of the differential equation f. The left side of relation 6

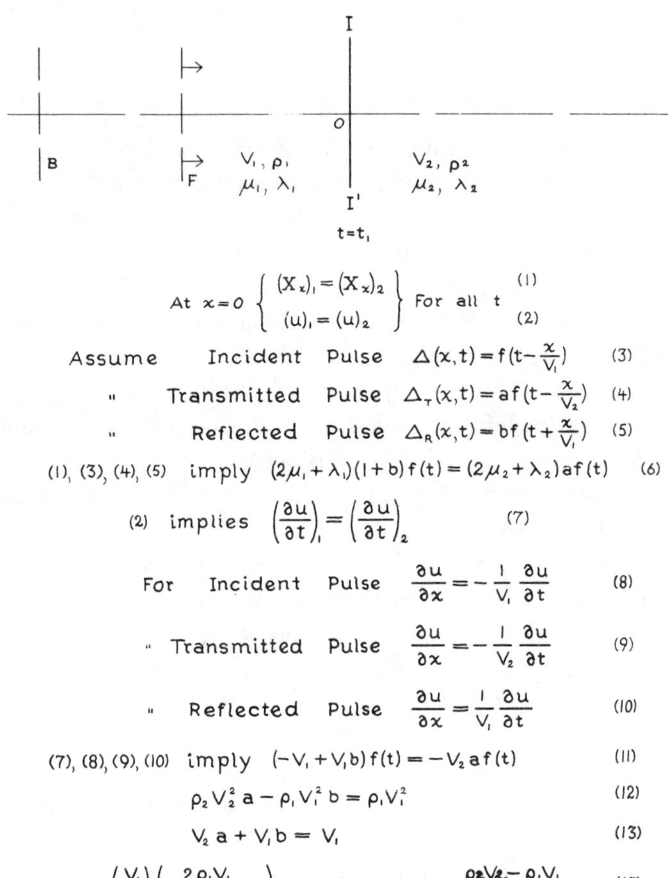

FIG. 14.11. Normal incidence reflection and transmission of a seismic pulse.

represents the left side of relation 1, and the right side represents the right side of relation 1. Now since relation 2 holds for all time, it is possible to differentiate it with respect to time and get relation 7. In other words, not only is the displacement continuous for all time, but the velocity of displacement is also continuous for all time at the boundary. We can then proceed to the consideration of the next three relations of Figure 14.11, namely, relations 8, 9, and 10. These relate the velocity of displacement to the dilatation. By combining relations 7 through 10 with relations 3, 4, and 5, we obtain relation 11. Relations 6 and 11 give two simultaneous equations from which we can calculate a and b. We take a time for which $f(t)$ is not zero and divide through by $f(t)$. When we do this, and take account of the relationship between the velocities, the elastic constants, and the densities, we obtain relations 12 and 13 which can be solved for a and b as shown in equations 14 and 15.

The value of a may be regarded as the coefficient of transmission and the value of b as the coefficient of reflection. That is to say, the larger the value of b the larger the amount of reflected energy, as indicated by relation 5. Thus we see that the amount of reflection depends entirely on the product of the density and the compressional wave velocity for each medium. If this product were the same in the two media the value of b would be zero and there would be no reflection. In practice, however, the value of b is likely to be of the order of one tenth for a reasonably good reflection, and it usually consists primarily of a contribution due to the variation of velocity. But sometimes a fairly large change in velocity is accompanied by an opposite change in density, so that the reflection is very weak.

We see from Figure 14.11, particularly equations 8 through 10 which relate the particle velocities to the dilatations for the various wave components, that we can easily compute the momenta of the various pulses by multiplying the element of volume with the corresponding density times velocity. The product of this multiplication can be summed or integrated over the whole pulse in each case. In this way we can compute the momentum carried by each pulse and it can easily be shown that in each process the momentum is conserved, not only for the whole pulse, but for any part of the pulse as well. In addition to calculating the momentum we can also calculate the energy, and it can be shown very readily that the energy is

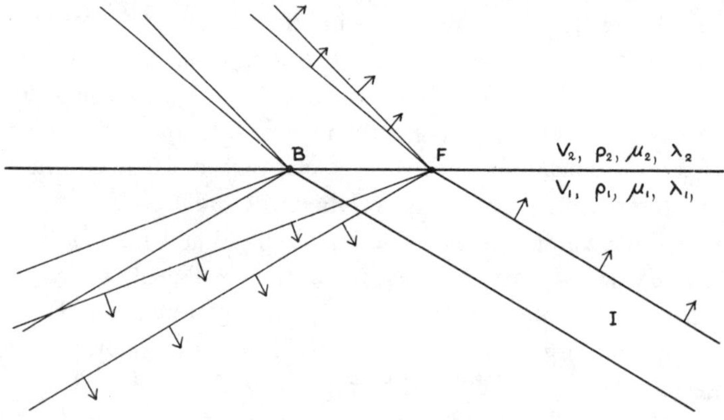

FIG. 14.12. Reflection and refraction of a seismic pulse at oblique incidence.

likewise conserved in the process of reflection and refraction described above. Thus the process described in Figure 14.11 meets all the mechanical requirements and its results car. therefore be regarded as having very high reliability.

14.3.1.1. General Incidence

When we go from normal incidence to general incidence, as illustrated in Figure 14.12, we do two things. In the first place we greatly increase the complexity of the problem. Naturally two more conditions are required corresponding to the continuity of tangential stress across the interface and the continuity of tangential displacement at the interface. These two conditions give rise to two additional relations, and we should expect to find two additional types of waves. This is indeed the case, because a reflected rotational wave and a refracted rotational wave are introduced by this process.

We shall not enter into the details of this deduction, but only note that it can be done along the same lines as were used in Figure 14.11, with the above additions. This computation gives relations that were previously obtained by C. G. Knott and Karl Zoeppritz.* These computations have attained a sort of classical authority over the years. It is true that there is nothing wrong with the mathematical deductions involved. The main question is whether this represents the type of physical problem that is of real interest in connection with an understanding of the reflection and refraction of seismic waves. In Figure 14.12 we note that, in addition to the incident pulse I, there are two reflected pulses and two refracted pulses joined at the front, F, and the back, B, on the interface. The figure shows an instantaneous situation. As time passes, points B and F move toward the right. Thus, although we attempt to consider a *transient* situation by considering an incident pulse I, in reality this is a *steady state* in which the whole system moves uniformly to the right.

If we wish to avoid this steady-state situation, we have to proceed somewhat as indicated in Figure 14.13, which shows the incident pulse I, and the interface passing through the back point, B, and the front point, F. Extending upward and to the left is the continuation of this incident pulse, shown in broken lines. If the incident pulse were not stopped along the interface between B and F, but continued, and if an infinite and homogeneous medium is assumed, no secondary reflected or re-

* C. G. Knott, *Philosophical Magazine*, 48:64 (1899); Karl Zoeppritz, *Gottingen Nachrichten*, p. 66 (1919).

fracted waves would be generated. It is possible to calculate the stress system required along the interface BF in order for the incident pulse to travel as if it extended out at the left to infinity. By applying this traveling stress distribution, we can keep the incident pulse from generating any secondary waves at all. Then if we want to create a transient situation, we can do so by suddenly or gradually removing this traveling stress distribution. When this is done, a process of reflection will

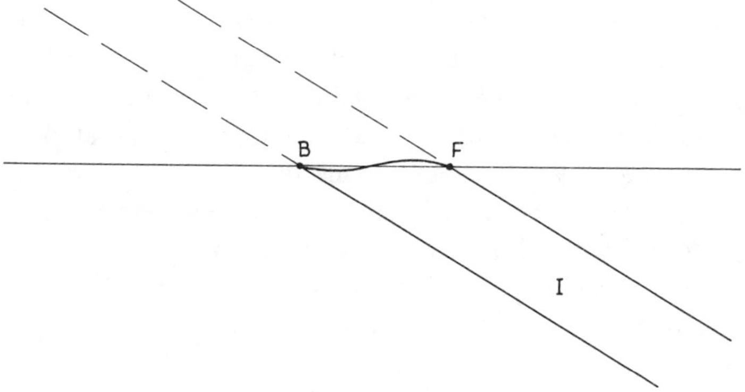

FIG. 14.13. A semiplane source for transient reflections.

begin. It is fairly evident that, in the early stages, this reflection process will not lead to a situation identical with that shown in Figure 14.12, although after a long period of time it may reach a stage that is indistinguishable from it. As we see, Figures 14.12 and 14.13 represent two kinds of physical problems. The problem represented in Figure 14.12 has been completely solved by Knott and Zoeppritz. That represented in Figure 14.13 has not been solved so far as the author knows. It seems probable that this problem has some aspects that come closer to the actual physical problem involved in seismic prospecting than that in Figure 14.12. Nevertheless, it must be admitted that neither of these figures represent the situation adequately.

14.3.2. The Point Source Problem

Let us consider the situation illustrated in Figure 14.14, which shows a source point S and a series of down-going rays

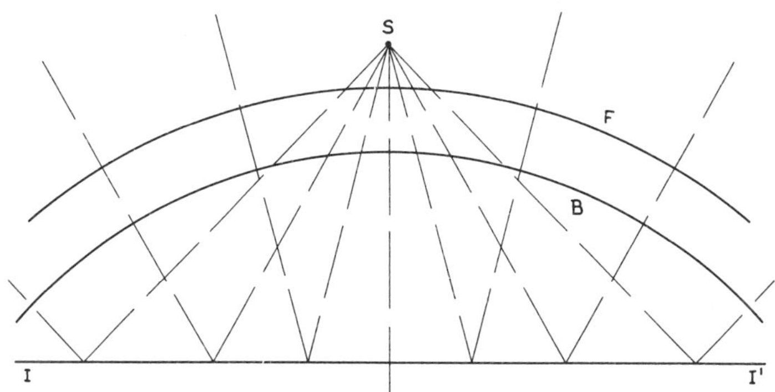

$$\frac{1}{r^2}\frac{\partial}{\partial r}\left(r^2\frac{\partial \Delta}{\partial r}\right)+\frac{1}{r^2 \sin\theta}\frac{\partial}{\partial \theta}\left(\sin\theta\frac{\partial \Delta}{\partial \theta}\right)=\frac{1}{V^2}\frac{\partial^2 \Delta}{\partial t^2} \tag{1}$$

$$\Delta(r,\theta,t)=\Delta_r(r,t)\,\Theta(\theta) \tag{2}$$

$$\left[\frac{\partial}{\partial r}\left(r^2\frac{\partial \Delta_r}{\partial r}\right)-\frac{r^2}{V^2}\frac{\partial^2 \Delta_r}{\partial t^2}\right]\Theta=\left[\frac{1}{\sin\theta}\frac{d}{d\theta}\left(\sin\theta\frac{d\Theta}{d\theta}\right)\right]\Delta_r \tag{3}$$

$$\left[\frac{\partial}{\partial r}\left(r^2\frac{\partial \Delta_r}{\partial r}\right)-\frac{r^2}{V^2}\frac{\partial^2 \Delta_r}{\partial t^2}\right]\Big/\Delta_r=\left[\frac{1}{\sin\theta}\frac{d}{d\theta}\left(\sin\theta\frac{d\Theta}{d\theta}\right)\right]\Big/\Theta=k \tag{4}$$

$$k=0 \quad \text{implies} \quad \frac{d}{d\theta}\left(\sin\theta\frac{d\Theta}{d\theta}\right)=0 \tag{5}$$

$$\sin\theta\frac{d\Theta}{d\theta}=c \longrightarrow d\Theta=\frac{c\,d\theta}{\sin\theta}$$

$$\Theta=c\log_e\tan\frac{\theta}{2}+c \tag{6}$$

$$\Delta_r=f(vt-r)/r \tag{7}$$

Fig. 14.14. Failure of wave equation to be satisfied for simple spherical reflection process.

striking the interface *II′* and being reflected back upward. Also shown is a front, *F*, and a back, *B*, for the reflected pulse. We now consider the possibility of a pulse reflection process in which the pulse shape changes only with the angle θ, away from normal. Without much trouble, we can show that such a pulse would have to satisfy relation 1. If the solution is to be of the type specified, it will have to be of the form shown in relation 2. If we substitute this in relation 1, we obtain an expression which we may write in the form shown in 3. It is then possible to divide both sides of this equation by $\Theta \cdot \Delta_r$, and obtain relation 4. The expression at the left of this relation is a function of r and t only; that in the middle is a function of θ only, and that on the right is a constant. It is evident that the first two expressions must be constant, since they depend upon different independent variables. If they varied with their variables they could not maintain their equality because these variables are independent. In order that we may continue to work with the pulse solution, we have to set $k=0$ to obtain the wave equation. This gives relation 5 which can be solved as shown in relation 6. The solution of our original equation is given in relation 7. It is therefore possible to obtain a solution of the equations of motion that will satisfy the specified conditions, providing relation 6 is satisfied. Unfortunately this relation is of the type that can never be satisfied in any actual case, because when θ is zero the tangent of $\theta/2$ is also zero and the logarithm is negative infinity. Thus the solution Θ has a negative infinite singularity at $\theta=0$. We could change the sign of the singularity but we could not remove it without making $c=0$. If we make $c=0$, we revert to our original type of solution, which is a pure spherical wave.

We thus find that a pulse of spherical shape with varying amplitude does not serve the purpose. It is this type of solution that has been presupposed as adequate for our purposes if we use the Knott and Zoeppritz type of solution to obtain a picture of the variation of reflected amplitude with the angle of incidence. We find that, although the picture of the process obtained by adapting the Knott and Zoeppritz results to the point source problem satisfies the conditions at the interface, it does not satisfy the equations of motion within the upper medium where a reflection is taking place. It is obvious, therefore, that something is wrong with this picture.

An adequate treatment of this problem requires a much more thorough procedure. We shall not discuss it in any detail here; instead we refer the reader to Cagniard's work.* His treatment covers the problem of a point source in a homogeneous medium separated from a second homogeneous medium by a plane interface. His method of treatment makes possible the discussion of pulses, and is admirably suited to the solution of this very difficult problem. Cagniard considers a pulse as his source, requires the four continuity conditions on the plane boundary, and also requires that his solutions satisfy the wave equations on the two sides of the boundary for compressional and rotational waves.

Love's *Mathematical Theory of Elasticity* presents a proof of a uniqueness theorem on pages 176-177. This theorem states that, given the initial conditions of a problem, such as that solved by Cagniard, the solution found—if any is found—is the correct one because it is impossible to have two different solutions of an initial value problem.

In view of this uniqueness theorem, it seems odd at first sight that Cagniard's treatment should in any way be different from earlier treatments. However, close study shows that it is indeed different. For example, studies similar in scope to Cagniard's were made by Lamb in 1904, by Sommerfeld in 1909, by Weyl in 1919, by Jeffreys in 1926, by Muskat in 1933, by Sakai in 1933, by Joos and Tetlow in 1939, and more recently (1947) by Fu. All these writers have approached the problem in a way essentially different from that used by Cagniard. They have considered an analysis of the pulse into its harmonic components. They have attempted to solve the problem for the

* L. Cagniard, *Reflection et Refraction des Ondes Seismiques Progressives*, pp. 1-255 (1939).

steady-state harmonic case and then, by means of a Fourier synthesis of the steady-state result, have presented a solution for the pulse case. In view of the uniqueness theorem, it may seem surprising that there has not been perfect agreement with regard to the results thus found. The reason for this lack of agreement can be found in a paper by P. S. Epstein,* in which he shows that a great deal depends on the mathematical representation of the source, several representations giving different results being possible. The question may be stated in terms of the physical problem that was solved by each author. It happened that *different physical problems were solved.* When the steady-state harmonic solutions are considered, it is necessary to specify the behavior of all disturbances in the neighborhood of the interface at very great distances from the source. It is the difficulty of specifying this on the basis of a physical intuition of the problem that makes this approach so difficult. In the first place, we simply do not know how the waves at the interface behave in the neighborhood of infinity. Yet we have to decide this if we are to state the physical problem completely. Cagniard gets around this difficulty by stating his problem in terms of an initial pulse. Thus he can require that everything be quiet beyond the pulse front, and this gives the necessary determination and uniqueness for the problem. Once the problem has been set up according to his procedure, mathematical care and rigor are all that is required to obtain the solution. Unfortunately, the mathematical difficulties involved in all the above studies are rather large, so we shall not attempt to review the methods here. All that should be understood is that Cagniard has stated the problem and solved it, apparently without any mistakes. The reader who is interested in the complete solution of this difficult problem is referred to Cagniard's book.

It will now be our purpose to consider some problems of

* *Proceedings of the National Academy of Sciences* (1947).

seismic pulse propagation which are more elementary. It is thought that methods should be devised to give the geophysicist a more direct intuitive grasp of the mechanics of seismic pulse propagation, and this is the purpose of the material immediately following.

14.4. DIFFRACTION

The reader's first contact with the phenomenon of diffraction was probably in the field of optics. As is well known, if a beam of light is allowed to pass through a large hole in a screen it travels through the hole in approximately straight lines in the form of a cylinder of light whose shape is roughly determined by the shape of the hole. If the hole is very small, this description is no longer adequate, for the light is spread laterally to a greater or less degree, the spread increasing as the hole grows smaller. Thus the light appears to bend around the edge of the hole and pass into the region of the shadow. This phenomenon in optics is called *diffraction.*

In the propagation of seismic pulses, diffraction is much less exceptional than it is in optics. The reason is that optics ordinarily deals with very high frequencies and very short wave lengths, short as compared with the dimensions of holes, slits, obstacles, etc. But seismic pulses are usually several hundred feet in length. Thus, the scale of the pulse compared with the scale of obstacles and irregularities is usually large. Hence diffraction phenomena assume primary importance in the transmission of seismic pulses.

The importance of diffraction phenomena is increased all the more because of the usual approach to the subject. The usual approach—the one that we have used—first discusses the propagation of plane and spherical pulses or, even worse, of plane and spherical harmonic waves. The simple properties of the propagation of such pulses are so easy to grasp that there is a serious danger of trying to use these properties for all types of seismic pulse propagation. The author has tried to warn against this in a recent paper* in which he shows that the simple outward cylindrical type of propagation is impossible.

In this section we consider briefly some of the exceptional cases and devise a means of picturing roughly what must happen in such cases.

Figure 14.15 shows a fault, the high side being on the left and the low side on the right. Above the high side is shown an idealized portion of a plane reflected pulse with a sharp edge directly over the edge of the fault. This is the type of propagation that is often pictured in seismic records illustrating the presence of faults. The records show sharp breaks corresponding to the shift in the fault. It is our purpose here to show that this situation must be very exceptional and not representative of the usual behavior of seismic waves from fault neighborhoods.

The first relation in Figure 14.15 is the wave equation for compressional waves in which there are two spatial dimensions, x and y. Here we suppose y to be the horizontal dimension and x the vertical dimension. Let us try to fit to the problem a solution of the type expressed in relation 2. This is the type of solution required for the simple reflected half plane pulse shown in the figure. Substitution of relation 2 in equation 1 yields relation 3, which can be integrated to obtain relation 4. Thus, if the solution is to be like 2, the lateral variation must follow relation 4. If $c=0$ we have the ordinary case of a plane wave that does not stop at the edge of the fault. If $c \neq 0$ we have a continuously varying plane wave that likewise does not stop at the edge of the fault. The simple solution pictured above is not possible.

Since we would like to form some idea of what does take place, we refer to Figure 14.16, which shows the same fault and

* C. Hewitt Dix, "Pulse Propagation in Two Spatial Dimensions," *Geophysics*, 15:447 (1950).

the same reflected pulse. However, the reflection is shown only in the first stage of the process; thus the thin layer above the reflector corresponds to the beginning of the reflection process. Except at the corner representing the top edge of the fault, the wave equation is satisfied, together with the conditions at the interface. Therefore, except for this corner region, this is a proper representation of the process in terms of relation 2 in

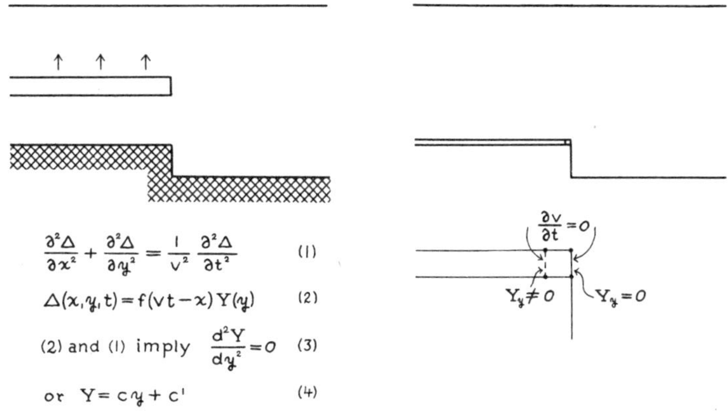

FIG. 14.15. Failure of wave equation to be satisfied at edge of fault for oversimplified reflection pulse.

FIG. 14.16. The beginning of a reflection process near the edge of a fault.

Figure 14.15. Let us see what happens at this corner. For this purpose we have enlarged the corner situation in the lower drawing. We see that at the corner we can consider a little cube along part of which the lateral stress is zero in the first approximation. But if we consider the part interior to the wave pulse we find that the lateral stress is not zero. Our assumption regarding the motion of the particles in the wave is that this motion is entirely vertical in the first approximation, so that in this approximation the lateral velocity $\partial v/\partial t = 0$ as indicated.

We now discuss what happens to a small cube when initially

the velocity in the lateral direction on the two faces is zero, the stress on one of the faces is zero, but the stress on the other face is not zero. For this, refer to Figure 14.17. At the left is

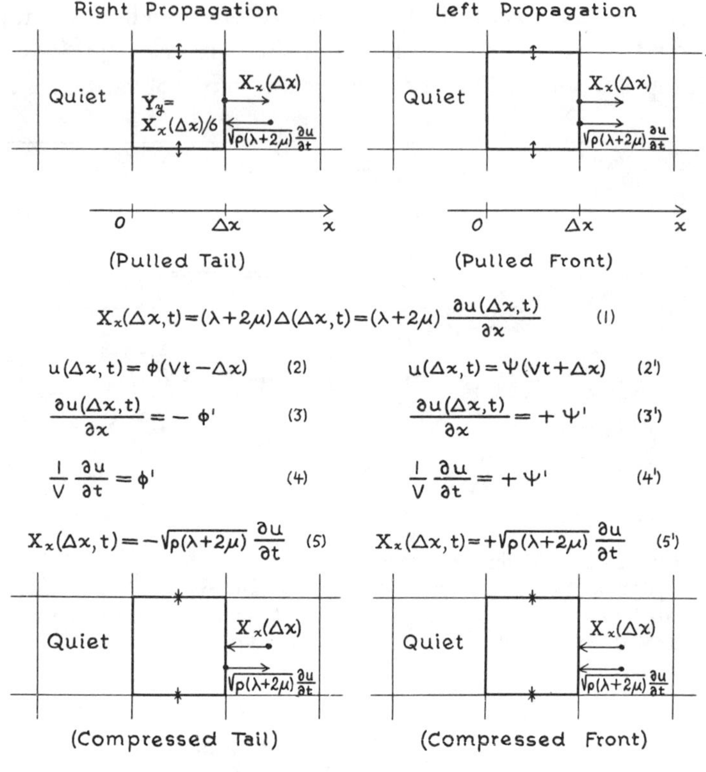

FIG. 14.17. Conditions for right and left propagation with pulled tail, compressed tail, pulled front, and compressed front.

shown the situation required for propagation toward the right; at the right is shown the situation required for propagation toward the left. In the upper left drawing, the medium is divided into a series of cubes; we suppose that the one at the left is quiet, as shown. We also suppose that the stress at a distance Δx from this quiet place is a stress toward the right, as shown. This is a tension on the material in the cube. We have added small tensions on the top and bottom faces that correspond to propagation for a medium in which Poisson's ratio is ¼. This particular specification of σ has no essential part in this description; it is introduced only to make the scale of the stresses definite. The vertical stresses in such a case prove to be one-sixth the horizontal stress. By using Hooke's law we obtain equation 1, which relates the stress to the dilatation. If the propagation is toward the right, the displacement is given in terms of a function shown in relation 2. In this the dilatation is given by relation 3. The velocity of displacement divided by the velocity of propagation is given in relation 4. Consequently, in a case of propagation toward the right, the relation between the stress and the velocity is given by equation 5. On the right edge of the cube being considered, the stress and velocity have *opposite* signs. This is shown in the diagram by the opposite direction of the corresponding arrows. In other words, the stress in this case is a tension toward the right, but the velocity is in a reverse direction toward the left. This is the particle velocity we refer to here. We might characterize this type of propagation as the pulled tail type, because in this case the pulse is moving toward the right and has a rarefaction at its rear. The stress is a tension and the velocity of particle displacement is toward recovery of its original undisplaced position.

The opposite situation (still right propagation of the pulse) is illustrated in the lower left drawing which may be called the compressed tail type. Here the stress is a compression and the velocity is oppositely directed, but it is still directed toward recovery of the position of the displaced particle.

Propagation toward the left is shown at the right of Figure 14.17. In the pulled front type at the upper right, the stress is a tension and the velocity is in the *same* direction as the stress.

This is illustrated more precisely in relations 2′ through 5′, which are analogous to relations 2 through 5 at the left. In this case the stress has the *same* sign as the particle velocity. We show that the tension and the velocity are in the direction necessary to increase the displacement.

The same is true in the *compressed front* type shown at the lower right. Here there is a compression on the right face of the cube under consideration and the velocity is represented by an arrow pointing in the *same* direction. In other words, the velocity is such as to increase the displacement that has already occurred.

These are the four simplest cases of propagation. It is recommended that the reader study these four situations and satisfy himself that the direction of propagation given in the headings is the correct one for these combinations of particle velocities and stresses or forces on the cube faces. Emphasis must be laid upon the fact that each of these situations represents a very precise relation between the stress and the particle velocity, and that, in general, none of these relations will be satisfied. When these relations are not satisfied we do not have propagation in just one direction, we have propagation in both directions.

Let us refer back to the lower part of Figure 14.16. We see that the lateral particle velocities are zero in first approximation, that the stress on the right face is zero, but that the stress on the left face is not zero. Thus none of the four conditions shown in Figure 14.17 is present. We must therefore conclude that energy will be transmitted laterally to both right and left. This is a property of diffraction at an edge that is well worth considering; in many cases it can be verified by direct observation of seismic records. The propagation of the diffracted part of the wave proves to be approximately what we should expect if a new source were located at the corner of the fault.

We shall not be very far off in our description of propagation, such as is illustrated in Figure 14.16, if we introduce a new source at the position of most serious irregularity—that is, at the upper corner of the fault—and describe the propagation in terms of the superposition of the waves from this new source and the regular reflected pulse. We must realize that this diffracted part receives its energy from the original reflected part. Furthermore, since it receives most of its energy near the edge of the reflected part, the greatest weakening in the reflected energy may be expected at this place. But this weakening is compensated for to a certain extent by a lateral flow of energy from the regular reflection on the left of the edge shown in Figure 14.16. It must be noted, however, that this flow is limited and that the reinforcement tends to travel backward along the pulse, since the lateral flow can only travel with the velocity of compressional waves.

The net result is a reflection in the neighborhood of the edge of a fault that has been weakened. Also its character has changed, that is, the shape of the pulse is altered. The pulse is also lengthened. Furthermore, to this must be added the disturbance of the diffracted part, which necessarily has to be superimposed on the regular reflection directly over the fault. This usually produces a rather confused situation that is difficult to untangle on a seismic record. Although it is difficult, it is not always impossible, especially if one has a fairly good picture of what is going on, such as that given in the foregoing discussion.

It will at once be realized that the edge of a fault is not the only irregularity that can cause a diffraction. A sudden change in the dip of a reflector can also give rise to diffractions that may be very disturbing. Another cause may be an erosional cut, perhaps due to an old buried stream bed. Another and very important cause is buried irregular deposits such as salt domes, reefs, sand bars, and the like. The alert seismologist will realize that these diffractions, although difficult to deal with, in many cases may form the most important pulses on the records he is studying. Often they appear to show a very high dip. In such a case, the dip of course has no significance. It only gives a measure of the apparent origin of the diffracted pulse, and thus it only locates the position of the structural irregularity in three-dimensional space.

That diffractions occur even in areas where the dip is very small, as in the gulf coastal region, furnishes the principal argument against using simplified plotting procedures in which reflections are plotted directly beneath shot points, or time sections are plotted. When such a procedure is used as a regular routine and no further refinements are considered, an interpretation of any diffraction pulses has been effectively eliminated. In many cases, this may mean elimination of an interpretation that has considerable economic value.

Although diffraction effects represent exceptional effects in most optical discussions, they represent ordinary, usual, to be expected, effects in the propagation of seismic pulses. It would be a great mistake to minimize the importance of diffraction effects because of their usual unimportance in most simple optical situations.

14.5. FOCI

In this section we consider what amounts to another diffraction effect. This one is due to the curvature of the reflector. Figure 14.18 shows a portion of a reflector which has two principal lines of curvature passing through a point of reflection P. One of these lines is C_1PC_1'; the other is C_2PC_2'. The curvature of these principal lines of curvature may not be quite the same. An attempt has been made to indicate *focal lines* corresponding to these at $L_1(P)$ and $L_2(P)$. For instance, in order to define $L_1(P)$, we pass a plane perpendicular to C_1PC_1', at point P_1' on the line. Then we pass another plane perpendicular to this line of curvature through point P_1. These two planes in general intersect in a line, $L_1(P)$. As P_1 and P_1' approach P, line $L_1(P)$ varies somewhat, but it approaches a definite line which we suppose to be the same as the line $L_1(P)$ shown. This, for example, would be the case if C_1PC_1' were a circular arc. The position of this so-called focal line, $L_1(P)$, depends upon the position of P on the surface. The same is true of the other focal line, $L_2(P)$. When the curva-

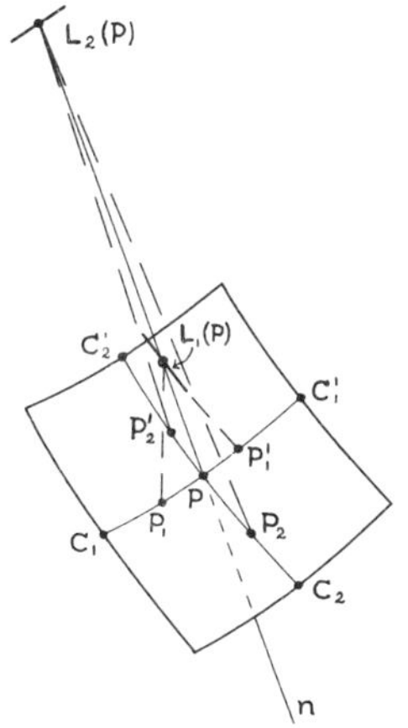

FIG. 14.18. Curvature of the reflector and lines of curvature.

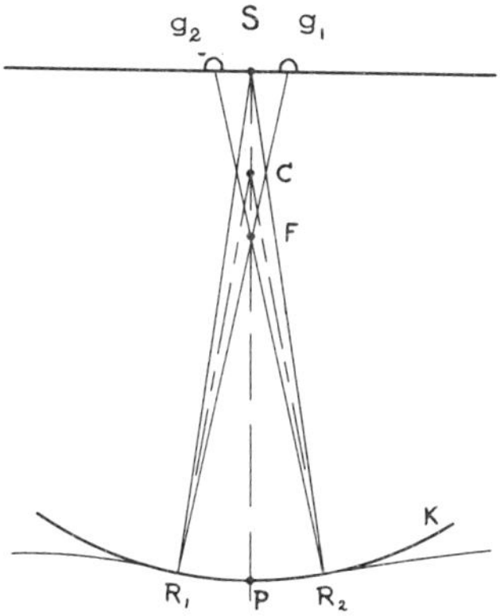

FIG. 14.19. Buried center of curvature and buried focus.

ture in the two principal directions of curvature is the same, the focal lines intersect in the *focal point*. But in a nonspherical surface these lines of curvature are usually separated, as in Figure 14.18.

In Figure 14.19 a shot point, S, appears at the top and a reflecting surface, R_1PR_2, is at the bottom. Rays travel from S down to R_1 and R_2. These two rays are reflected back, pass through a focal point or line, F, and return to the surface at geophone positions g_1 and g_2. This illustrates the *buried* focal point or focal line.

When the energy is concentrated at F and crosses through it, the order in which the energy is received at the surface is the opposite of that in which it is received at the reflector below. This reversal is of little importance unless we are en-

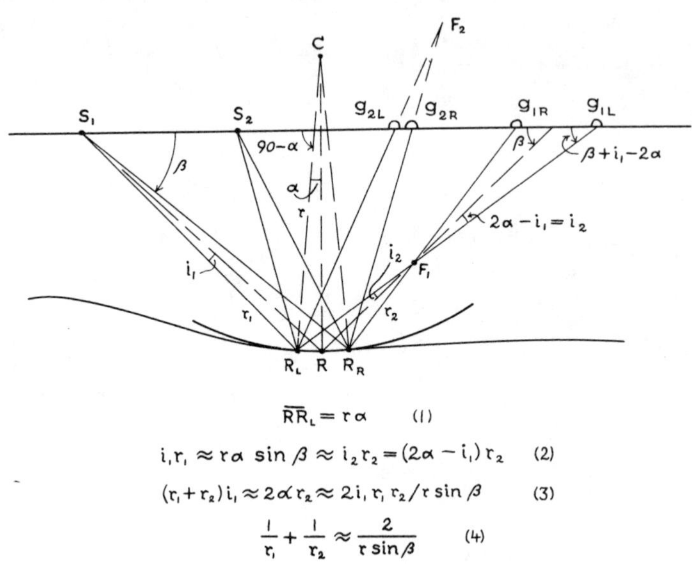

FIG. 14.20. Unburied center of curvature and buried or unburied focus.

gaged in continuous profiling, when we assume that this order has not been reversed. Therefore, in such work it is of some importance to know when there are buried foci.

The center of curvature of the reflector is the point, C. The focal point F will be buried if the center of curvature is buried under the shot point.

But this condition is not always required, as we may see in Figure 14.20. Consider, for example, the shot point, S_1, at the left. In this case we assume that the center of curvature of the reflector is well above the surface of the ground, at point C. We draw the rays down to two reflection points R_L and R_R and we find that the rays are reflected in such a way that they cross at the focal point, F_1. Thus, even when the center of curvature is well above the surface of the ground, a buried focal point may be obtained by offsetting the shot point.

Also illustrated is another case in which the shot point is not offset so far but is located at S_2. In this case the focal point F_2 proves to be above the surface of the ground. Thus, for a given concave upward curvature, there can be two kinds of situations. The focus can be buried, and the focus can be unburied, depending on the distance the shot point is offset from the geophones.

This situation is illustrated quantitatively in the formulas in Figure 14.20. The distance RR_L is given by $r\alpha$, r being the radius of curvature of the reflector and α the half angle shown in the figure. By using rays S_1R_L, S_1R, R_LF_1, and RF_1, we obtain relation 2. This relation is obtained by taking account of the fact that we can project RR_L in a direction perpendicular to ray r_1 and get the same length of projection as if we projected this distance on the line perpendicular to ray r_2. This is true because of the equality of the angle of reflection and the angle of incidence. Relation 3 is merely relation 2 rewritten, and relation 4 is a rearrangement of relation 3. When the angle, β, is 90° we have the regular relation between focal points and the radius of curvature. Relation 4 generalizes this reciprocal relationship when there is an offset, as in Figure 14.20, and gives a quantitative approach to the problem of the buried focus.

So far we have considered only the *geometrical* aspects of the buried focus problem. But this problem also has very important *diffraction* aspects. Consider the situation in Figure

14.21, in which is shown the shot point S on the surface, the center of curvature C, and the buried focus F. If we consider that the energy travels downward and is reflected upward, the energy will have a strong tendency to be concentrated as the pulse approaches focal point F. But this tendency toward concentration is associated with a tendency toward dispersing this energy laterally in both directions, toward both right and left. This dispersal of energy arises from the situation illustrated in Figure 14.17, and corresponds to an effect which may be considered to depend upon the *linear* differential equation governing the motion of the elastic medium. If this dispersion did not take place, the energy would be concentrated with infinite density at the focus, and consequently the linear differential equations would break down. But long before this happens the linear equations themselves take care of the spreading, the lateral decay of energy at the focus, so that no such infinite concentration ever occurs. A discussion of this effect on the basis of the linear equations has been given by Poincaré.* He shows that in the case of harmonic reflected waves, a change of phase of a quarter wave occurs in passing through each focal line. Thus, in passing through a focal point a change of phase of a half wave takes place, which is equivalent to a complete change of sign. In practice, only one focal line is usually passed through; hence the change that occurs is less simple

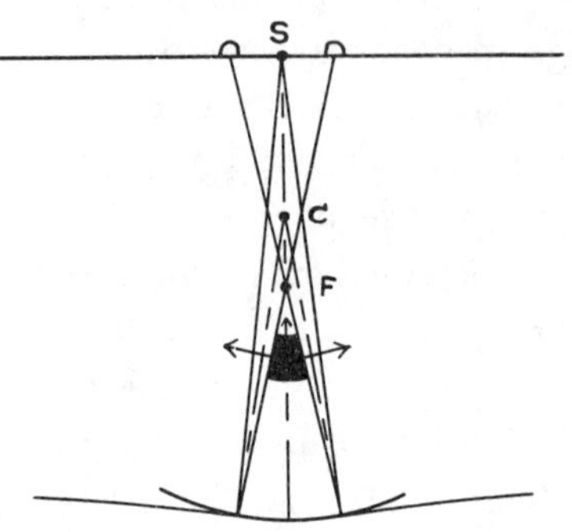

FIG. 14.21. Focal concentration and spreading of energy by diffraction.

* *Théorie Mathématique de la Lumière*, Tome 2, pp. 162-181 (1892).

than the complete change of sign deduced by Poincaré. This effect has been observed optically by several writers.*

The effect of the buried focus is complicated, when we remember that each focal line or focal point is a property of a particular point on the reflecting surface and that, as we pass from point to point, there is a large variation in the focal lines corresponding to these points. What we see is a synthesis of effects from a great many different places on the reflector; hence a localized depression giving rise to a buried focus may have little effect upon the resultant seismic reflected pulse that is observed. On the other hand, when the curvature effects are more widespread, they will be very important. It is extremely important that the interpreting seismologist be aware of the kind of complications that can be introduced by the burial of foci. The author recalls one case when a well was drilled on a small depression that was supposed to be a high. The existence of this depression was verified from several different places after it was first suspected. It had the effect, among other things, of making the reflections hard to work with, because it gave them an irregular character. Of course, the existence of buried foci becomes exceedingly important in case continuous profiling is attempted with continuous time ties. Failure to take account of their existence in this case usually leads to entirely erroneous results, as has been pointed out by Widess.†

14.6. SURFACE WAVES

We are now ready to consider very briefly one of the most difficult aspects of the propagation of seismic pulses, namely, the propagation of surface waves. It has been known for a long time that surface waves may be propagated along the

* M. Gouy, *Annales de Chimie et de Physique*, 6th Serie, 24:145-213 (1891); G. Sagnac, *Journal de Physique*, 4th Serie, 2:721-727 (1903); R. W. Wood, *Physical Optics*, p. 288 (1934).

† M. B. Widess, *Geophysics*, 8:93 (1943).

free surface of an elastic solid. The type of surface wave that apparently is most often encountered in the case of an explosive source is the Rayleigh wave. The theory of this wave was first stated by Lord Rayleigh* in 1887. Later on, his results were extended by R. Stoneley to waves along an interface between two elastic media, and these waves have been called Stoneley waves. Stoneley found that in order for such waves to be propagated, it is necessary that the shear moduli of the two substances be very close. This has been verified by Cagniard for the more general case when there is a point source. Stoneley's original discussion applied only to plane waves.

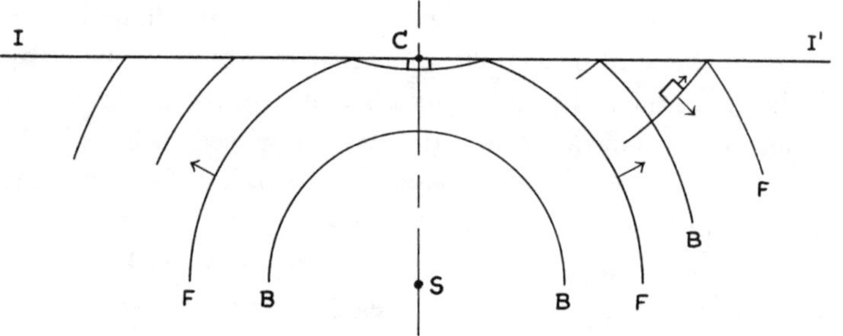

FIG. 14.22. Generation of surface waves.

None of the above-mentioned discussions of surface waves gives the slightest hint as to their *mode of generation.* Perhaps this information could be obtained from Cagniard's discussion if various stages were computed according to his method and examined very closely. However, it is possible to obtain a rough picture of the generation of surface waves without reference to Cagniard's work. The following discussion will provide such a picture.

Figure 14.22 shows a source S and an upward-traveling pulse with a front F and a back B. This pulse is pictured as

* *Proceedings of the London Mathematical Society,* 17:4 (1887).

just having struck the surface *II'*. The center of impact is at point *C*, which is the top of the perpendicular from *S* to the surface *II'*. We may suppose, with considerable justification, that the pulse is first reflected in the immediate neighborhood of *C*, in almost exact accordance with the reflection coefficient calculated in Section 14.3.1, since conditions here are almost the same as in the above section. We may then suppose that, as we move laterally from *C*, we introduce only a slight variation from the facts, if we assume that the reflection follows the laws of intensity deduced by Knott and Zoeppritz. But this deviation, though small in the immediate neighborhood of *C*, soon assumes major proportions as we go farther and farther away from *C* along the interface. If we consider an element of volume in the reflected pulse so that this element is symmetrically located with respect to *C* and consider the conditions for no lateral propagation, we find that the conditions are precisely met. If, on the other hand, we move away from *C* so that the element of volume is not symmetrically located with respect to this point, downward propagation no longer holds, but there are conditions for propagation laterally in a direction tangent to the reflected pulse front. That is to say, the conditions present are those we described for diffraction effects. These diffraction effects arise in the immediate neighborhood of the surface, immediately after reflection. It may be supposed that they are zero on the axial center line *SC*, but that they increase rather rapidly as we move away from this axis. It is also evident that their origin is in the immediate neighborhood of the surface. We thus have, from this diffraction effect, a kind of source that is of a secondary nature, is distributed symmetrically about the axis *SC*, and generates compressional waves traveling in practically all directions, including back toward the surface.

As the distance from *C* increases, in accordance with the relations of Knott and Zoeppritz rotational waves as well as compressional waves begin to be generated and these also give rise to secondary sources in the neighborhood of the surface which generate diffracted rotational waves. Thus, in the neighborhood of the surface, there are secondary sources due to diffraction effects of both rotational and compressional waves.

Since it is possible to have freely traveling waves of the Rayleigh type at such a free surface, the energy from these diffraction sources may be expected to be partially absorbed in waves of the Rayleigh mode.

It should be clearly understood that this is no proof that Rayleigh waves are generated in this manner. We may, however believe that if Rayleigh waves are to be generated, their source must be very similar to that described above. We can perhaps state this as follows: If the Rayleigh wave is not to be generated by secondary sources, these secondary sources must be distributed in a very particular manner because they occur primarily in the neighborhood of the interface. Since it is quite improbable that the actual distribution of secondary sources corresponds to a particular arrangement that prevents the generation of Rayleigh waves, we may suppose that our picture shows approximately the true generation process.

There is an aspect of Rayleigh wave propagation that perhaps should be emphasized in considering their generation. This is the fact that Rayleigh waves—at least the harmonic Rayleigh waves described originally by Lord Rayleigh—decay exponentially as we measure them farther and farther away from the surface, and the rate of exponential decay is inversely proportional to the wave length of the harmonic component. Thus very long or low-frequency Rayleigh waves decay very slowly, and very high-frequency waves decay very rapidly. This leads to other curious situations when we have a stratified medium, as we almost always do in seismic prospecting. For instance, if a weathered layer overlays a layer of higher velocity, the shorter components of the Rayleigh waves generated will travel more slowly and in the upper medium, whereas the longer waves will have greater penetration and travel more rapidly. Thus the velocity of transmission of Rayleigh waves is subject to what is called *dispersion,* that is, variation of the velocity with wave length. This has the effect of spreading out pulses. A little consideration will show that if we start with a short pulse and make a harmonic or a Fourier integral analysis of it and then insist on varying velocities for varying frequencies or wave lengths, the pulse generated will soon be no longer concentrated but spread out. However, another effect occurs, namely, any pulse that does travel does so with what is called *group velocity* rather than *phase velocity.** A familiar example of group velocity is seen in the transmission of waves on the surface of water. If a stone is dropped in a pond of still water, the waves travel outward in concentric circular rings. If we observe this motion we see that the outermost crest dies away and disappears successively as the group moves outward. Thus the crests travel faster than the group does.

Pekeris and others have shown that in a layered earth in which velocity increases with increasing depth, it is possible to have minimum or maximum group velocity corresponding to certain frequencies, depending upon the velocity distribution. The effect of these stationary values of group velocity is to concentrate the energy corresponding to these particular frequency neighborhoods in groups. Thus, after a certain amount of travel, it will be found that Rayleigh waves in a stratified medium tend to have certain characteristic frequencies, because these particular parts of the original pulse have become separated from the remaining parts. This coincides with the observation that Rayleigh waves show certain char-

* For a discussion of group velocity, refer to Rayleigh, *Theory of Sound,* vol. 2, pp. 297-302 (1878).

acteristic frequencies, not only in refraction seismograph records, but also in earthquake seismograms. Thus there is a tendency for energy to be concentrated in the Rayleigh wave mode with a narrow frequency range because energy can be transmitted not only in this mode but also in the group corresponding to a stationary value of the group velocity. This latter effect is usually pronounced on the observed records.

Let us consider for a moment the situation that occurs in the Stoneley waves. As was mentioned earlier, in order that these may exist, we must have only a small change in the rigidity modulus on the two sides of the interface. Such a small variation of rigidity moduli as this probably exists in nature only rarely.

On the other hand there is a case that is of some importance for seismic prospectors. Miller Quarles, Jr.,* has discussed faults that occur in the southern gulf coastal region of Texas. These faults are found in a section that is fairly uniform and the faults themselves have rather small displacements. We may suppose that in this case the fault interface itself represents an interface across which, at least in large part, the variation of the rigidity modulus is small. In some cases this variation may be small enough so that Stoneley waves associated with a fault interface may be generated by the diffraction process at the fault. If this were the case, these waves would travel down along the fault interface until they came to a place where they could not be transmitted. It is probable that they would be reflected at such a location. Stoneley waves would cease to be transmitted as soon as a level was reached where the rigidity moduli on the two sides varied too much. Under these circumstances we may suppose that Stoneley waves might be reflected and the energy be returned to the top of the fault and from there travel upward through the medium as if the top of the fault were approximately a new source. It is con-

* *Geophysics,* 7:123 (1942).

ceivable that this took place in the case described by Quarles. It is clear from his data that such a procedure is possible; moreover, this fits in very well with the hypothesis that Stoneley waves are generated and travel down in association with the fault face itself until they are reflected by the first major discontinuity. It would be of considerable interest to reëxamine Quarles' records in detail with this possibility in mind.

Of course the principal application of the idea of surface waves, perhaps of the Rayleigh type or some modification of this type due to stratification of the upper section, is to the *noise problem.* The noise, which has been called *ground roll* for many years in the seismic prospecting business, is probably primarily of the Rayleigh type. It is of course highly desirable to understand the properties of this kind of noise in some detail so that it may be eliminated. Since in many cases this ground roll tends to show characteristic frequencies that are much lower than those used for reflections, the ground roll can be removed by filtering. Filtering consists of cutting out these lower frequencies, or reducing them to such low levels that the ground roll becomes unobjectionable. This is the practical way of solving this problem.

The preceding sections describe certain properties of the propagation of seismic pulses. It must be admitted that the descriptions have often fallen far short of what might be considered adequate. As was pointed out in the introduction to this chapter, the principal reason for this is that the subject itself is extremely difficult. We have therefore approached each stage of the discussion with the knowledge that we could not present a really adequate description, but could give only a few bare suggestions to help the reader picture the possibilities. It is hoped that the development of experimental seismic modeling techniques, such as are being worked on vigorously at present by several organizations, will enable us to secure a much more precise picture of what happens than is possible at the present time.

Therefore we have referred to pulses somewhat loosely. But the ground motion and its record are different. The recording mechanism introduces distortion. Let us investigate a little further what we see when we look at a pulse on a seismic record.

14.7. COMMENTS TO THE SECOND EDITION

The problem of a wave passing through and around a buried focal point, as shown in Figure 14.21, should be clarified. We do this using Huygens' construction, as in Figure 14.23, where we show the reflection of a plane wave reflected

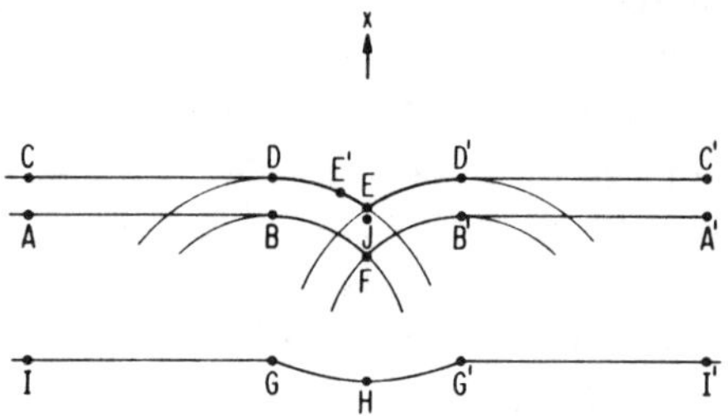

FIG. 14.23. The Effect of a Buried Focus on Phase Advance.

by a parabolic mirror, *GHG′*. The incident plane wave comes down and is reflected back up to *ABFB′A′* at the time when it just gets back to the focus, *F*. A little later, the pulse front reaches *CDED′C′*. But at that later time, the reflection from *H* arrives only at *J*, below *E*. It is in this sense that a wave shows an *advance* in passing "through" a focus. The advanced part does not really pass through the focus, it takes a *shorter* path. If you picture the vertical axis *EH* as an axis of 3-D symmetry, then *E* is a place of some energy magnification, as compared with *E′* beside it. However, *E* is not a focus in the ordinary sense.

CHAPTER 15

Recording

It is not the purpose of this chapter to discuss in detail the design of any instruments such as amplifiers, filters, geophones, recording cameras, and the like. Our only intention is to describe some of the general features of such equipment that are of particular value to the interpreter. In doing this, we shall have an entirely external viewpoint; that is, we shall not examine the mechanism of any of these instruments, but study them on the basis of their performance. This chapter is written primarily for seismologists, but it may be of some interest also to instrument operators and designers.

15.1. INDICIAL RECORD

In Figure 15.1 is shown a geophone, *G*, hung on a spring system of long period. The period of this spring system should be of the order of one second or so. Coming out of the geophone are two leads which go to the amplifier and then to the recorder. To the underside of the platform upon which the geophone rests is attached a small thread that supports a small mass. The mass is raised suddenly, as indicated by the hand, thus suddenly changing the force on the spring system including the geophone. This sudden change of force is shown in the first diagram below the hand. The hand raises the mass suddenly at the time $t = 0$ in the graph. This changes the force in the system by an amount mg. After this, of course, the system slowly recovers itself and oscillates with a natural period corresponding to the whole spring system. It is very important that this period be much longer than any period of interest in the recording mechanism. Raising the mass suddenly is approximately equivalent to *closing a switch* suddenly and applying a unit force to the system. The *unit step force* thus applied gives rise to a *response* that can be recorded in the recorder. This response we assume to be similar to that in the lowest diagram. This response is called the *indicial response for the force*, and it represents all that can be said about the characteristics of the equipment from an external point of view. The response function we have

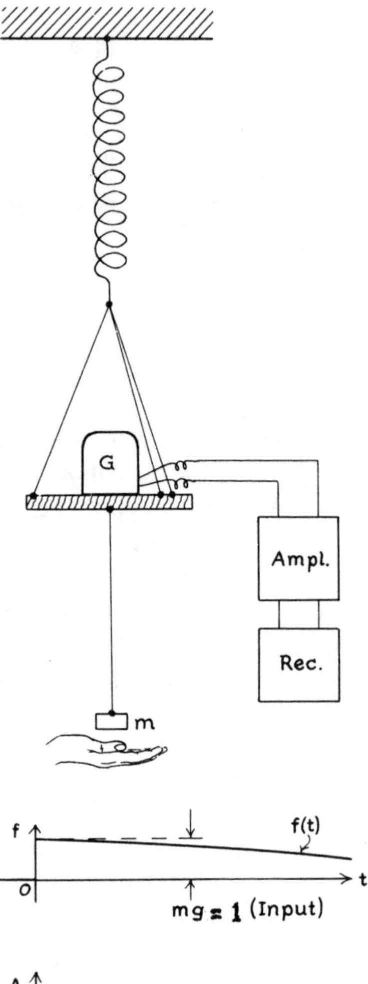

FIG. 15.1. Field or laboratory arrangement for measuring indicial response.

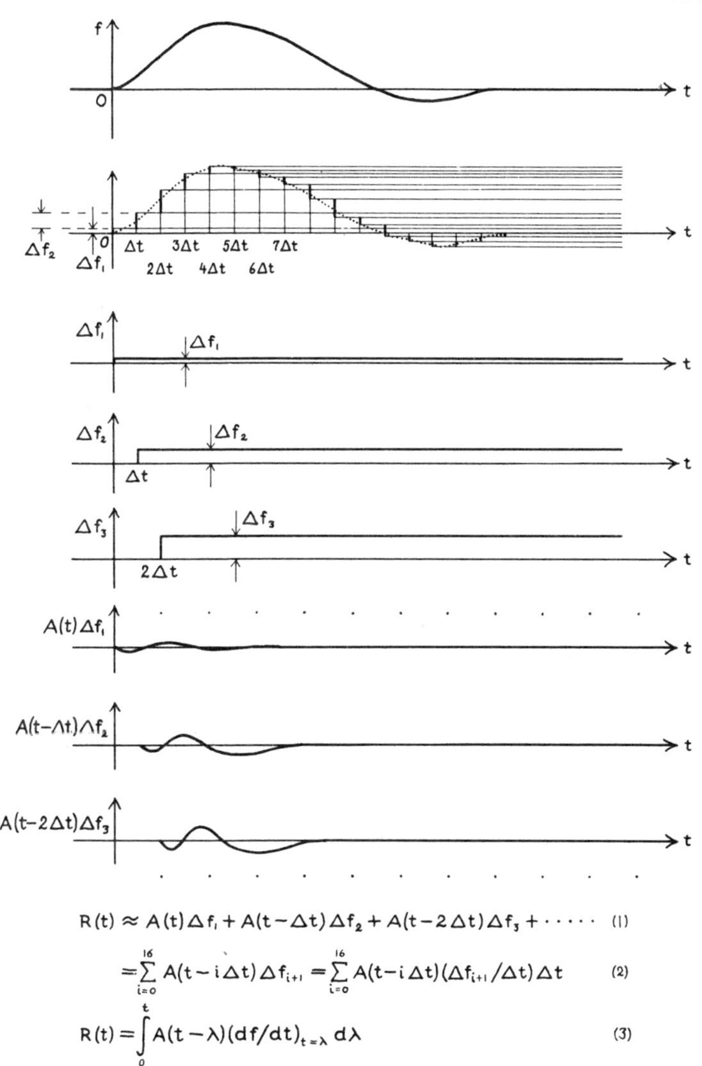

Fig. 15.2. Indicial response used to compute response to general input signal.

called $A(t)$. This response function is significant for our system largely because the system of geophone, amplifier, and recorder is approximately a *linear system*, so that if we double the input amplitude we double the response amplitude.

In the above analysis we follow a well-worn path which has been laid out by electrical engineers since Heaviside's time. Accounts of these methods may be found in Bush's *Operational Circuit Analysis* (1929) and in the later book by Gardner and Barnes, *Transients in Linear Systems I*. This particular application was suggested by R. G. Piety.*

The value of this particular type of analysis can be seen directly in Figure 15.2. We see here that if instead of a step function input, we have a *more general input* such as that at the top of the figure, we can compute from the indicial response the output for this more general input. The method is shown step by step graphically. The second drawing shows the smooth signal approximated by a series of steps. It is assumed that this kind of approximation does not introduce serious additional response terms that will not disappear as the length of the steps decreases. When we divide the general input function into a series of steps in this way, we suppose that each step corresponds to a small step function. We have drawn this step function by extending each step to the right to infinity. The first step is very small; it is represented by Δf_1. The second step is a little larger and is represented by Δf_2. The third drawing shows this first step alone; the fourth shows the second step alone, and the fifth shows the third step alone. The line of dots indicates that this sequence continues until the whole input pulse has been shown. When we step up, of course, we have *positive* input step functions, and when we step down we have *negative* input step functions. The sixth drawing represents the response due to the first input step function, the seventh represents the response due to the second

* *Geophysics*, 7:123 (1942).

input step function, the eighth represents the response due to the third input step function, and so on, as indicated by the dots. We assume that the total response is the sum of these parts. In other words, if we want to determine the response of the series of steps in the second drawing, we have only to add all the responses given in the sixth, seventh, eighth, etc., drawings. This addition is shown in relation 1. That this is an approximate expression is indicated by the approximate equality sign. In our example there are 16 time intervals in the total input time, so we have to add 16 terms as shown in equation 2. The relation at the right of it has been altered by dividing each term by Δt and multiplying each term by Δt. This does not change the numerical value but only the form of the expression. We can then allow Δt to approach zero—that is, the number of steps to approach infinity—and relation 2 then approaches relation 3, which expresses the response as a function of t in terms of the indicial response, A, and the input force f.

Several results can be obtained directly from relation 3. One of the most important is that the response $R(t)$ is generally of a length that is equal to the sum of the indicial response length and the input force length. This relationship gives us a means of estimating the length of the time through which a geophone is accelerated by a reflected pulse, for we can easily determine the indicial response and we know the apparent length of the pulse on the seismic record. It is, of course, true that there is seldom or never a pure case, without external noise, of a reflected pulse; but often we can estimate the minimum length of the pulse on a record and from it deduce the minimum length of the input force corresponding to this pulse.

Since reflected pulses are usually sharp and snappy, whereas noise is generally much more prolonged and more irregular in character, it would seem advantageous to design equipment so that the indicial response graph would correspond to a very short graph in time. Of course the shortest one would be the one corresponding to a sharp little peak just after the application of force. Such an instrument arrangement can be made, but it entails a kind of flat response characteristic that brings in a lot of extraneous noise of both high and low frequency. The high-frequency noise is generally due to the wind, and the low-frequency noise is generally due to ground roll or surface waves. A compromise has to be achieved and this has been done by most modern recording and amplifying equipment. It should be remembered, however, that in making this compromise a certain amount of information is lost by cutting down the highs and lows.

15.2. GROUND-TO-GEOPHONE COUPLING

Ground-to-geophone coupling has been treated in a paper by Washburn and Wiley,* to which the reader is referred for further information. Our purpose is to explain the points covered in it by means of an exaggerated and simplifying procedure. Figure 15.3 shows a geophone, *G*, resting upon sponge rubber, which in turn rests upon the ground. The geophone is connected to its amplifier and its recorder. It is evident that when the geophone rests upon sponge rubber, its response to upcoming pulses will be greatly altered. For example, if the geophone is pushed down into the rubber and suddenly released, it will oscillate on this rubber—in a damped fashion, to be sure, but nevertheless with a certain characteristic frequency. If this frequency is around 50 or 60 cycles—that is more or less in the midst of the main reflection frequency—the situation is exceedingly bad, for we have a means of introducing serious distortion right where we want the distortion to be minimized. A way of measuring this effect is illustrated in Figure 15.4, which is a modification of Figure 15.1 made by attaching a large mass below the supporting platform and

* H. W. Washburn and H. Wiley, *ibid.*, 6:116 (1941).

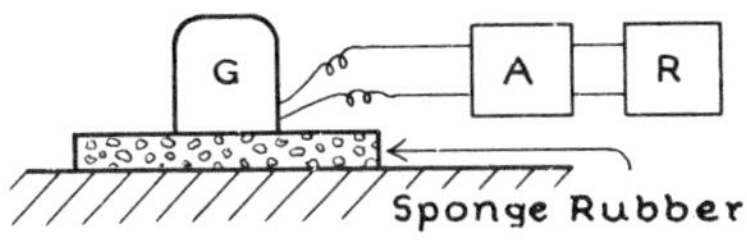

Fig. 15.3. The geophone-to-ground coupling problem exaggerated.

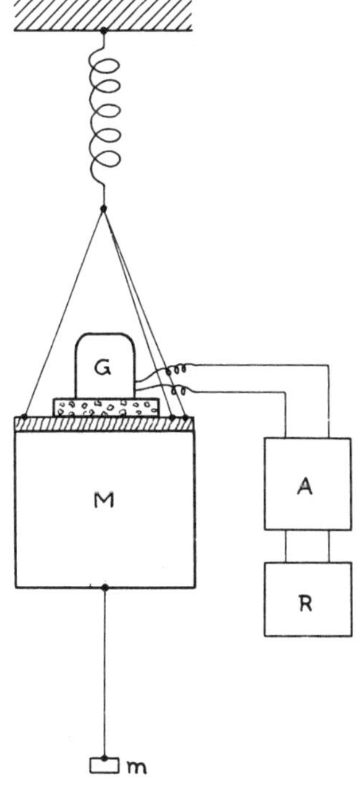

Fig. 15.4. Means of showing the geophone-to-ground coupling effect.

putting the sponge rubber between the platform and the geophone *G*. Such a change alters the dynamic picture considerably. The addition of the mass *M* has the effect of making the mass against which the geophone oscillates as large as it would be if the geophone were resting on sponge rubber on the ground. The effect of the change in the indicial response of the system can easily be observed in this manner.

Let us consider, for a moment, how the coupling might be altered by means of the *area of contact* between the geophone and the sponge rubber. At the left of Figure 15.5 is shown a geophone with a small contact; at the right the same geophone with a large contact, and in the middle an intermediate case. It is quite evident that if we assume that the platform which provides the contact is practically without mass so that the whole mass is the mass of the geophone itself, what corresponds to the *spring constant* in the oscillating system of the geophone coupled through the sponge rubber to the earth is proportional to the area of the base. Thus if the spring is stiffer and stiffer, that is, if this is a larger and larger area of contact, there is a larger and larger natural frequency for the oscillation of the geophone in contact with the rubber in contact with the ground. Hence, in order to secure a firm contact with the ground, so that the geophone will be capable of measuring certain features of the ground motion with a certain degree of reliability, the large contact at the right is necessary. The small contact is likely to be unsatisfactory. That this is true can be verified directly by experiment.

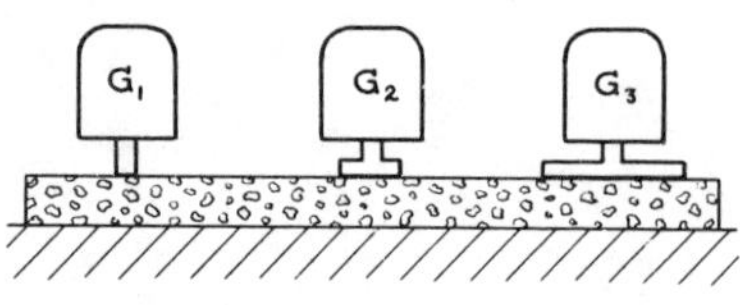

FIG. 15.5. The effect of the area at the base of the geophone on its coupling to the ground.

It is a matter of some interest, in this connection, that the shape of the geophone used in prospecting may determine the character of the geophone-to-ground coupling. For example, a tall geophone with a small base may be like that at the left of Figure 15.5; a squat large-based geophone may be like that at the right. A company that operates primarily with geophones like that at the left will probably bury them in order to secure better contact. If on the other hand they use geophones of the type illustrated at the right, they will probably never bury them except in extremely windy conditions. The above survey of the situation will be found to correspond quite well with actual facts. Since it costs a little more to bury geophones, it may be desirable to use geophones with a slightly broader base corresponding to the mass of the geophone.

If the reader is interested in making a more detailed investigation of the geophone-to-ground coupling problem, he will do well to use the methods outlined in the Gardner and Barnes book, mentioned above, for analyzing mechanical systems.

15.3. MIXING AND FILTERING. DECAY OF INFORMATION

It is common practice in certain regions and in certain companies to do a good deal of what is called *mixing*. In this the outputs from two geophones located at different places on the ground are fed into a common amplifier and appear summed on the output trace that is recorded. This mixing can be carried to such an extreme degree that all the geophone outputs are mixed and this mixed energy is fed into all the amplifiers. Such an extreme is obviously not desirable. The question is whether any mixing is desirable. The argument in favor of it claims that mixing tends to increase the amplitude of signals which come directly upward to the line of geophones as compared with random noise that travels more or less horizontally. This is indeed true, and for this reason mixing is sometimes valuable. On the other hand, it must be pointed out that mixing never actually increases the information that can be obtained from a record. On the contrary, it always decreases it, because it is always possible to mix a record that originally was unmixed. Even though this mixing may be very laborious, it can be done. But the reverse process of unmixing a record can never be done. Thus, mixing always reduces the amount of information available from a given record, but it sometimes seems to increase it by making the information a little easier to see.

There is another operation that is commonly carried out on records. This is the *filtering*, or cutting out, of both high and low frequencies, high frequencies to reduce *wind noise* and low frequencies to reduce *ground-roll noise*. Here again the total amount of information on the record is reduced by the filtering process, but at the same time the record is easier to read. Thus both mixing and filtering reduce the total content of information, but that which remains is usually seen more easily by the persons interpreting the record.

This is the case if only rather rough interpretations are being made. But anyone who is interested in details, in using information to its utmost, had better not either mix or filter very much. A good deal can be said for making the permanent record either on variable area or variable density film or on magnetic tape which can be played back later with various mixing and filtering arrangements. When such a record is made, the information is stored; but when the mixing and filtering are only done on a paper record, the information that has been cut out is lost forever (or until the record is retaken with less mixing and less filtering). The economics of this situation seem pretty obvious to the writer.

15.4. VOLUME CONTROL

Another alteration of the recording consists of a *volume* or *sensitivity control* arrangement. It is usually desired to bring in the first breaks on the record for weathering corrections. In order to do this, a high level of sensitivity is used on the early part of the record. Immediately afterward, the sensitivity is rapidly reduced to a low level, so that the energetic shallow reflection can be recorded. Following this, increasing the sensitivity allows deeper and deeper and thus weaker and weaker reflections to be recorded.

It is desirable that a record be kept of the variation of sensitivity. This can be done by a calibrated trace on the record measuring the sensitivity. When volume control is automatic, it is especially desirable that some measure of this control be recorded, for otherwise the information regarding the relative strengths of various reflections will be lost. It is the writer's opinion that this particular information is often of considerable value when used in connection with other bits of information in difficult cases.

15.5. COMMENTS TO THE SECOND EDITION

Referring to section 15.2 on ground-to-geophone coupling, I should mention a booklet by Heinz Rosemann, *Der Einfluss der Ankopplung das Seismometers an den Untergrund auf die Energieubertragung* Freiberger Forschungschefte, C64, 1959, Berlin, East Germany.

Now, this problem has been turned around, to consider the large vibrators used for Vibroseis * sources. The development and handling of these "chirp" sources has been of major importance over the past twenty years, becoming more important every day. For an adequate discussion of this work, see the 1978 book by K. H. Waters, *Reflection Seismology.* Mr. Waters was closely associated with the whole Vibroseis development within the Continental Oil Company for many years.

Referring to section 15.3 on mixing and filtering, the improvements have been great. Before reproducible recording (magnetic and digital), "mixing" meant stacking without any attempt to get coherence—a desperate attempt at improvement of the records.

Filtering has also been greatly improved by using the advantage of reproducible recording. The original unfiltered record is taped. Then if filtering is wanted, that may be done through the play-back of the original.

There is always some degradation of the signal whenever filtering is done. Filtering always subtracts energy from the signal and lengthens it. However, the signal may be redistributed in such a way that its energy density is much increased in some part of the response. For example, a chirp signal may start at 10 Hz and the frequency increase to, say, 50 Hz in 25 seconds, and then turn off. By cross correlating this long

* Trade name of Continental Oil Company

signal with itself, a new symmetrical signal is made with most of its energy concentrated near its middle in much less time than 25 seconds. This is the process used in the detection of chirps in the Vibroseis process.

REFERENCES

Claerbout, Jon F., *Fundamentals of Geophysical Data Processing.* New York: McGraw-Hill, 1976.

Crawford, J. M., W. E. N. D. Doty, and M. R. Lee, Continuous Signal Seismograph, *Geophysics, 25* (1960), 95–105.

Dix, C. H., Seismic Velocities from Surface Measurements, *Geophysics, 20,* (1955), 68–86.

Dohr, G., *Applied Geophysics: Introduction to Geophysical Prospecting.* New York: Wiley, 1974.

Hubral, P., and T. Krey, *Interval Velocities from Seismic Reflection Time Measurements.* Soc. Expl. Geoph. (1980) Tulsa.

Jeffreys, Sir Harold, *Theory of Probability.* Oxford Univ. Press (1939).

Martin, L. A., United States Patent #4,101,867.

Mayne, W. H., Common Reflection Point Horizontal Data Stacking Techniques, *Geophysics, 27* (1962), 927–938.

Peterson, R. A. and W. C. Walter, Pictorial Digital Atlas 1966, Seismic Energy Sources 1968, Seismography: The Writing of the Earth Waves 1970, Through the Kaleidoscope 1974, Seismic Imaging Atlas 1976, 1977, 1978. United Geophysical Company, Bendix. Pasadena, CA.

Robinson, E. A. and S. Treitel, *Geophysical Signal Analysis.* Edgewood Cliffs, N. J.: Prentice-Hall, 1980.

Sheriff, R. E. *A First Course in Geophysical Exploration and Interpretation.* Boston: IHRDC, 1978.

Telford, W. M., L. P. Geldart, R. E. Sheriff and D. A. Keys, *Applied Geophysics.* New York: Cambridge Univ. Press, 1976.

Waters, K. H., *Reflection Seismology.* New York: Wiley, 1978.

Index